A Treatise on Science, Technology and Society

A TREATISE ON SCIENCE, TECHNOLOGY AND SOCIETY

By

Dr. M.R. SHARMA

Director-cum-Principal
MIT College of Engineering and Management,
BANI, Hamirpur (HP)
Formerly
Director-cum-Principal
MG Institute of Engineering & Technology, Badhu, Mandi (HP)

UNIVERSITY SCIENCE PRESS
(An Imprint of Laxmi Publications Pvt. Ltd.)
BANGALORE • CHENNAI • COCHIN • GUWAHATI • HYDERABAD
JALANDHAR • KOLKATA • LUCKNOW • MUMBAI • RANCHI
NEW DELHI • BOSTON, USA

Published by :

UNIVERSITY SCIENCE PRESS

(*An Imprint of Laxmi Publications Pvt. Ltd.*)

113, Golden House, Daryaganj,
New Delhi-110002

Phone : 011-43 53 25 00
Fax : 011-43 53 25 28

www.laxmipublications.com
info@laxmipublications.com

Price : ₹ **250.00** *Only.*

First Edition : 2009, *Second Edition* : 2011
Reprint : 2013

OFFICES

© **Bangalore**	080-26 75 69 30	© **Chennai**	044-24 34 47 26
© **Cochin**	0484-237 70 04, 405 13 03	© **Guwahati**	0361-251 36 69, 251 38 81
© **Hyderabad**	040-24 65 23 33	© **Jalandhar**	0181-222 12 72
© **Kolkata**	033-22 27 43 84	© **Lucknow**	0522-220 99 16
© **Mumbai**	022-24 91 54 15, 24 92 78 69	© **Ranchi**	0651-220 44 64

UTS-9339-250-A TREATISE SC TECH SOC-SHA
Typesetted at : Shubham Composer, New Delhi.

C—5399/012/07
Printed at : Ajit Printing Press, Delhi

CONTENTS

Preface to First Edition

The book has been specially written to cater to the needs of first year B.Tech students of Himachal Pradesh who study the course in "Science, Technology and Society". There is a dearth of text-books on this topic. This book humbly attempts to bridge the gap to some extent. The book has been divided into 14 chapters.

How far this endeavour has been successful will be known by the response the book receives from the students and appreciation from the learned teachers. I would always gratefully welcome their valuable suggestions to improve the book.

The book has been written after consulting vast literature in the field of Science Technology and Society. I owe my gratitude to the distinguished authors who have been referred to at many places in the book. I acknowledge my indebtedness to all the authors.

In spite of my best efforts there may be some errors, omissions and misprints. Any suggestions and constructive criticism towards the improvement of this text will be highly appreciated.

I wish to thank the publishers for their support. I would like to thank Chairman, Bhuvneshwari Educational Society, Sundernagar for his support. As a personal note I would like to thank my wife Kamal and my loving children Anshul and Pranshul for sacrificing their share of time, which I devoted in writing the book.

—Author

Preface to Second Edition

The general response to first edition of the book has been encouraging. The author is very much grateful to his colleagues and friends for their help, constructive criticism, constant encouragement and suggestions. He is highly thankful to Sh. Karan Thakur, Lecturer for his valuable suggestions to improve the book. He is also highly grateful to Sh. Vikas Banial and Sh. Vikram Pathania Lecturers for their suggestions. In the light of their suggestions, the book has been revised thoroughly. Numerous additions have been made in various chapters to bring out up to date information on the subject. The following additions have been made;

1. A new portion on the Scientific Methods and Scientific Theory has been added in Chapter first.
2. A new topic on Jacques Ellul and his works has been added in Chapter three.
3. A new topic on Club of Rome - Limits to Growth has been added in Chapter four.
4. A new portion on features of industrial revolution and impacts of industrial revolution has been added
5. The topic of eco-friendly technologies has been enlarged.
6. A new section on 'Carbon credits' has been added.
7. A new portion on Nuclear weapons in India has been added.
8. Bibliography has been added at the end of book.
9. The latest question papers of B.Tech 1st year H.P University have been added at the end of book.

It is now believed that this edition of the book will meet the requirements of all the students preparing themselves in this subject for B. Tech. and other Degree Examinations of H. P. University and other universities.

Any errors, omissions and suggestions for the improvement of volume brought to our notice will be thankfully acknowledged and incorporated in the next edition.

Finally, author express his thanks to Publisher for bringing out this edition in a short time and pricing it so moderately in spite of heavy cost of paper and printing.

—**Author**

SYLLABUS FOR B. TECH. 1ST YEAR HIMACHAL PRADESH UNIVERSITY

SCIENCE TECHNOLOGY AND SOCIETY (HU-1002)

SECTION-A

1. Science, Technology and Engineering, as knowledge and as social and professional activities.
2. Inter-relationship of technology growth and social, economic and cultural growth: historical perspective.
3. Ancient, medieval and modern technology/Industrial revolution and its impact. The Indian Science and Technology.

SECTION-B

1. Social and Human critiques of technology: Mumford and Ellul
2. Rapid technological growth and depletion of resources. Reports of the club of Rome.
3. Energy crisis; renewable energy resources.
   Environmental degradation and pollution. Eco-friendly technologies. Environmental regulations. Environmental ethics.

SECTION-C

1. Technology and the arms race. The nuclear threat.
2. Appropriate technology movement Schumacher; later developments.
3. Technology and the developing nations. Problems of technology transfer. Technology assessment/impact analysis.
4. Human operator in Engineering projects and industries Problems of man machine interaction. Impact of assembly line and automation. Human centered technology.

SECTION-D

1. Industrial hazards and safety. Safety regulations. Safety Engineering.
2. Politics and technology. Authoritarian versus democratic control of technology. Social and ethical audit of industrial organizations.
3. Engineering profession. Ethical issues in Engineering practice, Conflicts between business demands and professional ideals. Social and Ethical responsibilities of the engineer. Codes of professional ethics. Whistle blowing and beyond. Case studies.

This book is dedicated
To
My mother

1

Science, Technology and Engineering as a Knowledge

1.1 INTRODUCTION

Science is an organised and ordered way of investigating and understanding the world which is essentially practical in nature. According to Science Manpower Project, "*science is a cumulative and endless series of empirical observations which result in the formation of concepts and theories. These theories and concepts can be subjected to modification in the light of further empirical observations. Science is both a body of knowledge and the process of acquiring and refining knowledge. Science has become an important institution in our day to day life. Thus, the starting place of any course should be what science is.*" What methodology allows us to know the past? To do so we need to place Science in a larger context of human knowledge and then explore precisely how does it function.

1.2 BELIEF AND RESEARCH KNOWLEDGE

What is knowledge? It is a statement about what you accept as sufficiently "***real***" to allow you to take action upon and thereby live your life. There are two kinds of knowledge. Both are equally valid, but stem from very different first assumptions.

1.2.1 Belief Knowledge

The essence of belief knowledge is that knowledge about the world is inherent and unique in each human being. It is attained by individual revelation. It is not open to testing, observation or qualification; it is based upon data that cannot be observed. By definition, these are "***supernatural***" data which are not amenable to investigation by research knowledge. Only belief knowledge can involve truth. Only with belief comes the absolute certainty that brooks no opposition and confers great personal strength. Only belief knowledge invokes value system. Herein lies the power of human spirit and some of its great creations, *e.g.*, purpose, aesthetics, morality, religion.

1.2.2 Research Knowledge

The essence of research knowledge is that the gathering of knowledge be not individual, but universal. It is group endeavor. To this end, research knowledge is based on a protocol to allow one to select between alternative observations or hypotheses.

1. It involves the requirements that all observations are repeatable by other person than the observer
2. And the requirement that all hypotheses are falsifiable. It means that they are open to test and that it is possible to prove the hypotheses wrong.
3. By definition, research knowledge involves natural phenomenon that are observable and measurable by humans.
4. By its very nature, this process forbids a scientist to say that "this is true" (although you will catch many scientists incorrectly speaking about "true" and "false" answers). Rather it can only be that *this is the most likely explanations in the light of existing knowledge.*
5. The closet this comes to belief knowledge is that the scientist does make a choice to believe in the *"research knowledge"* methodology of science.

Both modes generate "facts":

1. The defense of a fact in belief knowledge is the spirit of the individual who believes.
2. Its success is independent upon the quality of argumentation, or upon greater force.
3. A fact in research knowledge is defended by the methodology of observation and testing that allows one to choose one fact over another.

What's a research knowledge fact?

A fact is a mini-hypothesis, subject to testing by repeated observation etc.

Any observation is an electromagnetic stimulation that passes through several filters:

a. The receiving human brain which then

b. Interprets the received data and

c. Communicate it to other humans who in turn.

d. Process the data through their brains.

4. The mind receives data constantly. What it chooses to observe is conditioned by culture and influenced by individuality.
5. Perception of act thus varies, and all facts must be treated as hypotheses and treated constantly.
6. Some will indeed become dominant (the Earth is round, gravity is real), while others are open to testing.

This brings us to consider an oft-misunderstood aspect of the research protocol, the "**Hypothesis**".

1.2.3 Hypothesis

A hypothesis is an explanation for the pattern created by two or more facts. The Hypotheses should be based upon observations that have repeatedly been made by several individuals. It should make predictions about how the word works (an experiment) or has worked in the past (a test in the fossil record). It must be testable –that is, be open to being proven wrong.

No amount of data that agree with a hypotheses ever *"prove"* it to be true! They only fail to disprove it. One significant datum can disprove a hypothesis.

1.2.4 The Death of Hypothesis

A hypotheses dies for one of 3 reasons:

1. Data are found which contradict it
2. What it predict consistently fail
3. It is supplanted by new hypotheses which explains more of the data, or explains the same data more elegantly.

Hypotheses often rise and fall quickly. From the outside this appears unstable, but the content tumult is a sign of healthy science, and usually results in a "*spiral*" of advancing knowledge.

1.2.5 Multiple Working Hypotheses

More than one hypothesis will often explain the observations with equal elegance and strength of predictions. This may lead to the situation where two or more hypotheses have equal currency, and both coexist until one is found wanting.

1.2.6 The Hypotheses Triumphant—a "law" of Science.

Some hypotheses offer such strong predictions and withstand testing for such a long period of time that they become generally accepted, first as "*theories*" and then as "*laws*".

However, even these are not "absolute." A scientific law is just a "very strongly supported inference." We do not know that it will survive in light of new data or technology. The empirical science has several such laws (*e.g.*, gravity, thermodynamics). The historical science has no or few laws because of the nature of their data.

1.2.7 The Scientific Method and the Scientific Theory

The knowledge of science is acquired by following a rigorous procedure consisting of the following steps:

(*i*) Taking a large number of systematic observations.
(*ii*) Studying these observations and looking for their logical behaviour.
(*iii*) Suggesting some model to account for the behaviour observed.
(*iv*) Theoretical prediction of what is not actually observed, on the basis of the suggested model.

All the above steps taken together form the scientific method. It is extremely effective in gaining, organizing and applying new knowledge.

The set of laws so developed which account for the behaviour of physical systems constitute the scientific theory. A theory must stand the test of time and should be able to explain what may be observed from time to time. When any new observation measurement shows some discrepancy or deviation from the existing theory, then the theory has to be duly modified. When even the modification of the theory fails to account for the new observations, then the old theory has to be replaced by the new theory. The new observations leading to the new theory have of course to be checked and rechecked for their own accuracy. Sometimes, new experiments may have to be planned and executed to verify the predictions of a theory. Thus formulation of theories on the basis of experimental results and experimental verifications of theoretical predictions go hand in hand to further our knowledge of any science.

Some of the striking examples, where small discrepancies have led to new theories in Physical sciences are :

(*i*) The concept of flat earth was replaced by the concept of spherical earth, from the observations of distant ships in a sea.

(*ii*) Geocentric theory imagining earth to be at the centre of the universe was replaced by Heliocentric theory imagining sun to be stationary and all planets revolving around it.

(*iii*) The corpuscular theory of light given by Newton was replaced first by Huyghen's wave theory of light, which was replaced later by Planck's quantum theory of light. Finally dual theory of light was given by de-Broglie to account for all the phenomena observed in case of light.

(*iv*) Thus, in science, the approach is always '*open-minded*', in which no points of view are overlooked without logical reasons. In fact, all theories of science have to be updated, whenever required, so that they are consistent with all the known experimental data.

1.3 THE GENERAL CONCEPT OF SCIENCE

Every scientific activity is characterized by two partial activities. One is some form of ***observation/ perception***. It can take place directly through senses, somewhat more indirectly via some form of instrument, in one or another sense such as a microscope, a telescope or more indirectly via some detecting instrument like a Geiger counter or and X-ray apparatus. It surrounds and penetrates the observation/perception, a more or less conscious thought activity takes place as an introduction to the observation. It directs the attention, chooses the observation, steps somewhat back during the direct moment of observation.

The thought activity distinguishes between different parts of that which has been observed/ perceived, give them names or makes a more specific conceptual analysis of them, it may also quantify them and then relates them to each other, logically or mathematically. On the basis of above we can conclude, that;

- Science is a study of natural phenomena
- It is organised and systematized learning
- It is study of cumulative and ordered observations
- It is knowledge based on observation, experiment and interference
- Science is a process as well as the product of that knowledge.

Thus we can say that *Science is a systematic and organised body of knowledge, based on cumulative observations, experiments and inference and is involved in finding out the meaning of various phenomena of nature.*

1.3.1 A "Cultivated" Cut Concept of Science

But if you want to relate the concept to the rich flora of activities that are today termed "science" and get any help to see what they have in common, you have to specify the concept a step further.

If you look at what is today termed science, you find that only certain types of perception and certain types of conceptual formulations are permitted to use in connection with activities that in a more strict sense are characterized as scientific.

As far as **perceptions** are concerned, a number of different types of instrumental perceptions dominate. Different forms of more direct sense perceptions have a more ambiguous status. If you continue to perceptions of different forms of inner, psychic states, states of the soul, you have come to a type of perception with a very dubious status, to put it mildly, as something on what to base scientific knowledge. When you come to perceptions of a more spiritual nature, you have passed outside the border surrounding those types of perceptions that are discussed.

On the **conceptual** side, spatially oriented concepts of a mechanical character dominate. They should preferably relate to something that is quantifiable and it is very satisfying if the quantified perceptions (especially when one of the not exact sciences is concerned) have been chosen in a random way, exist in a great number and have to be put through a computer program to make it possible to describe the results with the help of a mathematical model, or to make it possible to point to more definite connections (significant correlations) between factors that you otherwise don't quite understand how they are related to each other.

1.4 TECHNOLOGY

In general, "technology" is the relationship that society has with its tools and crafts, and to what extent society can control its environment. The Merriam-Webster dictionary offers a definition of the term: *the practical application of knowledge especially in a particular area and a capability given by the practical application of knowledge.*

Technology can be most broadly defined as the entities, both material and immaterial, created by the application of mental and physical effort in order to achieve some value. In this usage, technology refers to tools and machines that may be used to solve real-world problems. It is a far-reaching term that may include simple tools, such as a crowbar or wooden spoon, or more complex machines, such as a space station or particle accelerator. Tools and machines need not be material; virtual technology, such as computer software and business methods, fall under this definition of technology.

The word "*technology*" can also be used to refer to a collection of techniques. In this context, it is the current state of humanity's knowledge of how to combine resources to produce desired products, to solve problems, fulfill needs, or satisfy wants; it includes technical methods, skills, processes, techniques, tools and raw materials. When combined with another term, such as "*medical technology*" or "*space technology*", it refers to the state of the respective field's knowledge and tools. "*State-of-the-art technology*" refers to the high technology available to humanity in any field.

Technology can be viewed as an activity that forms or changed culture. A modern example is the rise of communication technology, which as lessened barriers to human interaction and, as a result, has helped spawn new subcultures, the rise of cyberculture has, at its basis, the development of the Internet and the computer. Not all technology enhances culture in a creative way; technology can also help facilitate political oppression and war via tools such as guns. As a cultural activity, technology predates both science and engineering, each of which formalize some aspects of technological endeavor.

1.5 SCIENCE, ENGINEERING AND TECHNOLOGY

The distinction between science, engineering and technology is not always clear. Science is the reasoned investigation or study of phenomena, aimed at discovering enduring principles among elements of the phenomenal world by employing formal techniques such as the scientific method. Technologies are not usually exclusively products of science, because they have to satisfy requirements such as utility, usability and safety.

Engineering is the goal-oriented process of designing and building tools and systems to exploit natural phenomena for practical human means, using results and techniques from science. The development of technology may draw upon many fields of knowledge, including scientific, engineering, mathematical, linguistic, and historical knowledge, to achieve some practical result.

Technology is often a consequence of science and engineering-although technology as a human activity precedes the two fields. For example, science might study the flow of electrons in electrical conductors, by using already-existing tools and knowledge. This new-found knowledge may then be used by engineers to create new tools and machines, such as semiconductors, computers, and other forms of advanced technology. In this sense, scientists and engineers may both be considered technologists; the three fields are often considered as one for the purposes of research and reference.

The human race's use of technology began with the conversion of natural resources into simple tools. The pre-historical discovery of the ability to control fire increased the available sources of food and the invention of the wheel helped humans in traveling in and controlling their environment. Recent technological developments, including the printing press, the telephone, and the Internet, have lessened physical barriers to communication and allowed humans to interact on a global scale. However, not all technology has been used for peaceful purposes; the development of weapons of ever-increasing destructive power has progressed throughout history, from clubs to nuclear weapons.

1.6 EFFECT OF TECHNOLOGY ON SOCIETY

Technology has affected society and its surroundings in a number of ways. In may societies, technology has helped develop more advanced economies (including today's global economy) and has allowed the rise of a leisure class. Many technological processes produce unwanted by-products, known as pollution, and deplete natural resources, to the detriment of the Earth and its environment. Various implementations of technology influence the values of a society and new technology often raises new ethical questions. Examples include the rise of the notion of efficiency in terms of human productivity, a term originally applied only to machines, and the challenge of traditional norms.

Philosophical debates have arisen over the present and future use of technology in society, with disagreements over whether technology improves the human condition or worsens it. New-Luddism, anarcho-primitivism, and similar movements criticize the pervasiveness of technology in the modern world, claiming that it harms the environment and alienates people; proponents of beneficial to society and the human condition. Indeed, until recently, it was believed that the development of technology was restricted only to human beings, but recent scientific studies indicate that other primates and certain dolphin communities have developed simple tools and learned to pass their knowledge to other generations.

The use of technology has a great many effects; these may be separated into intended effects and unintended effects. Unintended effects are usually also unanticipated, and often unknown before the arrival of a new technology. Nevertheless, they are often as important as the intended effect.

The most subtle side effects of technology are often sociological. They are subtle because the side effects may go unnoticed unless carefully observed and studied. These may involve gradually occurring changes in the behavior of individuals, groups, institutions, and even entire societies.

1.7 ETHICAL IMPLICATIONS OF TECHNOLOGICAL DEVELOPMENT

Winston provides an excellent summary of the ethical implications of technological development and deployment. He states there are four major ethical implications:

- **Challenges Traditional Ethical Norms.** Because technology impacts relationships among individuals, it challenges how individuals deal with each other, even in ethical ways. One example of this is challenging the definition of *"human life"* as embodied by debates in the areas of abortion, euthanasia, capital punishment, etc., which all involves modern technological developments.
- **Creates an Aggregation of Effects.** One of the greatest problems with technology is that its detrimental effects are often small, but cumulative. Such is the case with the pollution from the burning of fossil fuels in automobiles. Each individual automobile creates a very small, almost negligible, amount of pollution, however the cumulative effect could possibly contribute to the global warming effect. Other examples include accumulations of chemical pollutants in the human body, urbanization effects on the environment, etc.
- **Changes the Distribution of Justice.** In essence, those with technology tend to have higher access to justice systems. Or, justice is not distributed equally to those with technology versus those without.
- **Provides Great Power.** Not only does technology amplify the ability, and hence the strength, of humans, it also provides a great strategic advantage to the human(s) who hold the greatest amount of technology. Consider the strategic advantage gained by having greater technological innovations in the military, pharmaceuticals, computers, etc. For example, Bill Gates has considerable influence (even outside of the computer industry) in the course of human affairs due to his successful implementation of computer technology.

1.8 EFFECT OF TECHNOLOGY ON LIFESTYLE

Technology, throughout history, has allowed people to complete more tasks in less time and with less energy. Many herald this as a way of making life easier. However, work has continued to be proportional to the amount of energy expended, rather than the quantitative amount of information or material processed. Technology has had profound effects on lifestyle throughout human history, and as the rate of progress increases, society must deal with both the good and bad implications.

In many ways, technology simplifies life:

- The rise of a leisure class
- A more informed society can make quicker responses to events and trends
- Sets the stage for more complex learning tasks
- Increases multi-tasking (although this may not be simplifying)
- Global networking
- Creates denser social circles
- Cheap price.

In other ways, technology complicates life:

- Pollution is a serious problem in a technologically advanced society. (From acid rain, to global warming, to Chernobyl and Bhopal)

- The increase in transportation technology has brought congestion in some areas.
- New forms of danger existing as a consequence of new forms of technology, such as the first generation of nuclear reactors.
- New forms of entertainment, such as video games and internet access could have possible social effects on areas such a academic performance.
- Increases probability of diseases and disorders, such as obesity.
- Social separation of singular human interaction. Technology has increased the need to talk to more people faster.

1.9 EFFECT OF TECHNOLOGY ON INSTITUTIONAL ISSUES

Technology often enables organizational and bureaucratic group structures that otherwise and heretofore were simply not possible. Examples of this might include:

- The rise of very large organizations: *e.g.*, governments, the military, health and social welfare institutions, supranational corporations.
- The commercialization of leisure: sports events, products, etc. The almost instantaneous dispersal of information (especially news) and entertainment around the world.

1.10 EFFECT OF TECHNOLOGY ON INTERNATIONAL ISSUES

Technology enables greater knowledge of international issues, values, and cultures. Due mostly to mass transportation and mass media, the world seems to be a much smaller place, due to the following, among others:

- Globalization of ideas
- Embedded ness of values
- Population growth and control
- Others.

1.11 EFFECT OF TECHNOLOGY ON ENVIRONMENT

Most modern technological processes produce unwanted byproducts in addition to the desired products, which is known as industrial wastes and pollution. While most material waste is re-used in the industrial process, many forms are released into the environment, with negative environmental side effects, such as pollution and lack of sustainability. Different social and political systems establish different balances between the values they place on additional goods versus the disvalues of waste products and pollution. Some technologies are designed specifically with the environment in mind, but most are designed first for economic or ergonomic effects. Historically, the value of a clean environment and more efficient productive processes has been the result of an increases in the wealth of society, because once people are able to provide for their basic needs, they are able to focus on less-tangible goods such as clean air and water.

The effects of technology on the environment are both obvious and subtle. The more obvious effects include the depletion of nonrenewable natural resources (such as petroleum, coal, ores), and the added pollution of air, water, and land. The more subtle effects include debates over long-term effects (*e.g.*, global warming, deforestation, natural habitat destruction, coastal wetland loss.)

Each wave of technology creates a set of waste previously unknown by human: *toxic waste, radioactive waste, electronic waste.*

One of the main problems is the lack of an effective way to remove these pollutants on a large scale expediently. In nature, organisms "recycle" the wastes of other organisms, for example, plants produce oxygen as a by-product of photosynthesis, oxygen-breathing organisms use oxygen to metabolize food, producing carbon dioxide as a by-product, which plants use in a process to make sugar, with oxygen as a waste in the first place. No such mechanism exists for the removal of technological wastes.

Humanity at the moment may be compared to a colony of bacteria in a Petri dish with a constant food supply, with no way to remove the wastes of their metabolism, the bacteria eventually poison themselves.

The atomic bombings of Hiroshima and Nagasaki are a tragic chapter in the history of the human race. These bombings not only demonstrate the cruelty that humanity can inflict upon itself, but they also foreshadow a terrifying future if we do not halt nuclear proliferation and embrace nuclear disarmament. It has been 61 years since nuclear weapons were first used in war. I wish I could say that the world has learned the lesson that the survivors, the hibakusha, have been trying to teach us since then: The lesson that humans and nuclear weapons cannot coexists.

Unfortunately, all evidence points to the contrary. The world is a far more dangerous place than it was 61 years ago. The world is a far more dangerous place than it was 10 or even 5 years ago. The likelihood that countries will seek nuclear weapons and the likelihood that countries will use nuclear weapons has increased.

To begin, nuclear weapons have become far more lethal since 1945. The 21 kiloton bomb used at Nagasaki is considered miniscule by modern standards. Today, there are thousands of missiles tipped with nuclear warheads hundreds of times more powerful. A full nuclear war would likely bring about the end of the human race. But, even the amount of suffering and destruction that would result from the detonation of just one of these nuclear weapons over a populated area is unprecedented. A book published in 2004, titled Whole World on Fire by Lynn Eden, details the heat and blast effects of a moderate-sized 300 kiloton weapon detonated over the Pentagon.

It would create a fireball more than a mile in diameter producing temperatures of more than 200 million degrees. Fahrenheit-about four to five times the temperature at the center of the sun.

In Pentagon City, asphalt and metal would melt, paint would burn. Offices and cars would explode into flames. The blast wave would create 750 mile per hour winds tossing burning cars into the air.

On the edge of the Potomac the fireball would be 5,000 times brighter than a desert sun at noon. It would melt the marble at the Lincoln and Jefferson memorials. Four seconds after detonation, these structures would collapse from the blast wave that followed.

For decades, the international community has tried to prevent countries from causing this level of destruction. The cornerstone of these efforts has always been international agreements that encourage arms control. *The most important of these is the nuclear Non-Proliferation Treaty (NPT). The NPT has three key provisions. It guarantees countries the right to use nuclear energy for peaceful means. It prohibits the passing of nuclear weapons technology to, and the development of nuclear weapons by, non-nuclear weapons states. Most important, it requires countries with nuclear weapons to negotiate nuclear disarmament*. The treaty establishes an effective framework discouraging more countries from developing nuclear weapons. When the treaty was signed in 1968, there were five nuclear weapons states. The NPT has not been perfect and currently there are nine, but without it there would likely be more.

Today, the international anti-nuclear framework set forth in the NPT is unraveling. There are countries such as Israel, India and Pakistan that have never signed the treaty, and developed large-scale programs. North Korea has broken away from the NPT in order to develop nuclear weapons. Iran may be in the early stages of a weapons program. These countries are endangering themselves and their neighbors, as are the original five nuclear weapons states.

One of the most dangerous recent developments is in Russia. Russia is currently building up its own nuclear arsenal, in significant part, in response to a US missile shield. Recent articles in the Nation and Foreign Policy magazines have argues that, given the state of Russia's infrastructure, such a build-up is extremely dangerous. The Russian government is not investing in proper safety mechanisms to prevent catastrophes such as accidental launches. There has already been a near miss. In 1995, the world came within minutes of nuclear Armageddon when the Russian early warning systems confused the launch of a Norwegian weather rocket with a preemptive nuclear attack by the United States. Boris Yeltsin had nuclear launch codes in front of him and would have retaliated had the mistake not been caught at the last minute. Russia's early warning system has only further deteriorated since then. There are massive holes in its detection capabilities. Russian commanders rely on antiquated radar rather than satellite technology to detect possible launches.

Together, the United States and Russia have 26, 300 nuclear weapons. They possess the ability to carry out precision nuclear strikes anywhere in the world. They have hundreds of nuclear missiles on hair-trigger alert, pointed at each other, that could be fired in a matter of minutes. An accidental nuclear launch by Russia and the retaliatory response by the United States would result in the deaths of millions of people.

But let's not forget the biggest nuclear player and the destabilizing effect it has on non-proliferation regimes. In 2002, the United States placed increased emphasis on the role that nuclear weapons play in its foreign policy. The Bush administration's Nuclear Posture Review states:

1. Nuclear weapons play a critical role in the defense capabilities of the United States, its allies, and friends.
2. Nuclear weapons can be used to achieve political or strategic goals.
3. US policy now supports preemptive attacks, possibly nuclear, on countries with Weapons of Mass Destruction or hardened targets.

The United States is relying now, more than ever before, on nuclear weapons. It also has lowered the threshold for the use of nuclear weapons.

In the past five years, the Bush administration has ignored many important international arms control treaties. It has failed to ratify the Comprehensive Test Ban Treaty, which prohibits all forms of nuclear testing. It has withdrawn from the Anti-Ballistic Missile Treaty. It has invested billions of dollars into a missile shield program; an action seen by many other countries as an aggressive gesture.

In violation of the Non-Proliferation Treaty, it has attempted to develop nuclear weapons that can be used more readily in combat, such as the *bunker buster*. It has also attempted to upgrade the US nuclear arsenal with the implementation of the Reliable Replacement Warhead Program. Most recently, it has negotiated a *deal* with India that allows the exchange of nuclear technology.

By steering around international treaties that encourage arms control, attempting to build new weapons and then seeking to use them for political or strategic goals, the United States is encouraging other countries to do the same. As more countries go down this road, the likelihood of nuclear weapons use will only increase.

Although the only way to ensure that nuclear weapons are never used again is through total disarmament, there are ways to stop proliferation and minimize the risk of nuclear weapons use. In order to be effective, these efforts must have the support of the United States. As the world's most powerful nation in possession of thousands of nuclear weapons, and as the only country that has used nuclear weapons as an instrument of war, the United States is ethically obligated to pursue non-proliferation, arms control and disarmament. This effort must begin with the following three steps:

1. Altering US current nuclear policy. The US must de-legitimize the idea that nuclear weapons are an effective way to achieve political or strategic goals by declaring that it will never be the first to use nuclear weapons in war.
2. Ratifying and complying with the provisions set forth in international treaties such as the Non-Proliferation Treaty, the Comprehensive Test Ban Treaty, the Anti-Ballistic Missile Treaty, the Strategic Arms Reduction Treaties, and the Strategic Arms Limitations Treaties that promote non-proliferation and disarmament.
3. Abandoning our policy of preemptive attacks, which further emboldens countries like Iran to pursue nuclear weapons.

Despite the dangers that we now face and despite all that needs to be done to make the world a safer place, historically, anti-nuclear activism in the United States has been incredibly effective. Anti-nuclear activism was a significant factor in bringing an end to nuclear testing in the US and around the world. Activism was influential in slowing the nuclear arms build up in the 1980s. History has shown that our government listens to the public about nuclear weapons. If the people of the United States work together to tell our government that the creation and use of unclear weapons is not acceptable, we can actually change nuclear policy to make the world safe. I strongly encourage you to find a way to get involved in anti-nuclear work.

Nuclear weapons are the most significant threat to the future of the human race. As long as they exist, no human being is safe. Today, more and more countries are adopting dangerous nuclear policies. It is imperative that we pressure our government to bring the world back from potential nuclear anarchy. Only then, can we prevent proliferation and prevent future Hiroshimas and Nagasakis.

QUESTIONS

1. What is Science? How research knowledge is generated?
2. What is hypothesis and also define multiple working hypotheses?
3. Define Science, Technology and Engineering.
4. What are the ethical implications of technological development?
5. What is the effect of technology on lifestyle of man, institutions and various groups?
6. What is the effect of technology on international issues?
7. What is the effect of technology on environment?

2

History and Development of Science and Technology

2.1 DEVELOPMENT OF SCIENCE AND TECHNOLOGY ABROAD

The day man tried to use fire for his day to day life, it was the beginning of science and technology. The Egyptians knew how to work on coloured glass. Egyptians were expert in picture writing and Sumerians knew the art of writing on clay tablets. The Babylonians developed a solar calendar. The people of Nile valley knew the art of Geometry and Trigonometry. The people of Egypt gave decimal system of numeration. In 3000 BC Edwin Smith discovered a surgical papyrus wherein 48 surgical cases are listed. In 2000 BC Code of Hammurabi was developed by the Babylonians. The great Greek Thates made his searches about what the world was made up of.

Between 450 BC and 1450 AD science and technology remained practically stagnant Egypt. Greece and Rome were no longer centers of science. During this period Arabic numerical developed. Al-Battani was a great astronomer. AI Hazen wrote the Theory of Optics wherein he discussed reflection, optics, refraction and vision marks. During this period Roger Becon gave the idea that alchemy was basis of all science. On the whole, during this period no distinction was made between science and magic.

The invention of printing press did not let knowledge remain the monopoly of the few. Science and technology now took birth again. Due to long voyages geographical facts came to light.

Galileo perfected the telescope and gave new knowledge in the field of mechanics and refraction. He also helped in the perfection of compound microscope and balance scale. Robert Hooke discovered the wave motion of light.

Period between 1765–1815 AD has been called the period of revolution of science and technology. James Watt and Mathew Bolten have made steam power for industry. Improved machinery and machine tool making came to be adopted on regular lines. Joseph Black discovered Carbon Dioxide and Joseph Priestly discovered Oxygen. Stephen Grey gave new knowledge about conductors and non conductors in the transmission of electricity.

Seventeenth century inventors also produced some mechanical contrivances such as the pendulum clock, the watch, the barometer, the air pump, and nautical instruments which were generally used and manufactured on a large scale during the 18th century.

Between 1815 and 1895 AD science and technology made rapid progress. John Dalton gave his atomic theory and William Prout gave the theory of atomic weight and complex atoms. In the field of electricity the names of Faraday, Andre Maire, Ampere, and George Simon Ohm deserve special care. Thomas Alva Edison gave the incandescent electrical lamp.

After 1895, age of science and technology came all over the world. In 1895 Roentgen announced the discovery of X-rays and Einstein published his theory of relativity. Many antibiotic drugs were invented. Submarines, gasoline engines, jet planes, rockets, helicopters, ballistic missiles and air-to-air, as-well-as ground-to-air missile and many other war weapons were manufactured. It was also discovered that Sun's energy was the result of nuclear transformation. Agriculture became highly scientific. Atomic energy was found revolutionary.

The man has reached moon and is trying to find out the possibility of human life and settlement on the planet. Several successful attempts made by big powers to put the man in orbit have succeeded and good data is available about the surface of the moon. In addition remarkable progress has been made in the field of nuclear technology which is now being used in and for peace. The progress of science in such branches as telemetry, flights, medicine and surgery has been extremely good.

2.1.1 Development of Science and Technology in India

(i) ***Pre and Proto Historic Period:***

It appears that during this period stone industry seems to have existed in the south. The people made core tools, particularly fire hand axes. These have affinities with similar core tools in Africa, Western Europe and Southern England. It was around 6000 BC that man learnt the technology of growing food crops and thereafter learnt the art of making well polished stone implements which have been found all over, but mostly in the North-West and Deccan.

(ii) ***Harappan Civilisation:***

The history of science and technology in India begins clearly with the Indus Valley civilization, which is also known as *Harappan civilization*. From the excavations it appears that pottery at that time was wheel made. A number of designs and colour combinations show that during these days production techniques had been of highly standardized degree and the potters ware familiar with the art of glazing. It appears that metal smiths of those days possessed sound knowledge of annealing of the metal. These people also used a number of minerals for ornamental, cosmetic and medicinal purposes. Their town planning was excellent. They had developed excellent sanitation and public hygiene system.

The people of Indus Valley Civilisation knew spinning and weaving. They used burnt bricks and mortar for construction work and evolved a system of measurement and computational techniques indispensable both for architectural and commercial purposes. Many places of equipment, such as bullock carts have provided prototypes for subsequent generations of Indians craftsman.

(*iii*) ***The Vedic Period*:**

The people had mathematical knowledge of high order and had full idea of several geometrical forms, squares and rectangles. Even the Greeks owe some of their mathematical knowledge to Indian mathematicians.

Aryans knew the extraction of metals such as gold, silver and copper and were able to make bronze. They were familiar with the tanning of leather. Woolen textiles were dyed. They also knew the use of such agricultural equipments as plough, sickle, selvage etc. They were aware of the operations of plouging, sowing, thrashing and grinding.

(*iv*) ***The Post Vedic Period*:**

During this period Ayurveda developed which employed all drugs, therapeutic procedures and surgery both for treatment and elimination of diseases. It appears that glass technology also made good progress because glass objects have been found in many parts of the country. Similarly progress was also made in so far as implements made of iron were concerned.

(*v*) ***Glorious Age of Science and Technology*:**

The period up to 1200 BC has been characterized by some scholars as glorious period in the field of science and technology for making great headway in different branches and producing scientific and technical texts of high standard.

It was during this period that mathematicians like Aryabhatta I, Bhaskara I, Brahmagupta, Ayrabhatta II, Sridhara and Bhaskara II developed new methamatical concepts. Aryabhatta had knowledge about extraction of square, cube roots, triangles, circles, arithmetical progression, summation of series and fractions. He initiated rules for the solution of intermediate equation.

Bhaskra II developed a method called Chackravala (Cyclic Method) for rational integral solution of the intermediate equation. He also found a geometric proof of the Phythagorean Theorem.

In the field of astronomy a new class of literature called *Sidhanta* was produced which was composed by Lata. Aryabhatta-I also gave the theory of the rotation of the earth as also that of epicycles, Sripati, Mujata and Bhaskara in this period recognized subject of study in the university of Nalanda, Srivikramasila and Valbhi.

Metal working received considerable attention during this period. Iron pillar at Delhi and copper statue of Buddha at Sultanganj in Bihar speak highly of the metal smiths of the classical age. The metal smiths also demonstrated their skill in making new appliances and other utilitarian devices. In fact metal working came to be regarded as an important art. Bronze working is an elegant style and ceramic wares of sophisticated nature are other signs of good technological practice. Glassy pottery was made with mica dust in graceful designs. During this period the art of jewelry making with precious metals and stones also started flourishing in a big way. Agriculture received royal patronage.

(*vi*) ***Science and Technology during Medieval India*:**

Technological developments took place in such fields as paper, gunpowder, empanelling, glass and metal working. In addition, certain types of new plants also appeared. Most of the scientific works which appeared in the medieval period were, however, of the nature of commentary of earlier treatises. No new development in the field of mathematics took place during this period. Medicine also remained the art of the individual rather than the science of all.

(vii) ***Science during 19th Century:***

19th century has been described the most important period in the history of modern science in India. It was during this period that there was a gradual expansion of science in geology, metrology, archaeology etc. on the one hand and establishment of such organizations as the Geological Survey, Botanical Survey, Archaeological Survey etc. on the other. Attention was also paid to scientific and technical education. In 1857, universities of Bombay, Calcutta and Madras were started. In 1784 Asiatic Society in Calcutta was founded which aroused public interest in scientific education. In 1818 the great Trigonometrical Survey of India came into existence and in 1820 Agricultural Society of India was formed. In 1876 the Indian Society for the Cultivation of Science was brought into existence. Through the efforts of V. Ramaswamy lyer in 1901, Indian Mathematical Society was founded. Geographical Survey of India under-took systematic geological work. It studied Indian earthquakes. P. N. Bose of this survey was the first to publish the study of micro section as an aid to the study of rocks. P. N. Datta discovered the vast deposits of manganese ore in the Bhandara and Chindwara river valley.

In 1902 a department of Mines Inspection was started. In 1875 Indian Meteorological Department was founded. But the most important feature of the 19th century was the growth of colleges and institutions for the development of scientific and technical education. Encouragement was given to the study of engineering and medicine. J. C. Bose established that metals have muscles and plants have hearts. Bose institute Calcutta is an immortal monument to his memory. P. C. Ray laid the foundation of Indian Chemical School.

In 1876 Mahendra Lal Sircar established the Indian Association for the Cultivation of Science. C.V. Raman's name is associated with this association. In 1895 foundation of Solar Physics Laboratory at Calcutta was laid. It started functioning in the year 1900. In 1896 Plague Research Laboratory was founded at Bombay. In 1900 P. C. Roy discovered a number of rare Indian minerals.

In 1913 the government issued Educational Policy Resolution and subsequently several new universities came into existence. Several research institutes were also set up which gave a new impetus to the development of science and technology. These included Indian Institute of Science, Bangalore; Bose-Research Institute Calcutta; Harcourt Butler Technological Institute, Kanpur, Indian institute of Mines; Dhanbad. The Mason College of Civil Engineering Roorkee and Imperial Agriculture Research Institute, New Delhi.

The man has reached moon and is trying to find out the possibilities of human life and settlement on other planets. Several successful attempts made by big powers to put the man in orbit have succeeded and good data is available about the surface of the moon. In addition remarkable progress has been made in the field of nuclear technology which is now being used in war and for peace. The progress of science in such branches as telecommunication, space flights, medicines, surgery has been extremely good. The peace of progress is likely to be fast in the days to come. Some major technological inventions and discoveries are listed on Table 2.1.

TABLE 2.1 Some Major Technological Inventions And Discoveries

Invention	*Name of Inventor*	*Country*	*Year*
1. Adding Machine	Blaise Pascal	France	1642
2. Aeroplane	Wright Brothers	USA	1902
3. Arc Lamp	C.F. Bush	USA	1879
4. Automobile	Nicolas Cugnot	France	1769
5. Ball Point Pen	John J. Loud	USA	1888
6. Carburettor	Gottlieb Daimler	Germany	1876
7. Electric Bulb	Edison	USA	1879
8. Diesel Engine	Rudlof Diesel	Germany	1896
9. Disk Brake	Dr. F. Lanchester	England	1902
10. Dynamo	Hypolite Pixil	France	1832
11. Electronic Computer	J.P. Brainered	USA	1942
12. Electro Magnetic	William Sturgeon	England	1824
13. Electro Motor	Zenobe Gramwe	Belgium	1873
14. Elevator	Elishagous	USA	1852
15. Helicopter	'd' Ascanio	Italy	1930
16. Locomotive	Richard Travithick	England	1804
17. Machine Gun	Richard Getling	USA	1861
18. Microphone	Alexander Graham Bell	USA	1876
19. Microscope	Galilio and Jansen	Netherland	1590
20. Radar	Dr. Albert H. Taylor	USA	1922
21. Telescope	Hans Lippershey	Netherland	1608
22. Revolver	Colt	USA	1835
23. Steam Engine	Thomas Neutomen	England	1712
24. Telephone	Graham Bell	USA	1876
25. Television	J.L. Baird	Scotland	1926
26. Welder	Elisha Thomson	USA	1877
27. Watch	Bortholomew Manfredi	Italy	1462

2.1.2 Modern Indian Science and Technology

The tradition of Science and Technology (S & T) in India is over 5000 years old. A renaissance was witnessed in the first half of the 20th century. The S&T infrastructure has grown up from about Rs. 10 million at the time of independence in 1947 to Rs. 30 billion. Significant achievements have been made in the areas of nuclear and space science, electronics and defense. The government is committed to making S&T an integral part of the socio-economic development of the country. India has the third largest scientific and technical manpower in the world; 162 universities award 4,000 doctorates and 35,000 post-graduate degrees and the Council of Scientific and Industrial Research runs 40 research laboratories which have made some significant achievements.

In the field of missile launch technology, India is among the five top nations of the world.

Science and technology, however, is used as an effective instrument of growth and change. It is being brought into the mainstream of economic planning in the sectors of agriculture, industry and services. The country's resources are used to derive the maximum output for the benefit of society and improvement in the quality of life. About 85 per cent of the funds for S&T come directly or indirectly from the Government. The S&T infrastructure in the country accounts for more than one per cent of the GNP. Science and Technology in India is entering a new frontier.

Atomic Energy

The prime objective of India's nuclear energy program is the development and use of nuclear energy for peaceful purposes such as power generation, applications in agriculture, medicine, industry, research and other areas.

India is today recognized as one of the countries most advanced in nuclear technology including production of source materials. The country is self-reliant and has mastered the expertise covering the complete nuclear cycle - from exploration and mining to power generation and waste management. Accelerators and research and power reactors are now designed and built indigenously. The sophisticated variable energy cyclotron at Calcutta and a medium energy heavy ion accelerator 'pelletron' set up recently at Mumbai are national research facilities in the frontier areas of science.

As part of its program of peaceful uses of atomic energy, India has also embarked on a program of nuclear power generation. Currently eight nuclear stations are producing 8 billion kilowatt of electricity. Four more nuclear power stations are planned. The new nuclear reactors are designed in India. The peaceful nuclear program also includes producing radioisotopes for use in agriculture, medicine, industry and research.

Space

The Indian Space Research Organization (ISRO), under the Department of Space (DOS), is responsible for research, development and operationalisation of space systems in the areas of satellite communications, remote sensing for resource survey, environmental monitoring, meteorological services etc. DOS is also the nodal agency for the Physical Research Laboratory which conducts research in the areas of space science, and the National Remote Sensing Agency which deploys modern remote sensing techniques for natural resource surveys and provides operational services to user agencies. India is the only third world country to develop its own remote sensing satellite.

India joined a select group of six nations on October 15, 1994, when the Polar Satellite Launch Vehicle (PSLV) successfully accomplished its mission of placing the 800-Kg remote sensing satellite, IRS-P2, in the intended orbit. Earlier in May, the fourth developmental flight of the Augmented Satellite Launch Vehicle (ASLV) achieved its mission by placing the 113 Kg SROSS-C2 scientific satellite in a near-earth orbit. India is well on its way to developing a Geosynchronous Satellite Launch Vehicle (GSLV) capable of putting 2000 Kg satellites into space. The Indian Space Research Organization (ISRO) is currently trying to develop an indigenous cryogenic engine GSLV. A GSLV model has already been tested in wind tunnel.

The INSAT series of satellite launched earlier are performing well and provide vital services for telecommunications, television, meteorology, disaster warning and distress detection. The latest INSAT series will include new features like Ku-band transponders and mobile satellite services transponders. The remote-sensing satellites, launched in 1988 and 1991, have already become the mainstay of the natural resource management system of the country. The projected launch of advanced remote sensing satellite will not only enhances the scope of their application, but will also offer commercial service to other countries.

The Indian achievement in the application of space-based remote sensing technology has led a US company to enter into an agreement for marketing the data from Indian satellites globally.

India's progress in space technology has attracted worldwide attention and demand, with leasing agreements for marketing of IRS data and supply of space hardware and services. India also believes in cooperation in space with agencies all over the world. A high-level UN team selected India for setting up a UN Center for Space Science and Technology Education. India is on the threshold of achieving self-reliance in the launch capability. It will be a befitting tribute to the father of the Indian space program, *Dr. Vikram Sarabhai, whose 80th birth anniversary was observed in August 1996.*

Electronics

The Department of Electronics plays the promotional role for the development and use of electronics for socio-economic development. Many initiatives have been taken for a balanced growth of the electronics industry. The basic thrust has been towards a general rationalization of the licensing policy with an emphasis on promotion rather than regulation, besides achieving economy of scale with up to date technology. A multi-pronged approach has been evolved for result-oriented R&D with special emphasis on micro- electronics, telematics, and high performance computing and software development.

Application of electronics in areas such as agriculture, health and service sectors has also been receiving special attention. For upgrading the quality of indigenously manufactured products, a series of test and development centers and regional laboratories have been set up. These centers for electronic design and technology help small and medium electronics units. A number of R&D projects have been initiated to meet the growing requirements of the industry.

Biotechnology

India has been the forerunner among the developing countries in promoting multi-disciplinary activities in this area, recognizing the practically unlimited possibility of their applications in increasing agricultural and industrial production, and in improving human and animal life. The nucleus of research in this area is the National Biotechnology Board, constituted in 1982. A Department of Biotechnology was created in 1986.

Recently, the Biotechnology Consortium India Ltd. was set up. It will play the role of catalyst in bridging the gap between Research and Development, Industrial and Financial Institutions. Some of the new initiatives taken include developing techniques for gene mapping, conservation of biodiversity and bioindicators research, special biotechnology programs for the benefit of the scheduled castes and scheduled tribes and activities in the area of plantation crops.

The areas, which have been receiving attention, are cattle herd improvement through embryo transfer technology, in vitro propagation of disease resistant plant varieties for obtaining higher yields, and development of vaccines for various diseases.

Council of Scientific and Industrial Research (CSIR)

CSIR was established in 1942, and is today the premier institution for scientific and industrial research. It has a network of 40 laboratories, two cooperative industrial research institutions and more than 100 extension and field centers. The Council's research programs are directed towards effective utilization of the country's natural resources and development of new processes and products for economic progress. It is now playing a leading role in the fulfillment of the technology missions evolved by the Government.

2.2 INTER-RELATIONSHIP OF TECHNOLOGY AND SOCIO-ECONOMIC GROWTH

2.2.1 Society and Technological Changes

We are living in an age of science and technology. The success of science and technology can be felt in each and every part of our day to day life. It has influenced our age-old joint family system, marriage system; religious out-lock and resulted in urbanization.

2.2.2 Effect of Technology on Family System

In all societies technology has influenced family system. Some of the important influences of technology on family system are as under;

(*i*) It has created trend for single family system.

(*ii*) In factories, women get employment according to their ability and capability. Thus it has made women economically self-sufficient.

(*iii*) Sense of independence has been created in women.

(*iv*) There is less care for the children. When the parents go out for work they have less time for their children.

(*v*) Narrow mindedness has been reduced.

(*vi*) Family members get less time to discuss family problems.

2.2.3 Effect of Technology on Religion

Some of the important influences of technology on religion are as under:

(*i*) Hold of religion on the people is relaxing. Orthodoxy is considerably replaced by wider outlook.

(*ii*) Importance of priest has decreased.

(*iii*) Caste rigidity has been considerably reduced. People of different castes work in the same outlook.

(*iv*) Technology has made it possible to have quick means of transportation and communication. Similarly it also becomes possible to print religious books at a very quick speed. Due to these factors it becomes possible to spread religious ideas and dogmas very quickly.

(*v*) People try to understand religion on scientific basis. The very existence of God is being challenged.

2.2.4 Effect of Technology on Urbanization

In all the societies there is a trend towards urbanization. People are migrating from villages to the cities and also getting themselves settled there. In some societies the rate of urbanization is quick whereas in others it is slow.

Some of the important influences of technology on urbanization are as under:

(*i*) Due to technological advancement new factories come up which have the capacity to employ more people. It results in the migration of people from villages to the cities.

(*ii*) Because of migration of people towards cities, problem of accommodation, and transportation crop up.

(*iii*) Crime rate in the cities has increase.

(*iv*) The standard of living of the people has increased.

2.2.5 Effect of Technology on Women

(*i*) Technology has very affected and influenced women both in the rural and urban areas.

(*ii*) Women are considered equal to man in all professions.

(*iii*) Women have started getting not only education but higher education as well. This stage of female literacy has considerably gone up.

(*iv*) Problem of dowry has been reduced because of the awareness of women.

2.2.6 Women in the Rural Areas

Technology has affected and influenced women not only in the urban sector, but in rural sector as well, some of the important effects are:

(*i*) Women in rural area are keen on providing education and health to their children.

(*ii*) All superstitious beliefs have reduced.

2.3 EFFECT OF TECHNOLOGY ON RURAL LIFE

Some of the influences of technology on rural life are as under:

(*i*) Due to migration less people are available for agriculture.

(*ii*) New technologies are being used in agriculture to get good yield.

(*iii*) Rural population is faced with the serious problem of brain drainage. The village people who get education are drained out to the cities. Everybody wants white collar job.

(*iv*) Joint family system is now being replaced by single family system.

(*v*) The villagers are considerably awakened. They do not accept anything blindly without judging its merit. Religious dogmas, superiority of priests and authority of elders no longer go unquestioned and unchallenged. These are tested on the criteria of truthfulness and justification.

2.4 EFFECT OF TECHNOLOGY ON SOCIAL LIFE

Some of the influences of technology on social life are as under:

(*i*) Technology results in the migration of people from villages to the cities.

(*ii*) It has created problem of unemployment.

(*iii*) It has resulted in social tensions.

(*iv*) Technology makes life very complex.

(*v*) It has made the world materialistic.

(*vi*) It has made religion a private affair.
(*vii*) Grip of family has now loosened.
(*viii*) Urbanization is the outcome of technology.
(*ix*) Technology has given a great setback to the caste system.
(*x*) Divorce rate has gone up.
(*xi*) People forget to live in harmony with nature.

Highlights:

- Technology has influenced our joint family system, marriage system, and religious outlook etc.
- Technology has created a trend for single family system.
- Technology has made women economically self-sufficient.
- Technology has considerably reduced the caste rigidity.
- Technology has influenced women both in the urban and rural sectors in more than one way.
- Women in the rural areas are becoming quite conscious.
- Urbanization is the outcome of technology.

2.5 INTER-RELATIONSHIP OF SOCIO-ECONOMIC AND SCIENTIFIC FACTORS

Society is a complex web and one aspect of development both favorably and reversely affects the other. Thus social factors effect economic development. A society which is tradition ridden and where people waste a lot of money performing social ceremonies and for the purpose go in debt can not economically advance. A socially less developed society with literate population and having poor health and no shelter cannot be expected to materially contribute to economic development and advancement of the society.

On the other hand economic development very significantly effects social development. The people develop materialistic outlook and sole aim of their life becomes earning of wealth. Social position and prestige of the person is measured in terms of his wealth. There is then less regard for old customs and traditions and respectively others goes down. The hold of religion on society becomes very well.

Not only this but economic development brings with it changes in living standards. The people can afford education for their children, nutritive food and also some shifter for the family. They can afford good living and also have comforts of life.

Scientific factors effect as well as get affected by economic and social factors. An orthodox society is not prepared to expose its problems to other societies by getting these scientifically analysed. Such a society is usually opposed to disturbing status quo. It believes that any change is likely to do more harm than good to the society. Not only this but every scientific invention and discovery is condemned and its introduction in the social system resisted. But on its turn when scientific knowledge is accepted by such a society, many of its problems are solved. Because of scientific outlook many superstitions and unscientific and irrational customs and traditions are given up and in their place new approaches are made for the study of social problems and rational solutions refund. Society then makes rapid advancement.

In the same way scientific factors effect economic advancement. When a society adopts the latest scientific means of production, then that goes up and society can meet its own needs. This reduces dependence on other nations. There is economic prosperity and living standard of the people goes up. New machines are set up which provide more employment and help in the eradication of poverty and unemployment.

2.6 ROLE OF SCIENCE AND TECHNOLOGY IN DEVELOPMENT

Science and technology plays a vital role in the development of our country. There is no aspect of human life including agriculture, rural development, health, energy, industry, commerce, transport, communication and defense which is not affected by science and technology.

1. **Agricultural Sector.** India's major population depends on agricultural sector. We have attained self sufficiency in food production. It is because of the application of scientific methods to agricultural sectors. Few of fields in which science and technology has helped in the development of agricultural sector is given below:
   (*i*) India has wide variety of soil. Soil erosion is one of the major problems of the country. Science has helped in checking it and has helped the cultivators by suggesting methods of saving soil from degradation.
   (*ii*) Science has helped in using water in a optimum way by using equipments like sprinkler system etc.
   (*iii*) Big dams have been built by using modern technology. These dams are used for storing water in sufficient quantity so that the some can be utilized for power generation as well as agricultural purpose.
   (*iv*) The scientists in India have developed better quality and variety of seeds.
   (*v*) The scientists in India have developed multi cropping system which is becoming popular in the country. Organic manures have been developed with a view to improve physical properties of soil and increasing its moisture holding capacity.
   (*vi*) Science and technology in India has done a lot in having good agricultural produce. This has been possible because of revolution in genetic engineering in plant varieties. Increases has been in wheat, rice, pulses, oil seeds, sugarcane etc.
   (*vii*) Animal science research in India has helped in the productivity of live stock.
2. **Rural Development.** Few of fields in which science and technology has helped in the rural development are given below:
   (*i*) Education is spread on the rural side.
   (*ii*) Advanced medical facilities have reached rural areas.
   (*iii*) General awareness has come in rural population because of availability of T.V. programme.
   (*iv*) Because of the efforts of scientists and technologists better road facilities have been provided. Field areas have been linked with market areas and transportation of goods has been made easy and quick.
   (*v*) Under DPAP (Drought Prone affected Areas Programme) scientists and technologists have achieved good results. Under this programme they are trying to conserve, develop, harness land and other resources, including rainfall, with a view to improve and restore ecological balance. They are also trying to minimize adverse effects of droughts on crops

and animals and integrate all activities so that drought situation does not arise.

(*vi*) Rural drinking water problem is being solved under National Drinking Water Mission started in 1986.

3. **Health Development.** Health is one such area to which scientists and technologists all over the world have very significantly contributed and have made serious researches and found out new medicines and vaccines for controlling fatal diseases. They have developed life saving drugs and also new techniques for diagnosing of diseases. They have too developed sophisticated instruments for testing various organs of the body for finding out which parts of the human body ail and in what way can ailments be reduced or removed. Because of their efforts;
   (*i*) Mortality rate has come down.
   (*ii*) Infant mortality has been checked.
   (*iii*) Life expectancy has gone up.
   (*iv*) Heart surgery has become quite common.
   (*v*) Deaths due to malaria, TB, leprosy, etc. has been reduced.
4. **In Industry.** Industry is most vitally affected by the growth of science and technology. It is science and technology which brought industrial revolution in Europe and because of which European nations became advanced nations of the world. Scientists and technologists invent and discover new tools, machinery and equipment, which result in increased production. It is because of their efforts that mass production becomes easy and machine begins to replace man. Not only this but cost of production goes down and consumer can purchase goods at cheap rates.

   Industry then begins to employ more persons and the people begin to migrate from the rural to the urban areas. Process of urbanization thus starts. Pressure on land gets reduces. New townships begin to develop.

   Scientists help the industry by developing roads and waterways and as such it becomes easy for the industrialists to quickly and safely transport their goods from one place to the other. Research activity becomes very brisk and scientists and technologists engage themselves in the task of new inventions and discoveries.

   But at the same time because of effort of scientists and technologists industrial growth results in atmospheric pollution, because of the smoke of chimney and use of poisonous gases. Serious slum problem is created. Industrial growth creates unemployment of many categories of persons.
5. **Energy Development.** Energy plays an important role and has deep impact on the development of a nation. Natural gas is emerging as an important source of commercial energy. About 30% of the country's output of LPG comes from this source. Nuclear energy is being used for the development of transport, communication, defense etc. In order to use solar energy, S Energy centre has been set up. India is today one of the largest manufacturer of Solar energy equipments.

2.7 RELATIONSHIP OF INDUSTRY WITH SOCIAL SCIENCE

Relationship of industry with other social sciences is very close and that is the reason that industry looks towards social scientists for development and survival.

1. **Political Science and Industry.** Political science deals with the study of the state and government. Industry cannot act in isolation from the government. Government policies are bound to act very effectively both positively as well as negatively on the development and growth on the one hand, and retardation on the other. A political scientist makes political predictions which are used by the industry.
2. **Industry and Economics.** Economics is the study of monetary transactions and dealings. Industry is so closely related to economics that we have a separate and distinct branch of the study of economics known as economics of Industry. Economics tells us the principles of sound spending and earning of wealth. Where and how to spend, what cuts should be imposed and when; how to ensure that the money spent in industry is properly rewarded are both the concern of economists as well as that of the industry.
3. **Sociology and Industry.** Sociology is the study of society. Today every individual is faced with certain social problems. There is use to the society, family and also the organization in which an individual is working. An industrial worker is fun of the society as a whole. The sociology tries to identify the problems of industrial workers. Similarly one finds that no industry can function satisfactorily without properly understanding needs and necessities of the society on the one hand and its workers on the other. If the industry does not help the social problems of its workers it is bound to suffer. Thus sociology and industry are very closely linked with each other.
4. **Industry and History.** History deals with what had happened in the past. The study of the past is very important for the growth of present day industry. Each industrialist must know the past history of its industry, in his own country and as it functioned in other countries of the world. IT must have historical background and knowledge of the factors and events which influenced the course of its working. He must also know the causes which were responsible for the success of one industry and the failure of the other.
5. **Psychology and Industry.** Most of the problems, including those of lock-outs and strikes, arise only because the workers are not psychologically tackled. Thus industry and psychology have close relationship with each other. Industry also takes the help of psychology even at the time of making recruitment of workers at different levels.
6. **Public Administration and Industry.** In our industry today one of the serious prob'ems is that relationship between the employers and the employees is not very healthy. Public Administration helps us in maintaining these healthy relations by avoiding strikes lock-outs, checking waste of man hours and less productions etc. A good personnel and administrative system is real backbone of industry.

2.8 ROLE OF SCIENCE AND TECHNOLOGY IN DEVELOPMENT

Interrelationship between Socio-economic and Scientific Factors: Scientific factors and socio-economic conditions are closely related to each other. As a cause, economic factors tries to know the level and type of technological needs, ability of society to procure them, resources availability for research and development and efforts and capacity to build scientific and technological institutions. There is no doubt that capital intensive technique ensure higher rate of production quantitatively and qualitatively. But poor countries are not able to buy these due to a number of reasons. They lack scientific and technical manpower too. On account of this reason they do not have good science and technical colleges. They do not have sufficient resources for building industries and certainly not for research and development. Similarly social conditions such as values, attitudes and motivation are necessary for promotion of science. Traditional societies are quite backward having no scientific temperament. In India political leaders, intellectuals and practicing scientists have often stressed the need to develop scientific temperament among the people which they find to be lacking due to various historical, ethnic and ethical forces active in society for centuries together.

It is, therefore agreed that science and technology play a very crucial role in development of a country. The successive five year plans have stressed the need towards scientific and technological maturity. Per capita income and standard of living depend heavily on technology. Some significant contributions of science and technology towards development are stated below:

1. Increased productivity qualitatively and quantitatively in both agricultural and industrial sector.
2. Upgrading of existing industries to make them domestically attractive as well as export competitive.
3. Generation of energy.
4. Advancement and improvement in defence capabilities.
5. Exploration and extraction of new natural resources.
6. Conservation of resources.
7. Development and growth of human resources including food, cloth, health, sanitation, hygience, nutrition and water supply.
8. Reduction in regional imbalance.
9. Employment generation and removal of poverty.
10. Development of innovation and discoveries.

Pt. Nehru in India was perhaps the first personality to recognize the need of science and technology to economic development. Recently National Committee on Science and technology and Science Advisory Council submitted 'A Approach Paper to a Perspective Plan for future. Role of Science and Technology in six issues as identified are;

(*a*) Control of population,
(*b*) Need for increased food production,
(*c*) Protection of the environment,
(*d*) Improving human health,
(*e*) Control of hyper-urbanization and
(*f*) Reducing unemployment.

2.9 QUALITY OF LIFE (QOL) AND PHYSICAL QUALITY OF LIFE (PQOL)

There are two issues related to development of process-quantitative aspect and qualitative aspect. Balance between the two is necessary for development QoL is quality aspect whereas PQoL is called quantitative aspect. In simple language quantity aspect called physical quality of (PQoL) measures development in terms of numerical expression that is the schools, colleges, institutions, universities, higher learning centres; number of students acquiring education and teachers imparting education, pass percentage in the field of education. Similarly in the field of health it is measured as number of medical colleges, hospitals, dispensaries, and number of doctors after patients. In the field of production number of goods produced both household and producer goods; quantity goods sold, number of machines installed and amount of wages and salaries paid to workers. If the number of these things increase it means the country has progressed and developed and development in terms of physical quantity. The change in number from time to time is depicted by tables and graphs so that people understand about the development related to quantitative aspect.

2.9.1 Measurement of QoL

QoL is an inclusive concept which covers all aspects of living including material satisfaction of vital need as well as more transcendental (prior knowledge) aspects of life such as personal development, self-realisation and a healthy eco-system.

Quality of life give importance not to numbers alone as mentioned above but the quality of things so produced. So it is immaterial to think as how many institutions and educational centres have been opened, what matters is the quality of students coming out or passing out from these institutions. Are they intelligent, disciplined having mental abilities and aptitudes? Whether their faculties and talents have been fully developed? Can they after completion of their education inspire other with their personal qualities? Similarly in the field of health it does not matter how many doctors do we have but what matter is the qualities of doctors produced. So the quality of life is related to such things as increase in expectancy of life, better food, skill formation etc. It is basically concerned with the degree of progress made towards achieving minimum levels of literacy, expectation of life at birth and reduction in infant mortality. Literacy enables people to participate in the process of development and take full advantages of cultural progress. Reduction in infant mortality relates to environmental cleanliness and availability of clean water. *Data of PQoL show that GNP per capita is not always a good sign of PQoL. In this connection example of Sri Lanka and Kerala is often quoted. Both the places are poor in terms of per capita income but they achieved high levels of literacy, life expectancy and reduction in infant mortality rate.*

2.9.2 Determinants of QoL

Recently, **United Nations (1990) has identified 10 major concerns of quality of life and they are;**

(*a*) Health
(*b*) Education
(*c*) Culture and time use
(*d*) Housing
(*e*) Environment
(*f*) Public safety
(*g*) Employment and working life
(*h*) Income, consumption of wealth
(*i*) Social welfare and
(*j*) People participation.

It is concluded that development process needs both quality and quantity to be rationally balanced.

2.10 IMPROVEMENT OF QUALITY OF LIFE

The quality of life can be improved through;

- Individual Efforts
- Small Group Activities
- Government Intervention.

Quality of Life-Individual Efforts: Quality of life is a new tool related to goal of development. This means improving physical quality life of individuals living in society. Achieving maximum levels of literacy, expectation of life at birth and reduction of infant mortality along with increasing material gains that is increasing per capita income and by that enhancing standard of living are a few features of quality of life. The quality of life can be improved through individual efforts because man is essentially creative by nature. Since the beginning of human civilization he has been making continuous efforts to make his life better and more meaningful. The urge for improving the quality of life can easily be evinced (made to know) present since time immemorial. Moreover, man is gifted with the quality of cooperation with fellow being to achieve aims to better living. This is the reason that man has grown from primitive nomadic stage to the modern stage. Though development towards quality of life is not uniform yet struggle and efforts made by man do not stop. If all individual in their individual capacity make efforts in improving their standard by doing more work and producing more than total improvement and development of nation can be achieved. Individual efforts pertaining to quality of life results in (*a*) adequate nutrition, including ready access to uncontaminated water supply (*b*) adequate housing (*c*) adequate clothing (*d*) access to prevention and curative medicine (*e*) security of life and property and (*f*) gainful employment.

Quality of life-Small group Activities: Every man possesses quality of being living together, may be in small or big groups. This is so because man fellow beings to achieve aims of better living. Small group activities so formed influence quality of life. Quality of life is both economic connotation and non-economic connotation. In economic connotation we include better material gains whereas in non-economic connotation non-material gains such as high character, morality, higher education and a number of other qualitative aspects are included. Small group activities with regard to quality of life include (*a*) organisation of food security/production and distribution systems; supply of clean water (*b*) housing scheme (*c*) organisation of production/distribution system for clothing (*d*) organization of network of medical facilities (*e*) security and police (*f*) employment schemes (*g*) mass education covering areas of nutrition, water use, child care and health care etc.

Quality of Life- Government Intervention: This means State and Central Governments interference with working system. For improving the overall quality of life, government intervention is important on account of the following reasons.

Firstly, government arranges a number of infrastructural facilities which producing sector particularly private sector cannot manage and arrange due to financial and resource constraints.

Secondly, to check adulteration in products and exploitation the intervention reduces inequalities of income and ensures even distribution of national income GDP (Gross Domestic Product). This is done by using effective fiscal and monetary measures.

Thirdly, government through public expenditure improves the quality of life.

Fourthly, planning is a great asset with government for development of culture upon which overall nation's quality of life depends.

2.11 INDUSTRIAL REVOLUTION

The **Industrial Revolution** was a period in the late 18th and early 19th centuries when major changes in agriculture, manufacturing, and transportation had a profound effect on socioeconomic and cultural conditions in Britain and subsequently spread throughout Europe and North America and eventually the world, a process that continues as industrialization. The onset of the Industrial Revolution marked a major turning point in human social history, comparable to the invention of farming or the rise of the first city-states; almost every aspect of daily life and human society was eventually influenced in some way.

In the later part of the 1700s the manual labour based economy of the Kingdom of Great Britain began to be replaced by one dominated by industry and the manufacture of machinery. It started with the mechanisation of the textile industries, the development of iron-making techniques and the increased use of refined coal. Once started it spread. Trade expansion was enabled by the introduction of canals, improved roads and railways. The introduction of steam power (fuelled primarily by coal) and powered machinery (mainly in textile manufacturing) underpinned the dramatic increases in production capacity. The development of all-metal machine tools in the first two decades of the 19th century facilitated the manufacture of more production machines for manufacturing in other industries. The effects spread throughout Western Europe and North America during the 19th century, eventually affecting most of the world. The impact of this change on society was enormous.

The first Industrial Revolution merged into the Second Industrial Revolution around 1850, when technological and economic progress gained momentum with the development of steam-powered ships, railways, and later in the nineteenth century with the internal combustion engine and electrical power generation.

The period of time covered by the Industrial Revolution varies with different historians. Eric Hobsbawn held that it 'broke out' in the 1780s and was not fully felt until the 1830s or 1840s, while T. S. Ashton held that it occurred roughly between 1760 and 1830. Some twentieth century historians such as John Clapham and Nicholas Crafts have argued that the process of economic and social change took place gradually and the term *revolution* is not a true description of what took place. This is still a subject of debate amongst historians.

GDP per capita was broadly stable before the Industrial Revolution and the emergence of the modern capitalist economy. The industrial revolution began an era of per-capita economic growth in capitalist economies.

Regional GDP/capita changed very little for most of human history before the industrial revolution. (The empty areas mean no data, not very low levels. There is data for the years 1, 1000, 1500, 1600, 1700, 1820, 1900, and 2003)

The causes of the Industrial Revolution were complicated and remain a topic for debate, with some historians feeling the Revolution as an outgrowth of social and institutional changes brought by the end of feudalism in Britain after the English Civil War in the 17th century. As national border controls became more effective, the spread of disease was lessened, therefore preventing the epidemics common in previous times. The percentage of children who lived past infancy rose significantly, leading to a larger workforce. The Enclosure movement and the British Agricultural Revolution made food production more efficient and less labour-intensive, forcing the surplus population who could no longer find employment in agriculture into cottage industry, for example weaving, and in the longer term into the cities and the newly developed factories. The colonial expansion of the 17th century with the accompanying development of international trade, creation of financial markets and accumulation of capital are also cited as factors, as *is the scientific revolution of the 17th century.*

Technological innovation was the heart of the industrial revolution and the key enabling technology was the invention and improvement of the steam engine.

The historian, Lewis Mumford has proposed that the Industrial Revolution had its origins in the early middle ages, much earlier than most estimates. He explains that the model for standardised mass production was the printing press and that *the archetypal model for the [industrial era] was the clock*. He also cites the monastic emphasis on order and time-keeping, as well as the fact that Mediaeval cities had at their centre a church with bell ringing at regular intervals as being necessary precursors to a greater synchronisation necessary for later, more physical manifestations such as the steam engine.

The presence of a large domestic market should also be considered an important driver of the Industrial Revolution, particularly explaining why it occurred in Britain. In other nations, such as France, markets were split up by local regions, which often imposed tolls and tariffs on goods traded amongst them.

Governments' grant of limited monopolies to inventors under a developing patent system (the Statute of Monopolies 1623) is considered an influential factor. The effects of patents, both good and ill, on the development of industrialisation are clearly illustrated in the history of the steam engine, the key enabling technology. In return for publicly revealing the workings of an invention the patent system rewards inventors by allowing, *e.g*, James Watt to monopolise the production of the first steam engines, thereby enabling inventors and increasing the pace of technological development. However, monopolies bring with them their own inefficiencies which may counterbalance, or even overbalance, the beneficial effects of publicizing ingenuity and rewarding inventors. Watt's monopoly may have prevented other inventors, such as Richard Trevithick, William Murdoch or Jonathan Hornblower, from introducing improved steam engines thereby retarding the industrial revolution by up to 20 years.

"What caused the Industrial Revolution?" remains one of the most important unanswered questions in social science.

2.11.1 Causes for Occurrence in Europe

European 17th century colonial expansion, international trade, and creation of financial markets produced a new legal and financial environment, one which supported and enabled 18th century industrial growth.

One question of active interest to historians is why the industrial revolution occurred in Europe and not in other parts of the world in the 18th century, particularly China, India, and the Middle East, or at other times like in Classical Antiquity or the Middle Ages. Numerous factors have been suggested, including ecology, government, and culture.

Benjamin Elman argues that China was in a high level equilibrium trap in which the non-industrial methods were efficient enough to prevent use of industrial methods with high costs of capital. Kenneth Pomeranz, in the *Great Divergence*, argues that Europe and China were remarkably similar in 1700, and that the crucial differences which created the Industrial Revolution in Europe were sources of coal near manufacturing centres, and raw materials such as food and wood from the New World, which allowed Europe to expand economically in a way that China could not.

However, most historians contest the assertion that Europe and China were roughly equal because modern estimates of per capita income on Western Europe in the late 18th century are of roughly 1,500 dollars in purchasing power parity (and Britain had a per capita income of nearly 2,000 dollars) whereas China, by comparison, had only 450 dollars. Also, the average interest rate was about 5% in Britain and over 30% in China, which illustrates how capital was much more abundant in Britain; capital that was available for investment.

Some historians such as David Landes and Max Weber credit the different belief systems in China and Europe with dictating where the revolution occurred. The religion and beliefs of Europe were largely products of Judaeo-Christianity, and Greek thought. Conversely, Chinese society was founded on men like Confucius, Mencius, Han Feizi (Legalism), Lao Tzu (Taoism), and Buddha (Buddhism). The key difference between these belief systems was that those from Europe focused on the individual, while Chinese beliefs centred around relationships between people. The family unit was more important than the individual for the large majority of Chinese history, and this may have played a role in why the Industrial Revolution took much longer to occur in China. There was the additional difference as to whether people looked backwards to a reputedly glorious past for answers to their questions or looked hopefully to the future. Furthermore, Western European peoples had experienced the Renaissance, Reformation and Enlightenment; other parts of the world had not had a similar intellectual breakout, a condition that holds true even into the 21st century.

Regarding India, the Marxist historian Rajani Palme Dutt said: *The capital to finance the Industrial Revolution in India instead went into financing the Industrial Revolution in England.* In contrast to China, India was split up into many competing kingdoms, with the three major ones being the Marathas, Sikhs and the Mughals. In addition, the economy was highly dependent on two sectors—agriculture of subsistence and cotton, and technical innovation was non-existent. The vast amounts of wealth were stored away in palace treasuries, and as such, were easily moved to Britain.

The debate about the start of the Industrial Revolution also concerns the massive lead that Great Britain had over other countries. Some have stressed the importance of natural or financial resources that Britain received from its many overseas colonies or that profits from the British slave trade between Africa and the Caribbean helped fuel industrial investment. It has been pointed out, however, that slavery provided only 5% of the British national income during the years of the Industrial Revolution.

Alternatively, the greater liberalisation of trade from a large merchant base may have allowed Britain to produce and use emerging scientific and technological developments more effectively than countries with stronger monarchies, particularly China and Russia. Britain emerged from the Napoleonic Wars as the only European nation not ravaged by financial plunder and economic collapse, and possessing the only merchant fleet of any useful size (European merchant fleets having been destroyed during the war by the Royal Navy). Britain's extensive exporting cottage industries also ensured markets were already available for many early forms of manufactured goods. The conflict resulted in most British warfare being conducted overseas, reducing the devastating effects of territorial conquest that affected much of Europe. This was further aided by Britain's geographical position — an island separated from the rest of mainland Europe.

Another theory is that Britain was able to succeed in the Industrial Revolution due to the availability of key resources it possessed. It had a dense population for its small geographical size. Enclosure of common land and the related Agricultural Revolution made a supply of this labour readily available. There was also a local coincidence of natural resources in the North of England, the English Midlands, South Wales and the Scottish Lowlands. Local supplies of coal, iron, lead, copper, tin, limestone and water power, resulted in excellent conditions for the development and expansion of industry. Also, the damp, mild weather conditions of the North West of England provided ideal conditions for the spinning of cotton, providing a natural starting point for the birth of the textiles industry.

The stable political situation in Britain from around 1688, and British society's greater receptiveness to change (compared with other European countries) can also be said to be factors favouring the Industrial Revolution. In large part due to the Enclosure movement, the peasantry was destroyed as significant source of resistance to industrialisation, and the landed upper classes developed commercial interests that made them pioneers in removing obstacles to the growth of capitalism.

2.11.2 Protestant Work Ethic

Another theory is that the British advance was due to the presence of an entrepreneurial class which believed in progress, technology and hard work. The existence of this class is often linked to the Protestant work ethic and the particular status of dissenting Protestant sects, such as the Quakers, Baptists and Presbyterians that had flourished with the English Civil War. *Reinforcement of confidence in the rule of law*, which followed establishment of the prototype of constitutional monarchy in Britain in the Glorious Revolution of 1688, and the emergence of a stable financial market there based on the management of the national debt by the Bank of England, contributed to the capacity for, and interest in, private financial investment in industrial ventures.

Dissenters found themselves barred or discouraged from almost all public offices, as well as education at England's only two universities at the time (although dissenters were still free to study at Scotland's four universities). When the restoration of the monarchy took place and membership in the official Anglican church became mandatory due to the Test Act, they thereupon became active in banking, manufacturing and education. The Unitarians, in particular, were very involved in education, by running Dissenting Academies, where, in contrast to the **universities of Oxford** and **Cambridge** and schools such as **Eton** and **Harrow**, much attention was given to mathematics and the sciences —areas of scholarship vital to the development of manufacturing technologies.

Historians sometimes consider this social factor to be extremely important, along with the nature of the national economies involved. While members of these sects were excluded from certain circles of the government, they were considered fellow Protestants, to a limited extent, by many in the middle class, such as traditional financiers or other businessmen. Given this relative tolerance and the supply of capital, the natural outlet for the more enterprising members of these sects would be to seek new opportunities in the technologies created in the wake of the Scientific revolution of the 17th century.

2.11.3 Innovations

The commencement of the Industrial Revolution is closely linked to a small number of innovations, made in the second half of the 18th century:

- **Textiles** - Cotton spinning using Richard Arkwright's water frame, James Hargreaves's Spinning Jenny, and Samuel Crompton's Spinning Mule (a combination of the Spinning Jenny and the Water Frame). This was patented in 1769 and so came out of patent in 1783. The end of the patent was rapidly followed by the erection of many cotton mills. Similar technology was subsequently applied to spinning worsted yarn for various textiles and flax for linen.
- **Steam power** - The improved steam engine invented by James Watt was initially mainly used for pumping out mines, but from the 1780s was applied to power machines. This enabled rapid development of efficient semi-automated factories on a previously unimaginable scale in places where waterpower was not available.
- **Iron founding** - In the Iron industry, coke was finally applied to all stages of iron smelting, replacing charcoal. This had been achieved much earlier for lead and copper as well as for

producing pig iron in a blast furnace, but the second stage in the production of bar iron depended on the use of potting and stamping (for which a patent expired in 1786) or puddling (patented by Henry Cort in 1783 and 1784.)

These represent three 'leading sectors', in which there were key innovations, which allowed the economic take off by which the Industrial Revolution is usually defined. This is not to belittle many other inventions, particularly in the textile industry. Without some earlier ones, such as spinning jenny and flying shuttle in the textile industry and the smelting of pig iron with coke, these achievements might have been impossible. Later inventions such as the power loom and Richard Trevithick's high pressure steam engine were also important in the growing industrialisation of Britain. The application of steam engines to powering cotton mills and ironworks enabled these to be built in places that were most convenient because other resources were available, rather than where there was water to power a mill.

In the textile sector, such mills became the model for the organisation of human labour in factories, epitomised by Cottonopolis, the name given to the vast collection of cotton mills, factories and administration offices based in Manchester. The assembly line system greatly improved efficiency, both in this and other industries. With a series of men trained to do a single task on a product, then having it moved along to the next worker, the number of finished goods also rose significantly.

2.11.4 Transfer of Knowledge

Knowledge of new innovation was spread by several means. Workers who were trained in the technique might move to another employer or might be poached. A common method was for someone to make a study tour, gathering information where he could. During the whole of the Industrial Revolution and for the century before, all European countries and America engaged in study-touring; some nations, like Sweden and France, even trained civil servants or technicians to undertake it as a matter of state policy. In other countries, notably Britain and America, this practice was carried out by individual manufacturers anxious to improve their own methods. Study tours were common then, as now, as was the keeping of travel diaries. Records made by industrialists and technicians of the period are an incomparable source of information about their methods.

Another means for the spread of innovation was by the network of informal philosophical societies, like the Lunar Society of Birmingham, in which members met to discuss 'natural philosophy' (*i.e.* science) and often its application to manufacturing. The Lunar Society flourished from 1765 to 1809, and it has been said of them, They were, if you like, *the revolutionary committee of that most far reaching of all the eighteenth century revolutions, the Industrial Revolution* other such societies published volumes of proceedings and transactions. For example, the London-based Royal Society of Arts published an illustrated volume of new inventions, as well as papers about them in its annual *Transactions*.

There were publications describing technology. Encyclopedias such as Harris's *Lexicon technicum* (1704) and Dr Abraham Rees's *Cyclopaedia* (1802–1819) contain much of value. *Cyclopaedia* contains an enormous amount of information about the science and technology of the first half of the Industrial Revolution, very well illustrated by fine engravings. Foreign printed sources such as the *Descriptions des Arts et Métiers* and Diderot's *Encyclopédie* explained foreign methods with fine engraved plates.

Periodical publications about manufacturing and technology began to appear in the last decade of the 18th century, and many regularly included notice of the latest patents. Foreign periodicals, such as the Annales des Mines, published accounts of travels made by French engineers who observed British methods on study tours.

2.11.5 Social Effects

In terms of social structure, the Industrial Revolution witnessed the triumph of a middle class of industrialists and businessmen over a landed class of nobility and gentry.

Ordinary working people found increased opportunities for employment in the new mills and factories, but these were often under strict working conditions with long hours of labour dominated by a pace set by machines. However, harsh working conditions were prevalent long before the industrial revolution took place as well. Pre-industrial society was very static and often cruel—child labour, dirty living conditions and long working hours were just as prevalent before the Industrial Revolution.

Industrialization led to the creation of the factory. Arguably the first was John Lombe's water-powered silk mill at Derby, operational by 1721. However, the rise of the factory came somewhat later when cotton spinning was mechanized.

The factory system was largely responsible for the rise of the modern city, as large numbers of workers migrated into the cities in search of employment in the factories. Nowhere was this better illustrated than the mills and associated industries of Manchester, nicknamed "Cottonopolis", and arguably the world's first industrial city. For much of the 19th century, production was done in small mills, which were typically water-powered and built to serve local needs. Later each factory would have its own steam engine and a tall chimney to give an efficient draft through its boiler.

The transition to industrialisation was not without difficulty. For example, a group of English workers known as Luddites formed to protest against industrialisation and sometimes sabotaged factories.

In other industries the transition to factory production was not so divisive. Some industrialists themselves tried to improve factory and living conditions for their workers. One of the earliest such reformers was Robert Owen, known for his pioneering efforts in improving conditions for workers at the New Lanark mills, and often regarded as one of the key thinkers of the early socialist movement.

By 1746, an integrated brass mill was working at Warmley near Bristol. Raw material went in at one end, was smelted into brass and was turned into pans, pins, wire, and other goods. Housing was provided for workers on site. Josiah Wedgwood and Matthew Boulton were other prominent early industrialists, who employed the factory system.

2.11.6 Child Labour

The Industrial Revolution led to a population increase, but the chance of surviving childhood didn't improve throughout the industrial revolution. There was still limited opportunity for education, and children were expected to work. Employers could pay a child less than an adult even though their productivity was comparable; there was no need for strength to operate an industrial machine, and since the industrial system was completely new there were no experienced adult labourers. This made child labour the labour of choice for manufacturing in the early phases of the industrial revolution.

Child labour had existed before the Industrial Revolution, but with the increase in population and education it became more visible. Before the passing of laws protecting children, many were forced to work in terrible conditions for much lower pay than their elders.

Reports were written detailing some of the abuses, particularly in the coal mines and textile factories and these helped to popularise the children's plight. The public outcry, especially among the upper and middle classes, helped stir change in the young workers' welfare.

Politicians and the government tried to limit child labour by law, but factory owners resisted; some felt that they were aiding the poor by giving their children money to buy food to avoid starvation, and others simply welcomed the cheap labour. **In 1833 and 1844, the first general laws against child labour, the Factory Acts, were passed in England:** Children younger than nine were not allowed to work, children were not permitted to work at night, and the work day of youth under the age of 18 was limited to twelve hours. Factory inspectors supervised the execution of the law. About ten years later, the employment of children and women in mining was forbidden. These laws decreased the number of child labourers; however, child labour remained in Europe up to the 20th century.

Living conditions during the Industrial Revolution varied from the splendour of the homes of the owners to the squalor of the lives of the workers. Cliffe Castle, Keighley, is a good example of how the newly rich chose to live. This is a large home modelled loosely on a castle with towers and garden walls. The home is very large and was surrounded by a massive garden, the Cliffe Castle is now open to the public as a museum.

Poor people lived in very small houses in cramped streets. These homes would share toilet facilities, have open sewers and would be at risk of damp. Disease was spread through a contaminated water supply. Conditions did improve during the 19th century as public health acts were introduced covering things such as sewage, hygiene and making some boundaries upon the construction of homes. Not everybody lived in homes like these. The Industrial Revolution created a larger middle class of professionals such as lawyers and doctors. The conditions for the poor improved over the course of the 19th century because of government and local plans which led to cities becoming cleaner places, but life had not been easy for the poor before industrialisation. However, as a result of the Revolution, huge numbers of the working class died due to disease spreading through the cramped living conditions. Chest diseases from the mines, cholera from polluted water and typhoid were also extremely common, as was smallpox. Accidents in factories with child and female workers were regular. Dickens' novels perhaps best illustrate this; even some government officials were horrified by what they saw. Strikes and riots by workers were also relatively common.

2.11.7 Luddites

The rapid industrialisation of the English economy cost many craft workers their jobs. The textile industry in particular industrialised early, and many weavers found themselves suddenly unemployed since they could no longer compete with machines which only required relatively limited (and unskilled) labour to produce more cloth than a single weaver. Many such unemployed workers, weavers and others, turned their animosity towards the machines that had taken their jobs and began destroying factories and machinery. These attackers became known as Luddites, supposedly followers of Ned Ludd, a folklore figure. The first attacks of the Luddite movement began in 1811. The Luddites rapidly gained popularity, and the British government had to take drastic measures to protect industry.

2.11.8 Organisation of Labour

The Industrial Revolution concentrated labour into mills, factories and mines, thus facilitating the organisation of *combinations* or trade unions to help advance the interests of working people. The power of a union could demand better terms by withdrawing all labour and causing a consequent cessation of production. Employers had to decide between giving in to the union demands at a cost to themselves or suffer the cost of the lost production. Skilled workers were hard to replace, and these were the first groups to successfully advance their conditions through this kind of bargaining.

The main method the unions used to effect change was strike action. Strikes were painful events for both sides, the unions and the management. In England, the Combination Act forbade workers to form any kind of trade union from 1799 until it was repeated in 1824. Even after this, unions were still severely restricted.

In the 1830s and 1840s the Chartist movement was the first large scale organised working class political movement which campaigned for political equality and social justice. Its *Charter* of reforms received over three million signatures but was rejected by Parliament without consideration.

Working people also formed friendly societies and co-operative societies as mutual support groups against times of economic hardship. Enlighted industrialists, such as Robert Owen also supported these organisations to improve the conditions of the working class.

Unions slowly overcame the legal restrictions on the right to strike. In 1842, a General Strike involving cotton workers and colliers was organised through the Chartist movement which stopped production across Great Britain.

Eventually effective political organisation for working people was achieved through the trades unions who, after the extensions of the franchise in 1867 and 1885, began to support socialist political parties that later merged to became the British Labour Party.

2.11.9 Other Effects

The application of steam power to the industrial processes of printing supported a massive expansion of newspaper and popular book publishing, which reinforced rising literacy and demands for mass political participation.

During the Industrial Revolution, the life expectancy of children increased dramatically. The percentage of the children born in London who died before the age of five decreased from 74.5% in 1730 – 1749 to 31.8% in 1810 – 1829. Besides, there was a significant increase in worker wages during the period 1813-1913.

2.12 KEY FEATURES OF INDUSTRIAL REVOLUTION

1. The Industrial Revolution came first to England
2. Machinery first came to cotton and then to woolen industry
3. The development in Silk industry was due to the influx of the Huguenot refugees.
4. The use of coke instead of charcoal for smelting iron resulted in increase in production tremendously.
5. The Factory System progressed slowly
6. The causes that assisted the Industrial Revolution in England were the scarcity of population, availability of coal and iron, availability of plenty of capital and a good banking system political security, her geographic position, free population and extensive markets which she had at home and abroad.

Results of the Industrial Revolution

It increased the national wealth. It raised the standard of living. It assisted the development of trade unions and brought about the Parliamentary Reforms of the 19th Century. But it destroyed the cottage industries and brought about the differentiation deplorable. There was over-crowding in factory towns. These led to the all-round decay of the labour. Labour was also displaced by machinery. It also brought about the conflict between labour and capital.

Impacts of Industrial Revolution in Europe

1. Industrial Revolution changed the social structure of society.
2. The people moved from villages to towns and cities to work in factories.
3. The machines capable of huge outputs made small hand weavers redundant.
4. There was unprecedented growth and profit that occurred during industrial revolution.
5. The relentless pursuit of money caused the ultimate breakdown of the family unit.
6. The children received very little education, had stunted growth and were sickly.
7. The working families were forced to live in slums with little sanitation and infant mortality sky rocketed.

Impacts of Industrial Revolution on Rest of the World

- The quick industrialization across Europe during the 19th century led to a great increase in goods produced as well as a demand for raw materials.
- This demand, coupled with increased nationalist pride, led nations to seek colonies abroad in which to produce and trade goods.
- The main expansion for the European colonial powers occurred in Africa. By 1914, the entire continent with the exception of Liberia and Abyssinia were controlled by European nations.
- England also took control of India and Hong Kong during this period of expansion. By the beginning of World War I, England had an empire which stretched across every continent in the world. Vast amounts of natural resources were extracted from these colonies, which aided the British industrial effort but left many of the nations bankrupt.
- In short, industrialization in Europe had far reaching consequences for the rest of the world. While it made Britain the ultimate power for over a century, it can be argued that its rule over the world caused conflict and internal strife which continues to this day.

QUESTIONS

1. State relation of industry with social science.
2. What is the role of science and technology in development?
3. What is the inter-relationship between social, economic and scientific factors?
4. State inter-relationship of technology and Socio-economic growth.
5. Give a brief historical account of development of science and technology in India.
6. List some of the technical innovations that led to Industrial revolution.
7. Give negative after effects of Industrial revolution.
8. Discuss Industrial revolution and its effects.

3

HUMAN CRITIQUES OF TECHNOLOGY

3.1 INTRODUCTION

Critique of technology is a theory which criticizes technology for alleged negative impact under capitalist conditions. Proponents of this theory argue that under capitalist conditions technology is or becomes a means of domination, control and exploitation, or more generally something which threatens the survival of humanity. In a wider sense a skeptical attitude towards technology which is not necessarily fully theoretically developed can also be related to the critique of technology.

In Europe skeptical attitudes towards technology became more prominent in the 1970s, *e.g.*, in the antinuclear movements. In the eyes of some critics, the Chernobyl disaster in 1986 seemed to confirm the impossibility of controlling large scale technology. The Internet euphoria of the 1990s favored more positive attitudes toward technology. Nevertheless fears about the increased possibilities of technological surveillance are still widespread.

Prominent authors elaborating a critique of technology are, *e.g., Gunther Anders, Jacques Ellul and Lewis Mumford*. In a wider sense some writings of Martin Heidegger and the critical history of technology (David F. Noble) are also part of the critique of technology.

The critique of technology is a variant of the critique of progress which dates back at least to Rousseau.

3.2 CRITIQUE OF TECHNOLOGY

The critique of technology overlaps with the philosophy of technology but whereas the latter tries to establish itself as an academic discipline the critique of technology is basically a political project, not limited to academia. It features prominently in *neomarxism (Herbert Marcuse), ecofeminism (Vandana Shiva) and in post development (Ivan Illich)*.

3.3 LEWIS MUMFORD AND HIS WORKS

Mumford was born in Flushing, New York and graduated from Stuyvesant High School in 1912. He studied at the City College of New York and the New School for Social Research, but became ill with tuberculosis and never finished his degree. In 1919 he became associate editor of *The Dial*, an influential modernist literary journal. He later worked for *The New Yorker* where he wrote architectural criticism and commentary on urban issues.

Mumford's earliest books in the field of literary criticism have had a lasting impact on contemporary American literary criticism. *The Golden Day* contributed to a resurgence in scholarly research on the work of 1850's American transcendentalist authors an *Herman Melville: A Study of His Life and Vision* effectively launched a revival in the study of the work of Herman Melville. Soon after, with the book *The Brown Decades*, he began to establish himself as an authority in US architecture and urban life, which he interpreted in a social context.

In his early writings on urban life, Mumford was optimistic about human abilities and wrote that the human race would use electricity and mass communication to built a better world for all humankind. He would later take a more pessimistic stance. His early architectural criticism also helped to bring wider public recognition to the work of Henry Hobson Richardson, Louis Sullivan and Frank Lloyd Wright.

3.3.1 Ideas

Mumford believed that what defined humanity (what set human beings apart from other animals) was not primarily our use of tools (technology) but our use of language (symbols). He was convinced that the sharing of information and ideas amongst participants of primitive societies was completely natural to early humanity, and had obviously been the foundation of society as it became more sophisticated and complex. He had hopes for a continuation of this process of information "pooling" in the world as humanity moved into the future.

Mumford's choice of the word "technics" throughout his work was deliberate. For Mumford, technology is one part of technics. Technics refers to the interplay of a social milieu and technological innovation-the "wishes, habits, ideas, goals" as well as "industrial processes." As Mumford writes at the beginning of *Technics and Civilization*, "other civilizations reached a high degree of technical proficiency without, apparently, being profoundly influenced by the methods and aims of technics."

3.3.2 Megatechnics

In *The Myth of the Machine: teachnics and human development* (1967), Mumford criticizes the modern trend of technology, which emphasizes constant, unrestricted expansion, production, and replacement. He explains that these goals work against technical perfection, durability, social efficiency, and overall human satisfaction. Modern technology-which he calls 'megatechnics'-evades producing lasting, quality products by using devices such as consumer credit, installment buying, non-functioning and defective designs, built-in fragility, and frequent superficial "fashion" changes. "Without constant enticement by advertising", he explains, "production would slow down and level off to normal replacement demand. Otherwise many products could reach a plateau of efficient design which would call for only minimal changes from year to year."

He uses his own refrigerator as an example, explaining that it "has been in service for nineteen years, with only a single minor repair: an admirable job. Both automatic refrigerators for daily use and deepfreeze preservation are inventions of permanent value... if biotechnic criteria were heeded, rather than those of market analysts and fashion experts, an equally good product might come forth from Detroit, with an equally long prospect of continued use."

3.3.3 Biotechnics

Mumford describes an organic model of technology, or *biotechnics,* as a contrast to megatechnics. Organic systems direct themselves to "qualitative richness, amplitude, spaciousness, and freedom from quantitative pressures and crowding. Self-regulation, self-correction, and self-propulsion are as much an integral property of organisms as nutrition, reproduction, growth, and repair." Biotechnics models life in seeking balance, wholeness, and completeness.

3.3.4 Polytechnics Versus Monotechnics

A key idea, introduced in *Technics* and *Civilization* (1934) was that technology was twofold:

- *Polytechnic,* which enlists many different modes of technology, providing a complex framework to solve human problems.
- *Monotechnic* which is technology only for its own sake, which oppresses humanity as it moves along its own trajectory.

Mumford commonly criticized modern America's transportation networks as being 'monotechnic' in their reliance on cars. Automobiles become obstacles for other modes of transportation, such as walking, bicycle and public transit, because the roads they use consume so much space and are such a danger to people. Mumford explains that the thousands of maimed and dead each year as a result of automobile accidents are a "ritual sacrifice" the American society makes because of its extreme reliance on highway transport.

3.3.5 Megamachines

Mumford also refers to large hierarchical organizations as *megamachines*-a machine using humans as its components. These organizations comprise Mumford's stage theory of civilization. The most recent Megamachine manifests itself, according to Mumford, in modern technocratic nuclear powers-Mumford used the examples of the Soviet and US power complexes represented by the Kremlin and the Pentagon, respectively. The builders of the Pyramids, the Roman Empire and the armies of the World Wars are prior examples.

3.3.6 Features of Megamachines

He explains that meticulous attention to accounting and standardization, and elevation of military leaders to divine status are spontaneous features of megamachines throughout the history. He cites such examples as the repetitive nature of Egyptian paintings which feature enlarged Pharaohs and public display of enlarged portraits of dictators such as Mao Tse Tung and Joseph Stalin. He also cites the overwhelming prevalence of quantitative accounting records among surviving historical fragments from ancient Egypt to Nazi Germany.

Necessary to the construction of these megamachines is an enormous bureaucracy of human which act as "servo—units", working without ethical involvement. Technological improvements such as remote control by satellite or radio, instant global communication, and assembly line organizations dampen psychological barriers against the end result of their actions, according to Mumford. An example which he uses is that of Adolf Eichmann, the Nazi official who conducted logistics behind the Holocaust. Mumford collectively refers to people willing to carry out placidly the extreme goals of these megamachines as "*Eichmanns*".

3.3.7 The Clock as Herald of the Industrial Revolution

One of the better-known studies of Mumford is of the way the mechanical clock was developed by monks in the Middle Ages and subsequently adopted by the rest of society. He viewed this device as the key invention of the whole Industrial Revolution, contrary to the common view of the steam engine holding the prime position, writing: "*The clock is a piece of machinery whose 'product' is seconds and minutes*".

3.3.8 Urban Civilization

In his influential book *The City in History*, which won the National Book Award, Mumford explores the development of urban civilizations. Harshly critical of urban sprawl, Mumford argues that the structure of modern cities is partially responsible for many social problems seen in western society. While pessimistic in tone, Mumford argues that urban planning should emphasize an organic relationship between people and their living spaces.

Mumford uses the example of the medieval city as the basis for the "ideal city", and claims that the modern city is too close to the Roman city (the sprawling megalopolis) which ended in collapse; if the modern city carries on in the same vein, Mumford argues, then it will meet the same fate as the Roman city.

Mumford wrote critically of urban culture believing the city is "a product of earth... a fact of nature... man's method of expression". Further Mumford recognized the crises facing urban culture, distrusting of the growing finance industry, political structures, fearful that a local community culture was not being fostered by these institutions. Mumford feared 'metropolitan finance', urbanization, politics and alienation.

3.4 JACQUES ELLUL AND HIS WORKS

Jacques Ellul (January 6, 1912-May 19, 1994) was a French philosopher, Law Professor, sociologist, theologian and Christian anarchist. He wrote several books about the "technological society", and about Christianity and politics, such as *Anarchy and Christianity* (1991)—arguing that anarchism and Christianity are socially following the same goal.

He is a philosopher who approached technology from a deterministic viewpoint. Ellul, Professor at the University of Bordeaux, authored some 40 books and hundreds of articles over his lifetime. The dominant theme of his writings has been the threat to human freedom and Christian faith created by modern technology. His constant concern has been the emergence of a "technological tyranny" over humanity. As a philosopher and theologian, he further explored the religiosity of the technological society.

Life

Ellul was born in Bordeaux, France, and was educated at the universities of Bordeaux and Paris. In World War II he was a leader in the French resistance. For his efforts to save Jews he was awarded the title *Righteous among the Nations* by Yad Vashem </wiki/Yad_Vashem> in 1981. He was also prominent in the worldwide Ecumenical movement, although he later became sharply critical of the movement for what he felt were indiscriminate endorsements of political establishments, primarily of the Left. However, he was no friendlier in his assessment of those of the Right, either; he fashioned an explicitly anti-political stance as an alternative to both.

Ellul studied Marx and became a prolific exegete of his theories. Ellul became a Christian at age 22. The influence of these ideologies, which are generally considered to be in opposition to one another, has alternately earned him devoted followers and vicious enemies. In large measure and especially in those of his books concerned with theological matters, Ellul restates the viewpoints held by the great Protestant theologian Karl Barth, who was a leader of the resistance against the German state church in World War II. Barth's polar dialectic of the Word of God, in which the gospel both judges and renews the world, helped to shape Ellul's theological perspective. But Ellul went beyond Barth in one particular observation: *That which desacralizes a given reality, itself in turn becomes the new sacred reality.*

The sacred is then, as classically defined, the object of both hope and fear, both fascination and dread. Once nature was the all-encompassing environment and power upon which human beings were dependent in life and death, and so was experienced as sacred. The Reformation desacralized the church in the name of the Bible, and the Bible became the sacred book.

Science (through Charles Darwin's theory of evolution) and reason (higher criticism and liberal theology) desacralized the scriptures; subsequent decades have seen science, particularly those in the applied categories amenable to the aims of collective economic production (whether capitalist, socialist, or communist), elevated to the position of sacred in Western culture.

Ellul argued, today it is the technological society that modern-day humans generally hold sacred. Ellul defined technique as *the totality of methods rationally arrived at, and having absolute efficiency (for a given stage of development) in every field of human activity*. Thus, it is not the society of machines as such, but the society of "efficient techniques" which is the focus of Ellul's sociological analysis.

Modern technology has become a total phenomenon for civilization, the defining force of a new social order in which efficiency is no longer an option but a necessity imposed on all human activity

"It is useless to think that a distinction can be made between technique and its use, according to Ellul, for techniques have specific social and psychological consequences independent of human desires. There can be no room for moral considerations in their use"

"Not even the moral conversion of the technicians could make a difference. At best, they would cease to be good technicians." In the end, technique has only one principle, **efficient ordering.**"

Philosophy

What many consider to be Ellul's most important work, **The Technological Society** (1964) was originally titled: *La Technique: L'enjeu du siècle, "the stake of the century."* In it, Ellul set forth seven characteristics of modern technology.

The characteristics of technique which serve to make efficiency a necessity are:

- Rationality,
- Artificiality,
- Automatism of technical choice,
- Self-augmentation,
- Monism,
- Universalism,
- Autonomy.

The rationality of technique enforces logical and mechanical organization through division of labour, the setting of production standard, etc. It creates an artificial system which "eliminates or subordinates the natural world."

Regarding technology, instead of it being subservient to humanity, "human beings have to adapt to it, and accept total change". As an example, Ellul offered the diminished value of the humanities to a technological society. As people begin to question the value of learning ancient languages and history, they question those things which, on the surface, do little to advance their financial and technical state. According to Ellul, this misplaced emphasis is one of the problems with modern education.

This, according to Ellul, produces a situation where an incredible stress is placed on information in our schools. The focus in those schools is to prepare young people to enter the world of information, able to handle computers, but knowing only the reasoning, the language, the combinations, and the connections between computers. This movement is invading the whole intellectual domain and also that of conscience.

Ellul's commitment to scrutinize technological development is expressed as such:

> *...what is at issue here is evaluating the danger of what might happen to our humanity in the present half-century, and distinguishing between what we want to keep and what we are ready to lose, between what we can welcome as legitimate human development and what we should reject with our last ounce of strength as dehumanization. I cannot think that choices of this kind are unimportant.*

Theology

Although a son of the French Reformed (Huguenot) tradition (a small minority in his native land) and thus a spiritual heir of thinkers like John Calvin and Ulrich-Zwingli, Ellul departed substantially from Reformed doctrinal traditions, but unlike other European Protestant thinkers, utterly rejected the influence of philosophical-idealism or romanticism upon his beliefs about God and human faith. In articulating his theological ideas, he mainly drew upon the corpus of works by the Swiss-German professor Karl Barth and the critiques of European state Christianity made by Soren-Kierkegaard. This made him one of the more ardent expositors of neo-orthodoxy, which was in decline elsewhere in the Western theological scene during Ellul's heyday. Much like Barth, Ellul had no use for either liberal (to him dominated by Enlightenment notions about the goodness of humanity and thus rendered puerile by its naïveté) or orthodox (*e.g.*, fundamentalism or scholastic Calvinism, both of which to him refuse to acknowledge the radical freedom of God and humanity) Protestantism and maintained a roughly anti-Catholic view of the Bible, theology, and the churches. The latter was a standpoint shaped, again, by his membership in a tradition historically persecuted by Catholic clergy and state officials.

Ellul espouses views on salvation, the sovereignty of God, and ethical action that appear to take a deliberately contrarian stance toward established, "mainstream" opinion. For instance, in the book *What I Believe,* he declared himself to be a Christian Universalist, writing "that all people from the beginning of time are saved by God in Jesus Christ, that they have all been recipients of his grace no matter what they have done." Ellul formulated this stance not from any liberal or humanistic sympathies, but in the main from an extremely high view of God's transcendence, that God is totally free to do what God pleases. Any attempts to modify that freedom from merely human standards of righteousness and justice amount to sin, to putting oneself in God's place, which is precisely what Adam and Eve sought to do in the creation stories in Genesis. This highly unusual juxtaposition of original sin and universal salvation has repelled liberal and conservative critics and commentators alike, who charge that such views amount to *antinomianism*, denying that God's laws are binding upon human beings. In most of his theologically-oriented writings, Ellul effectively dismisses those charges as stemming from a radical confusion between religions as human phenomena and the unique claims of the Christian faith, which are not predicated upon human achievement or moral integrity whatsoever.

Political Standpoint

Ellul identified himself as a "Christian Anarchist." For him, this meant that nation-states should neither be praised nor feared, but merely ignored. To him, human government is irrelevant in that the law contained in Scripture is sufficient and exclusive. That is, being a Christian means pledging absolute allegiance to Christ, which makes other laws redundant (at their very best) or counter to the Law of God. Despite the initial attraction of some evangelicals to his thinking because of his high view of Biblical texts (*i.e.*, generally eschewing the historical-critical method), this position eventually alienated the few followers he had among conservative Protestants (this despite the fact that a U.S. evangelical school, Wheaton College, houses numerous manuscripts of his). Later, he would attract a following among adherents of more ethically compatible traditions such as the *Anabaptists*, the *Mennonites*, and the *house church movement*. Similar political ideas to Ellul's appear in the writings of a corresponding friend of his, *the American William Stringfellow.*

On Media and Propaganda

Ellul saw the power of the media as another example of technology exerting control over human destiny. As a mechanism of change, the media are almost invariably manipulated by special interests, whether of the market or the state. Using the term propaganda to address both political and commercial communication, Ellul wrote:

> *It is the emergence of mass media which makes possible the use of propaganda techniques on a societal scale. The orchestration of press, radio and television to create a continuous, lasting and total environment renders the influence of propaganda virtually unnoticed precisely because it creates a constant environment. Mass media provides the essential link between the individual and the demands of the technological society.*

In all of this, Ellul continued to place his understanding of technology and its proper role in this present society in a context that recognizes a faith in the eternal. This allowed Ellul to propose a more explicit alternative to the technology of the technician than those provided by some of his contemporaries, such as Heidegger.

To throw this wager or secular faith into the boldest possible relief, Ellul places it in dialectical contrast with Biblical faith. As a dialectical contrast to **La Technique**, for instance, Ellul writes **Sans feu ni lieu** (published in 1975, although written much earlier). Whereas technology is the attempt of human beings to create their home in this world, the Bible denies that people, the children of a Creator God, can ever be truly at home here.

Ellul adhered to the maxim "Think globally, act locally" throughout his life. He often said that he was born in Bordeaux by chance, but that it was by choice that he spent almost all his academic career there. After a long illness, he died in his house in Pessac, just a mile or two from the University of Bordeaux campus, surrounded by those closest to him. Not long before his death, the treatment for this illness illustrated to him once again one of his favourite themes - the ambivalence of technological progress.

QUESTIONS

1. Define critiques of technology and who are its prominent authors.
2. Who is Lewis Mumford and what are his main ideas about humanity, technology and civilization?
3. What do you mean by Megatechnics and Biotechnics?
4. Define Monotechnics and Polytechnics.
5. Give views of Lewis Mumford on Urban Civilization.
6. Who is Jacques Ellul and what are his main ideas about philosophy, Theology, political standpoint and on media?

4

TECHNOLOGICAL GROWTH AND DEPLETION OF NATURAL RESOURCES

4.1 INTRODUCTION

Things and services provided to man by nature are known as natural resources. Natural resources include both living and non-living resources on land as well as in ocean. **Living resources** include forests, wild life and creatures living in water or on land. **Non-living resources** include air, water, soil and mineral deposits. Some of these resources are found in abundance while others in limited quantity in nature. Depending upon their abundance and availability, natural resources may be of two types:

1. Renewable resources
2. Non-renewable resources

1. **Renewable resources:** The in exhaustive resources which can be regenerated within a given span of time are known as renewable resources, *e.g.*, forests, wild life, aquatic life, water, solar energy, etc.

   Thus, it may be said that renewable resources undergo natural multiplication but the rate of multiplication may very for each resources.
2. **Non-renewable resources:** Natural resources which do not have the capability of recycling and hence cannot be regenerated are known as non-renewable resources, *e.g.*, Fossil fuels like coal, petroleum, mineral deposits, etc., Renewable resources can become non-renewable if used without proper management.

   Main natural resources are discussed below;

4.2 LAND RESOURCES

Soils are an integral component of our environment and together with water constitute the most important natural resource of our nation. Land is perhaps the most precious resource to us provided by nature as entire ecological systems and all the terrestrial plants and animals reside on it. Land constitutes about 29.22% of the total earth's surface and covers 14,960 million hectares (Mha) of area.

With the exception of a small proportion (0.4%) of this land, most of the agricultural land has a number of physical and climatic constraints which limit productive potential. Land is used by living beings for different purposes, as for agriculture, forestry, etc., but its utilization for these purposes differs from area to area. In India, a rough idea of relative distribution of land use for different purposes is given here,

Agriculture	44%
Forests	11–14%
Pastures and grazing land	4%
Housing, agroforestry, industries, roads, reservoirs, etc.	8%
Barren land	14%
Cultivable wastelands	15%

Land has many functions of vital importance in life :

1. **The Production Function:** It is the basis for many life support systems through the production of biomass that provides food, fodder, fibre, fuel, timber and other biotic materials for human use, either directly or through animal husbandry including aquaculture and inland and coastal fishery.
2. **The Biotic Environmental Function :** Land is the basis of terrestrial biodiversity by providing the biological habitats and gene reserves for plants, animals and micro-organisms, above and below ground.
3. **The Climate Regulative Function:** Land and its uses are a source and sink of greenhouse gases and form a co-determinant of the global energy balance-reflection, absorption and transformation of radiative energy of the sun and of the global hydrological cycle.
4. **The Hydrologic Function:** Land regulates the storage and flow of surface and ground water resources and influences its quality.
5. **The Storage Function:** Land is a storehouse of raw materials and minerals for human use.
6. **The Waste and Pollution Control Function:** Land has a receptive, filtering, buffering and transforming function for hazardous compounds.
7. **The Living Space Function:** Land provides the physical basis for human settlements, industrial plants and social activities, such as sports and recreation.
8. **The Heritage Function:** Land is a medium to store and protect the evidence of the cultural history of mankind and a source of information on past climatic conditions and past land uses.
9. **The Connective Space Function:** Land provides space for the transport of people, inputs and produce and for the movement of plants and animals between discrete areas of natural ecosystems.

The suitability of land for these functions varies greatly over the world. Landscape units, as natural resources units have a dynamism of their own but human influences affect this dynamism to a great extent, in space and time. The quality of the land for one or more functions may be improved (for instance, through erosion control measures), but more often than not the land has been or is being degraded by human activities.

Soil, the uppermost layer of land is its most important part as it,

1. Supports the whole life system.
2. Stores water.

3. Provides food and fodder in the form of vegetation.
4. Provides mechanical support to the plants.
5. Has a capability of ion exchange, making the micro and macronutrients available for the growth of plants and animals.
6. Helps in the decomposition and mineralization of organic matter and regeneration of nutrients.

The word *'soil'* has been derived from the Latin word *'Solum'* meaning upper crust of the earth. It is composed of sand, silt, clay, organic and mineral nutrients along with air and moisture but its composition goes on changing constantly. Soil formed by the decomposition of rocks and minerals. They type of soil varies from place to place. Each type of soil is different with respect to its productivity.

Fertility of soil depends upon:

1. Extent of organic matter in it.
2. Its capacity to retain water and oxygen.
3. Its texture, structure, pH and microbial properties.

4.2.1 Land Degradation

Land is an indispensable resource for most of the essential human activities. It provides the basis for agriculture and forest production, water catchments, recreation and settlement. The range of uses that can be made of land for human needs is limited by environmental factors including climate, topography and soil characteristics and is to a large extent determined by demographic, socioeconomic, cultural and political factors, such as population density, land tenure, markets, institutions and agricultural policies.

In most of the developing countries, the needs and demands of rapidly increasing population have been the principal driving force in the allocation of land resources to various kinds of uses with food production as the primary land use. Population pressure and an increased competition among different land users have emphasized the need for more effective land-use planning and policies.

Rational and sustainable land use is an issue of great concern to governments and future generations. An integrated approach to planning and management of land resources is a key factor to implement solutions which will ensure that land is allocated to uses providing the greatest sustainable benefit. The increasing human population in several developing countries is laying pressure on the finite land resources, at the risk over-exploitation and land degradation. Sectoral and single objective approaches used to alleviate this situation have frequently not been effective. An integrated approach is required that should involve all stakeholders, accommodate the qualities and limitations of each land unit and produce viable land use options.

The deterioration or loss of fertility or productivity of soil is known as land degradation. Land degradation is the result of natural disasters as well as of human activities.

It occurs through the natural and man-made processes of wind erosion, water erosion and water-logging and has been identified as one of the priority concerns in India. The result of such degradation is the loss of invaluable nutrients and lower food grain production. Poor land use practices and management are responsible for the rapid land degradation in India.

The rate of land degradation may continue unabated or even increase under conditions of any human-induced global climatic changes, but this cannot be automatically assumed. Land degradation can be controlled, redressed or even reversed if the land is used wisely, if all the functions of land are taken into account and if short-term vested interests of privileged groups are replaced by long-term enlightened interest of all segments of mankind, globally, nationally and locally.

4.2.2 Causes of Land Degradation

Main factors responsible for the problem of land degradation are :

1. **Soil erosion:** Land degradation is a composite term, loosely defined as a sustained loss in the quality and the productivity of the land. As land degradation progresses, efforts by land users to secure a living become increasingly precarious and uneconomic. The most common indication of land degradation is soil erosion. The wearing away of soil, *i.e.* the loss or removal of upper layer of soil by the action of wind, water or human activities is known as soil erosion.

   Factors affecting soil erosion : The amount of erosion varies considerably from one field to another depending on soil type, slope of the field, drainage pattern and crop management practices and the effects of erosion also very. Areas with deep organic loams are better able to sustain erosion without loss of productivity than are areas where top soils are shallower. However, the extent of soil erosion mainly depends upon following factors :

   (*i*) Distribution, intensity and amount of rainfall. Heavy rainfall for a particular period of time causes soil erosion.

   (*ii*) Steep slope of the ground.

   (*iii*) Bare land is easily eroded than land covered with vegetation.

4.2.3 Causes of Soil Erosion

Climate affects soil erosion and the chemical and biological deterioration of soil. The state of the soil (texture, structure and chemical and biological properties) is a major factor, particularly in the sub-humid zones where the influence of climatic factors is less marked. Soil mismanagement including:

(*i*) Uncontrolled grazing by cattle

(*ii*) Faulty methods of surface drainage

(*iii*) Wrong cultivation practices

(*iv*) Removal of forest litter, etc.

are the major causes leading to soil erosion.

Types of soil erosion: Depending upon its mode of occurrence, soil erosion may be of two types:

(*i*) Normal erosion or geologic erosion.

(*ii*) Accelerated erosion.

(*i*) Normal erosion or geologic erosion: This type of erosion is caused by the removal of top soil by natural processes.

(*ii*) Accelerated erosion: This type of erosion is caused by human activities. The rate of this type of erosion is much faster than that of the previous one.

4.2.4 Effect of Soil Erosion

Erosion affects productivity as it removes the surface soil, containing most of the organic matter, plant nutrients and fine soil particles, which help to retain water and nutrients in the root zone where they are available to plants. The subsoil that remain behind tend to be less fertile, less absorbent of water and less able to retain moisture, fertilizers and other plant nutrients.

The effects of erosion are also felt elsewhere in the environment. A recent study estimated the off-site cost of cropland erosion in the United States to be in the range of a billion dollar per year. Eroded soil clogs streams, rivers, lakes and reservoirs, resulting in increased flooding, decreased reservoir capacity and destruction of habitats for many species of fish and other aquatic life. The eroded soils contain nutrients and other chemicals that are beneficial on farm fields, but can impair water quality when carried away by erosion. As a result, drinking water supplies may contain nitrates or organic chemicals in concentrations that exceed public health standards or surface water may become clogged with excessive plant growth from the added nutrients.

4.2.5 Prevention of Soil Erosion

Up to a certain extent, increased fertility and irrigation may compensate for the lower soil productivity. Long-term loss of farmland productivity and damage to the environment from eroded sediments, therefore, often are overlooked in the need for short-term economic gains. Over the past 50 years, the negative effects of soil erosion on farm productivity have been masked by improved technology and increased use of fertilizers and pesticides. In natural ecosystems, soil fertility is maintained by the diverse contribution and recycling of nutrients by a wide range of plant and animal species. However, following steps may be taken to control soil erosion:

(*i*) **Conservational Till Farming or no-Till-Farming :** It involves the use of tilling machines which make slits in the unplugged soil and inject seeds, fertilizers, herbicides and a little waste to allow the crop to grow without making any competition with weeds.

(*ii*) **Contour Farming:** It is a process in which crops are grown in rows on slopes which helps in holding the soil and preventing it from running off with water.

(*iii*) **Terracing:** Terracing retains water for crops, thus checking running off of soil.

(*iv*) **Strip Cropping:** In this method, crop strips are grown alternate to the strips of grasses which retain the run off coming from cropped soil.

(*v*) **Alley cropping or Agro Forestry:** In this, crops are planted between rows of trees or shrubs.

4.2.6 Salination

Increase in the concentration of soluble salts in the soil is known as salination. It is result of poor drainage of over irrigation and flood waters. High concentration of salts in soil leads to its poor productivity, as salination. It is a result of poor drainage of over irrigation and flood waters. High concentration of salts in soil leads to its poor productivity.

4.2.7 Water Logging

Water logging is a condition in which accumulation of water results from poor drainage system ultimately leading to the formation of a continuous column of water table. In water logged soils, availability of oxygen is decreased which results in the reduced fertility of soil.

4.2.8 Desertification

Slow process of land degradation ultimately leading to desert formation is known as desertification. It is the degradation of the land in arid, semi-arid and sub-humid dry areas caused by climatic changes and human activities. It is accompanied by a reduction in the nutritional potential of the land and depletion in surface and ground water resources. Desertification not only occurs in natural deserts, but can also take place on land which is prone to desertification processes. Desertification reduces the productivity of land. Depending upon the extent of loss of productivity of land, desertification may be categorized into three groups:

(*i*) Moderate desertification causing 10–25% drop in productivity.
(*ii*) Severe desertification causing 25–50% drop in productivity.
(*iii*) Very severe desertification causing more than 50% drop in fertility.

The seriousness of desertification depends on factors which vary from one region, country or year to another. These factors include:

1. The severity of the climatic conditions during the period under consideration (particularly in terms of annual rainfall).
2. Population pressure and the standard of living of the people concerned.
3. The level of the country's development and the quality of preventive measures prevailing there.

Characteristics of Desertification:

(*i*) Devegetation
(*ii*) Depletion of ground water
(*iii*) Salination
(*iv*) Sever soil erosion.

4.2.9 Causes of Desertification

Desertification may be due to natural climatic changes or due to abusive and improper use of land. Climatic changes are actually both a consequence and a cause of desertification. The destruction of the natural grass and woody vegetation cover in dry areas affects the topsoil temperature and atmospheric humidity and consequently influences the movement of atmospheric masses and rainfall. The major causes of desertification are :

(*i*) Deforestation
(*ii*) Overgrazing of land by cattle
(*iii*) Mining and quarrying

4.2.10 Effects of Desertification

Desertification is a world-wide phenomenon which causes the earth's ecosystems to deteriorate. It affects about two-thirds of the countries of the world and one-third of the earth's surface, on which one billion people live.

4.2.11 Shifting or Jhum Cultivation

Jhum cultivation leads to complete destruction of forests resulting in soil erosion and other associated problems. This is more popular in tropical and sub-tropical regions of Africa. Asia and Islands of Pacific Ocean. In India, it is common in North-East and Orissa.

4.2.12 Urbanisation

Urbanisation, *i.e.*, developmental activities, as such industry settlement, etc., has reduced the productive land.

4.2.13 Landslides

Certain human activities, as construction of roads, canals, dams, etc., lead to landslides especially in hilly areas. (For details see unit-IV).

4.2.14 Soil Pollution

Agents of soil pollution badly affect the physical, chemical and biological properties of soil.

4.3 WATER RESOURCES

Water may be considered as a synonym for life. It is one of the five fundamental components of the universe; namely *Aakash (Space), Vayu (Air), Tej (Fire), Jal (Water) and Prithvi (Earth).* It is a vital and highly valued resource. It is the most common substance found in the body of a living organism. The evolution of human civilization has been closely related with the availability of this natural and replenish able resource. It is the singly key factor which:

- Determines the environment
- Regulated the health of soil
- Makes nutrients available to the plants.

Besides these basic activities, a number of functions are performed by water. Hence, if it is said that life revolves around water, it will not be a hyperbola.

Water is the most abundant substance in our environment. It is the major constituent of hydrosphere that consists of oceans, seas, rivers, streams, glaciers, lakes, reservoirs etc. About 70% of earth's surface is covered under water. About 97% of total water is present in oceans and 2% in glaciers and ice caps. It means that 90% of total water remains unavailable for human beings either to the presence of excessive salt contents or being in frozen state. Thus, only 1% water is available for use in the form of fresh water. Different sources of fresh water are rivers, streams, lakes and reservoirs.

Water resources are not distributed uniformly across the globe. India is a land of rivers. There are 14 major, 44 medium and 55 minor river basins in the country. Major rivers have a catchment area of 20, 000 square kilometers or above, medium between 2,000 and 20,000 square kilometers and minor systems have a catchement area of 2,000 square kilometers or less. But, in some dry areas of Rajasthan, every inch of land contributes water through runoff to these rivers.

4.3.1 Uses of Water

Water is essential for life. It is required by human beings as well as by animals and plants. It is impossible to survive without water. It is the most important basic need for life. We have been blessed by nature with this invaluable resource as earth is the only planet where water exists in liquid form and can be utilized in different ways. Human beings depend upon this indispensable natural resource for almost every physiological and developmental activity. Some important uses of water are:

1. Drinking
2. Irrigation
3. Bathing
4. Transportation of nutrients.
5. Regulation of temperature

6. Disposal of wastes from industries
7. As coolant for thermal power plants etc.

Man requires water for every activity as water acts as a medium in which all the living processes take place.

Population and environment of any nation are dependent on the wise use of its water resources, including the protection of water quality and availability. Its availability greatly influences the prosperity of people and development potential in addition to improved health status of the people of the area. In fact, food security and human health, energy supplies and industrial production are all dependent on water to a large degree. Water also plays an important role in regulating global climate. Major uses of water are discussed here in detail.

It is estimated the 70% of world wide water used for the irrigation, 18% world wide other purpose. These include drinking water, bathing, cooking, sanitation and gardening.

Another important use of water is in the production of electricity. For the first time water was used to generate electricity in 1882 on the Fox river, in the USA, which produced enough power to light two paper mills and a house.

Water has a lot of recreational value also.

Recreational usage is non-consumptive. However, it may reduce the availability of water for other users at specific times and places.

Use of water for environmental purpose is small but in growing percentage. Environmental water usage includes artificial wetlands, artificial lakes intended to create wildlife habitat, fish ladders around dams and water releases from reservoirs timed to help fish spawn.

4.3.2 Over-utilisation of Surface and Ground Water

Although water is a renewable resource but it is also a finite one. It is abundant on earth but is not available in sufficient amount as required. Requirement and use of water is continuously increasing but the requirements are not met properly and this results in fresh water crisis. Fresh water crisis is not merely due to lack of water as provided by nature but it is also a result of:

1. Increasing human population.
2. Rapid development.
3. Destruction of fresh water by some anthropogenic activities.
4. Overuse of water for drinking, irrigation and domestic purposes.
5. Improper use of water.
6. Wastage of water.

Freshwater available for use in nature is in two forms:

1. Surface water
2. Ground water.

1. **Surface Water:** Surface water is water in a river lake in a fresh water Worland. The water among through pros potation in the form of rainfall and snow is known as surface water. This type of water neither percolates into the ground nor returns back to the atmosphere via transpiration or evaporation rather it gathers in the form of streams, ponds, lakes or reservoirs. This type of water in widely used for agricultural, domestic and industrial purposes.

2. **Ground Water:** Sub-Surface water or ground water is fresh water located in the pore space of soil particles. Ground water is the major source of fresh water. Ground water sources are easy to develop. Ground water is clear and colourless and is the major source of fresh water for agricultural and domestic use.

Ground water is considered to be almost pure, but it may be contaminated by different contaminants. In ground water sources, there are found some aquifers which hold water. An aquifer is a layer of sediment or rock, which is highly permeable and contains water. Depending upon their position, aquifers may be of two types:

(*i*) **Unconfined Aquifers:** These are overlaid by permeable earth material and are recharged by water seeping down from above in the form of rainfall and melting of snow.

(*ii*) **Confined Aquifers:** These are in the form of sandwich between two impermeable layers of rocks or sediments and are recharged in those areas where the aquifer intersects the land surface.

Demand for water outgrows supply in a number of countries and the quality of that supply is rapidly declining. 450 million people in 31 countries already face serious shortage of water. These shortages occur almost exclusively in developing countries, which are ill-equipped to adopt the policy and technology measures needed to meet the crisis. By the year 2025, one-third of the world's population is expected to face severe and chronic water shortage. Water shortage and poor water management can lead to loss of biodiversity and agricultural production, increase in malnutrition and disease, reduced economic growth, social instability, and conflict.

Overuse of ground water may cause several problems as:

(*i*) A heavily pumped well can lower the water table leading to the drying of shallower wells.

(*ii*) Heavy pumping may cause depletion of aquifer.

(*iii*) Overuse of fresh water reservoirs causes accumulation of salt water in aquifers.

4.3.3 Drought

Definition: People use water in different ways, thus it may not be defined universally. However, there are four different ways by which drought can be defined.

1. **Meteorologically,** it is a measure of departure of precipitation from normal. Due to difference in climatic conditions, what is considered a drought in one region may not be a drought in another region.
2. **Agriculturally,** it refers to a situation when the amount of moisture in soil no longer meets the needs of a particular crop.
3. **Hydrologically,** it occurs when surface and sub-surface water suppliers are below normal.
4. **Socioeconomically,** it refers to the situation that occurs when physical water shortage begins to affect people. When the word *"drought"* is used commonly, the most often intended definition is meteorological drought. However, when the word is used by urban planers, it is more frequently in the sense of hydrologic drought.

Drought is an insidious hazard of nature. It is a prolonged, abnormally dry period when there is not enough water for users' normal needs. It originates from a deficiency of precipitation over an extended period of time, usually a season or more. The severity of drought mainly depends upon climatic factors such as high temperature, strong wind and low relative humidity. Besides, there are other factors that affect the severity of a drought. Some of these are :

- the season in which a drought occurs.
- delay in the start of the rainy season.
- no rain during critical stages of crop growth.
- the growth stage of the crops when a drought begins.
- high temperature.
- strong winds.
- low humidity.

The condition of low rainfall leading to lack of water is called drought.

It is a common natural disaster occurring in the arid and semiarid regions. In these areas, the rainfall is below normal. Drought usually occur in hot dry areas of land. In most cases the area is dry because there is very little rainfall. Droughts usually start in the east and move to the west.

4.3.3.1 Causes of Drought

Drought may result :

1. From natural reasons, as less rainfall, more evaporation, etc.
2. Over grazing.
3. Deforestation.
4. Mining.
5. Desertification.

4.3.3.2 Consequences of Drought

Drought should not be viewed merely as a physical phenomenon or natural event. Its impacts on society result from the interplay between a natural event (less precipitation than expected resulting from natural climatic variation) and the demand that people place on water supply. Some adverse effects arising due to drought are :

1. **Effect on Agriculture:** Agriculture suffers first and most severely from drought. Drought disrupts cropping programs, reduces breeding stock and threatens permanent erosion of the capital and resource base of farming enterprises. This deficiency results in water shortage for some activity, group or environmental sector. It largely causes damage to crop farmers, such as
    - Crop losses
    - Loss of income because of reduced crop yields.
    - Damage to crops from wildlife looking for food.
    - Increased irrigation and other water development costs.
    - Higher fees for transporting produce.
2. **Effect on Environment:** The risk of serious environmental damage, particularly through vegetation loss and soil erosion, has long term implications fro the sustainability of our agricultural industry. Drought causes a lot of damage to the environment. This makes it harder to have successful crops in the future. If drought is not managed properly, it could become a permanent feature. Water quality suffers and toxic algae outbreaks may occur, plants and animals are also threatened. Bushfires and dust storms often increase during dry period.
3. **Effect on Economy :** Drought can cause a wide range of economic effects to both mankind and its community. Drought can affect crop farmers, livestock farmers and other parts of the

economy. Drought causes other economic impacts such as :

- Unemployment.
- Decreased land prices.
- Higher interest rates for loans.
- Decrease in village population.

4. **Effect on Forestry and Water Resources :** In forestry, drought can result in:
   - Wild fires.
   - Tree and plant diseases.
   - Insect infestation.
   - Less productivity of forest land.
   - Loss of trees, especially young ones.
   - Land degradation.
5. **Effect on Water Resources :** Drought also affects water resources in that :
   - Ponds, tanks and other water sources dry up.
   - Shallow wells dry up.
   - Water sources become polluted when percentage poisonous elements such as arsenic, fluoride and nitrate increases.
   - Salt content increases in deep wells.
6. **Effect on Health :** Drought can affect health in many ways, such as :
   - Malnutrition because of high cost of food and polluted water sources.
   - Faster spread of diseases because of more animals around people, decreased sewage flow and more pollutants.
   - Mental and physical stress such as anxiety, depression and domestic violence.
   - More people die from disease, heat stress and suicides.
   - Public safety is at risk from forest and range fires.
   - Higher infant mortality.
   - Women are overburdened because they have to do extra work in relief programs as well as their everyday work.
   - Reduced recreation.

Control : Drought cannot be checked completely but may be controlled to a certain extent. Some important measures include carefully selected mixed cropping, understanding the ecological requirements, etc.

It is important to monitor drought as the drought conditions continue. Drought can be monitored by using information on rainfall, weather, crop conditions, water availability and other factors. The easiest way to get this information is from weather forecasts. A weather forecast is a prediction or early warning of what the weather will be over a period of time. Forecasts are divided into three categories:

Short range : predicts upto 3 days.

Medium range : predicts from 3 to 10 days.

Long range : predicts from 10 to 30 days.

4.3.3.3 Famous Droughts

Major droughts which have occurred during 18th and 19th century till now all over the world are summarized in the following table :

Year	*Place*	*Destruction*
18th & 19th Century	Cape Verde	Death of over 100,000 people due to starvation
1900	India	Death of 250,000 to 3.25 million people
1921–22	Soviet Union (Ukraine & Volga)	250,000 to 5 million perished from drought
1928–30	North-west China	Femine resulted in over 3 million deaths
1932–34	Ukraine, Kuban & North Caucasus	5 to 10 million perished
1931–38	Great Plains of USA	Cars, homes and farms destroyed in dust storms
1930–37	United States	Severe impact on the U.S., refuting in all the districts of the American Great Plains being depopulated
1936	Sichuan, China	Worst drought in modern history, 34 million farmers were displaced and 5 million people starved
1941	Sichuan, China	Death of 2.5 million people
1965–67	India	Death of over 1.5 million people
1968–74	Sahel	Death of between 100,000–250,000 people; disruption of millions of lives and collapse of agricultural bases of five countries
1968–72	Iran	Death of approximately 4,000 people
1972–74	Africa	Country's financial business went down
1973–74	Ethiopia	Military coup, which overthrew the Government of Haile Selassie
1976	Britain	Water rationing was greatly needed in the larger cities
1982–83	Australia	60% of Australian sheep & cattle died
1983	Bolivia	80% of normal harvest of staple crop was lost. Total agricultural loss was at U.S.$417.2 million & $277.7 million in crop losses and loss of 200,000 cattle, 3,000,000 sheep and 660,000 lames and alpacas
1984–85	Ethiopia	Regime change to Marxist Government

4.3.3.4 Conflicts over Water

As demand for water hits the limits of finite supply, potential conflicts are brewing between nations that share transboundary freshwater reserves. Conflicts are brewing now over rivers and river basins shared by many countries around the world. This isns not unexpected given the fact that more than 200 bodies of water are shared by two or more countries.

Water is present in nature in abundance. But water has been a major issue of conflicts among different countries and different states even in one country. These issues arise mainly in those areas where a water source is shared by two or more than two countries or states. Some of the important disputes caused by this water sharing world over are:

1. Water conflict in the Middle East making issue on three river basins, the Jordan, the Tigris-Euphrates and the Nile. These three rivers are shared by different countries of the Middle East, as Ethiopia, Sudan, Egypt and Turkey and each of these countries is trying its best to have the maximum hold on them.
2. The Indus Water Treaty dealing with the conflict over Indus, Jhelum, Satluj, Ravi and Beas between India and Pakistan.
3. The Cauvery water dispute between Tamilnadu and Karnataka.
4. The Satluj-Yamuna link canal dispute between Punjab and Haryana.

A tussle is simmering in South Asia's Ganges-Brahmaputra Basin, where Bangladesh, India and Nepal dispute over the best use of water. India and Nepal want to exploit the basin's huge hydroelectric power-generating potential, whereas Bangladesh wants the water managed in such a way as to minimize flooding during monsoon and water shortage during dry months.

Between nations, intergovernmental activities to settle conflicts over shared bodies of water have had mixed success. A 1977 agreement between Bangladesh and India allocated 63% of the dry season flow of the Ganges at the Bangladesh and India border to Bangladesh. However, in 1988, India did not renew the agreement and on several occasions has cut the flow of the Ganges at the Farraka barrage just inside the border to less than one third of the amount agreed upon in 1977.

4.4 DAMS

One of the fundamental requirements for socio-economic development in the world is the availability of adequate quantity of water with the appropriate quality. In the past, the main sources of domestic and industrial water have been aquifers. Today, many of these are overused and their rate of recharge is far less than what is extracted. Their supply must be augmented with additional water from reservoirs. Properly planned, designed and constructed and maintained dams contribute significantly toward fulfilling our water supply requirements.

Water is an important and invaluable natural resource that is depleting continuously. Thus, it is necessary to save this precious gift supplied by nature to us so that it can be used for various purposes. Different methods have been adopted for this objective and dam is one of the latest advancements in this field. The first dam was built in 1890 and now there are 45,000 dams all over the world, out of which 4291 are in India alone. Dams may be considered as the symbol of national development of a particular country.

4.4.1 Benefits of Dams

1. They act as a potential for economic progress and growth of the country.
2. They help in checking floods.
3. They are used for generating electricity.
4. They reduce water sand power shortage.
5. They supply water to areas having scarcity supply of water.
6. They provide drinking water in remote areas.
7. They ensure an year-round water supply.
8. They help in soil protection.
9. Multipurpose river valley projects are used.
   (*i*) For navigation.
   (*ii*) To develop fish nurseries.
   (*iii*) For irrigation
   (*iv*) For Hydroelectric power production

Hence, dams are a vital part of a Nation's infrastructure. Dams provide a range of economic, environmental and social benefits, including hydroelectric power, river navigation, water supply, wildlife habitat, waste management, flood control, and recreation. Important uses of dams are discussed here in detail:

1. **Generation of Hydroelectricity:** The availability of energy is essential for the socio-economic development of a nation. It is advantageous to use energy that is clean, efficient, dependable and renewable. Hydropower meets all these requirements. In countries, where a vast amount of development is still awaited, good conditions often exist for renewable energy sources. The technically most advanced and economical source of renewable energy is hydropower.
2. **Flood Control:** Dams and reservoirs can be effectively used to regulate river level and flooding lldlf of the dam by temporally the flood volume and releasing it later. The most effective method of flood control is accomplished by a number of multipurpose dams strategically located in a river basin.
3. **Recreation:** The attraction of reservoirs for tourists is often a significant benefit, in addition to other purposes of a dam. This is very significant in areas where natural surface water is scarce or non-existent.

4.4.2 Problems of Dams

In spite of different advantage, a number of disadvantages are also associated with big dams.

1. Loss of water through evaporation and seepage is found in some dams which decrease their capacity to store water.
2. Evaporation of excessive water from dams leads to salinization of rivers making them unusable for people.
3. Enormous weight of water borne by dams may produce cracks in its walls resulting in flooding.
4. Submerging of large areas of land.
5. Resettlement and rehabilitation of displaced people.
6. Loss of forests, flora and fauna.
7. Displacement of tribal people.

8. Stagnation of water near reservoirs.
9. Breeding of vectors.
10. Spread of diseases.
11. Growth of aquatic weeds.

4.4.3 Dams and Their Effects on Forests and Tribal People

Dams are the most important means of water storage in a country and also contribute to the economy of the nation. Dams have been built for thousands of years to manage flood waters, to harness water as hydropower, to supply water for drinking or for industry or for irrigation of fields. By 1950, Governments, or in some countries the private sector, had built a large number of dams with the increase in population and growth of national economies. At least 45,000 large dams have been built as a response to meet an energy or water need. In India 4291 dams are there which have enormous benefits in the development of the nation. Today nearly half of the world's rivers have at least one large dam each. But, it is important to note that these dams also have some adverse effects on environment, especially on forests.

A dam affects not only a river, but the land and the people and animals that depend upon that river. The impact of large dams on ecosystem, biodiversity and downstream livelihood is well known.

It has been estimated that a large dam leads to :

1. The loss of forests and wildlife habitat.
2. The loss of species population.
3. The degradation of upstream catchments' areas due to inundation of the reservoir areas.
4. The loss of the services of downstream floodplains, wetlands and riverine, estuarine and adjacent marine ecosystems.
5. The loss of aquatic biodiversity, upstream and downstream fisheries.
6. Increase in water borne diseases.
7. Floods, droughts and landslides.
8. Destruction of downstream ecosystems.
9. Environmental damage.
10. Coastal erosion.
11. Earthquakes.

The construction of dams not only destroys the forest but often displaces tens of thousands of people destroying both their land and culture. Indigenous and tribal people and vulnerable ethnic minorities have suffered disproportionate levels of displacement and negative impacts on livelihood, culture and spiritual existence. The major outcome is the displacement of people living in that area. Affected population living near reservoirs as well as displaced people and downstream communities face averse health and problem of livelihood resulting from environmental changes and social disruption. Further, among affected communities, gender gaps have widened and women have frequently borne a disproportionate share of the social costs and are often discriminated in the sharing of benefits.

4.5 FOREST RESOURCES

As a natural resource, forests provide an important economic base for employment, tourism and recreation and support a diverse forest products industry. Forests have acquired increasing importance in the recent past for their role not only in meeting the material requirements but also for their ecological and environmental functions. Therefore, accurate and updated information on forest resources is of paramount importance. Some of the important parameters used for measuring forest resources are forest cover, growing stock, annual increment, species composition, biodiversity, non-timer forest products, etc.

Forests are an important component of our environment and economy and are essential for ecological balance of an area. Forests are important renewable natural resource. Different varieties of trees dominate the forest ecosystem. Forests play an important role in the economic development of a country. Besides economy, they are also linked with our culture and civilization. India has a very rich fauna and flora, most of which is present in forests. Forests vary in their composition and density.

Further, the distribution of forests is not uniform worldwide and has been changing continuously by human activities. Climate is probably the most important determinant of vegetation patterns globally and has significant influence on the distribution, structure and ecology of forests. It has been estimated that about 30% of land in the world is occupied by forests. Forests are not uniformly distributed among continents or countries. According to a world survey conducted in the early 1970s by Reidar Person, about one-fifth of the earth's land is covered by closed forests (where tree crowns cover 20 percent of more of the ground when viewed from above). Roughly another 12 percent of the land is covered by open woodlands (where scattered trees provide a crown cover of 5–19 percent). India is a mega-biodiversity country where forests account for about 20% (64 million ha) of the geographical area. In India, 20% land is forested out of which about 12% is dense forest and 8% is open forest.

4.5.1 Uses of Forests

People began life on this planet as forest dwellers. They were food gatherers and depended on the forests for all their needs; food, clothing and shelter. They gradually became food growers, clearing a small patch in the forest to grow food. Even today people depend on the forests for paper, timber, fuel wood, medicine and fodder. Forests provide many ecological and socioeconomic benefits. Ecological value not only encompasses forest attributes, such as biodiversity and wildlife habitat, but also includes processes that are critical for protecting our air, water and soil. From the perspective of human use, forests have tremendous aesthetic appeal, recreational use and economic value. In addition to the many natural resource based industries and activities, such as fisheries and hunting that are associated or influenced by forest lands, the forestry and wood products industry is the fifth largest in the country. Forests are of great use in the life and prosperity of human being. The main points of importance of forests are summarized here:

1. Forests check air pollution. They produce oxygen during photosynthesis, making it available for respiration. Further, they can absorb a number of toxic gases keeping the air pure and suitable for breathing.
2. Forests bind soil particles tightly by their roots and thus prevent soil erosion.
3. They save the hill slopes from landslide.

4. In desert areas, forests reduce wind erosion by reducing wind velocity.
5. They play an important role in enhancing the quality of environment by influencing the life supporting system.
6. They provide protection to wild life.
7. They help in balancing the gaseous cycles in the atmosphere. Further, they also reduce global warming by absorbing CO_2 during photosynthesis, the main greenhouse gas.
8. They increase the local rainfall and water holding capacity of soil.
9. They maintain soil fertility.
10. They regulate earth's temperature regimes.
11. They reduce flooding.
12. Forests contribute to the economic development of a country by providing a number of useful goods.
13. The chief product obtained from forests is wood which is widely used as,
    (*i*) Fuel.
    (*ii*) Raw material for various industries, viz. pulp and paper, board, timber for furniture, windows, doors, etc.
    (*iii*) In packaging articles.
14. Besides wood and timber, forests also provide resins, fruits, nuts, herbs, medicinal plants, pharmaceuticals, oil, spices and syrups.
15. Bamboos obtained from forests are used in roofing, walling, flooring, matting and basket industry.
16. Canes are used for making furniture, ropes, walking sticks, umbrella handles and sports goods.
17. Different types of oils (sandal wood, rosha grass, khas, etc.) are extracted from trees, which find use in the manufacture of cosmetics, soaps, pharmaceuticals, tobacco and confectionery.
18. Forest insects are a source of lac, honey, wax and silk.
19. A number of drugs, insecticides and poisons are obtained from forests.
20. Forests also have aesthetic and touristic value.

4.5.2 Over Exploitation of Forests

Since the very beginning human beings depend upon forests for their numerous requirements. With increasing population and developing civilization, this demand for forest products has increased continuously and has resulted in the rapid depletion of forests has come in the form of deforestation.

The reckless felling of trees by human beings is known as deforestation. Deforestation is the permanent destruction of indigenous forests and woodlands. It is a slow-onset disaster that may contribute to other cataclysmic disasters.

4.5.3 Rate of Deforestation

The rate of deforestation is increasing tremendously, however actual rate of deforestation is difficult to determine. The rate of deforestation varies from region to region. Different surveying efforts have been made to know about the current rate of deforestation since long. The Food and Agriculture Organization (FAO) estimates that 53,000 square miles of tropical forests were destroyed each year during the 1980s. Recent research results show that in the Brazilian Amazon, the rate of deforestation was around 6200 square miles per year from 1978–1986, but fell to 4800 square miles per year from 1986–1993.

It has been estimated that if the current rate of deforestation continues, the world's rain forests will vanish within 100 years causing unknown effects on global climate and eliminating the majority of plant and animal species from the planet. Scientists are currently analyzing rate of deforestation for the current decade as well as studying how deforestation changes from year to year. Man cuts trees for his own interest and adds to the ecological crisis.

4.5.4 Causes of Deforestation

The causes of deforestation are very complex. Deforestation occurs in many ways. The spread of agriculture, firewood collection and unregulated timber harvesting are the principal apparent causes of forest losses. It is most frequently caused by human beings taking care of immediate needs while not caring for the long-term effects of their actions.

Deforestation occurs in many ways which may be summarises as follows :

1. Jhum cultivation (Shifting cultivation) is the major cause of deforestation. In this process, due to low productivity of tropical forest soils, farms move to new sites after a few years leaving behind the previous land as such. These left lands called forest fallows may be converted back into forests with time if left untouched but this actually does not happen because due to increasing population farmers return back to these fallows within a very short interval of time for agriculture.
2. Overgrazing of forests by livestock.
3. Conversion of forests into pastures for domestic animals.
4. Cutting of trees for wood required for different purposes.
5. Clearing of forests for hydroelectric projects, big dams, road construction, mining, etc.
6. Clearing of forests for agriculture to meet increasing food demand.
7. Development of cash crops and cattle ranching, both of which earn money for tropical countries.
8. Commercial logging (which supplies the world market with woods such as meranti, teak, mahogany and ebony) destroys trees thereby opening up forests for agriculture.
9. Felling of trees for firewood and building material, the heavy lopping of foliage for fodder and heavy browsing of saplings by domestic animals like goats.

Commercial logging and spread of farming in one form or the other are by far, the major causes of outright forest loss today. Much of the blame for tropical forest destruction is often laid on the shoulders of shifting cultivators those who slash and burn a clearing in the forest, grow crops for a few years until soil fertility dissipates, and then move on the clear a new patch.

4.5.5 Consequences of Deforestation

High rate of deforestation has various direct and indirect effects, both on living beings and on environment. Some of them are as follows:

1. In deforested areas, especially on slopes, soil gets washed away with water leading to soil erosion.

2. In rain lacking areas, deforestation accompanied by strong winds results in desertification.
3. Decrease in rainfall is an important consequence of deforestation.
4. Loss of fertile land.
5. Due to decrease in rainfall, the climatic temperature of the deforested area is also changed.
6. Decrease in rainfall also causes a decrease in water table.
7. Loss of certain species of flora and fauna depending upon forests.
8. Disturbance in ecological balance.
9. Loss of plants of medicinal importance.
10. Environmental changes including air pollution and global warming.
11. Lack of wood and timber.
12. Flooding problems.
13. Loss of shelter for rural and tribal people.
14. Landslides may occur in hilly areas due to deforestation.
15. Alteration of local and global climate through disruption of :
    (*i*) **The carbon cycle:** Deforestation increase the amount of carbon dioxide (CO_2) and other trace gases in the atmosphere. It is well known that green plants take up carbon dioxide from atmosphere for photosynthesis, thus forests act as a major store of carbon. By cutting or burning of forests this entrapped carbon is released in the atmosphere as carbon dioxide leading to an increase in the concentration of carbon dioxide which is the major contributor to green house effect.
    (*ii*) **The water cycle:** Trees absorb ground water up through their roots and release it into the atmosphere via transpiration. With removal of the forests nothing is left there to hold as much water leading to a drier climate.
16. It results in the extinction of species which depend on the forests for survival. Forests are the abode of more than half of all species on our planet. As the habitat of these species is destroyed, the number of species declines.
17. Forest biodiversity is being lost due to rapid deforestation, fragmentation and degradation of all forest types. According to FAO, there has been an annual net loss of 9.4 million hectare (0.22 percent annually) forest since 1990, of which major part was of natural forests in the tropics.

4.5.6 Timber Extraction

Trees are cut down for procuring timber of economic importance. Along with the large trees, a number of small trees and creeper plants are also cut down which depend upon these large trees. Increasing industrial use of timber for human consumption has become the major cause of forest destruction. Various factors are responsible for the extraction of wood from forest. Wood extraction has increased over time from the early colonial period to 1947. With the beginning of the forest department in 1984 and passing of the Forest Act in 1865, state control over forests was strengthened and most of the forests were made available for exploitation leading to the rapid growth of extraction of timber. By the 1870s, the demand for timber further rose and the forest department began searching for more timber producing areas and started building roads to reach there. Hence forests were surveyed to estimate their resources. Once potential timber producing areas were identified and made approachable, timber extraction grew rapidly. Timber extraction started increasing rapidly from the mid 19th century and peaked in the early twentieth century. Timber extraction has different bad effects on different aspects of life. Some of them are:

1. Degradation of forests.
2. Soil erosion.
3. Floods.
4. Climatic changes.
5. Loss of biodiversity.
6. Loss of flora and fauna found in that particular area.
7. Exploitation of tribal societies.

Along with the forests containing timber, Non-Timber Forest Products (NTFP) are also extensively extracted from Indian forests and have an immense role in rural and forest economy.

NTFP harvests badly affect plant populations as :

1. It affects the growth and reproductive capacity of the individual plant.
2. It affects the demographic structure and dynamics of a particular plant community.

The direct effect of intensive forest produce collection include:

1. Decline in productivity
2. Decline in density
3. Regeneration of the targeted plant species depending on the utilized part of the plant.

4.5.7 Forest Conservation

Forest biodiversity may be the richest of all the terrestrial systems. During the last 8000 years about 45 percent of the Earth's original forest cover has disappeared. Some of the important illegal activities in India and abroad contribute to the following outcomes.

1. Reduction in the income of local communities from legal and improved forest management and thereby giving rise to further illegal activities.
2. Creation of a cycle of bad governance, where corrupt politicians. Government officials and private individuals support bad governance in order to maintain their profits.
3. Contribution to extensive deforestation and the destruction of biological diversity.
4. Increase in the risk of forest fires and distort forest product markets.

Sustainable use of our forest resources requires planned management in such a way that the needs of today are met without hurting the opportunity of future generations to meet their needs.

4.6 MINING

Mining is any activity that attempts to extract minerals (which are valuable and locatable) from their natural setting. Mining may be of two types:

1. **Surface mining,** involving extraction from deep deposits.
2. **Sub-surface mining,** involving extraction from deep deposits.

As per a recent Supreme Court order, the law of Ministry of Environment & Forests and the Department of Mines & Geology (DMG), mining in forest areas is illegal. Yet, mining is quite common in a number of forest regions. The well known examples are the Chandewadi and Ilavadabe forests. In an area of about 1200 acres of forest land in the Ilvadabe forests, mechanized sand mining is being pursued for several years. In this forest area, a minimum of 400 trucks haul sand every day.

4.6.1 Impacts of Mining

Forest protection, biodiversity conservation and prevention of devastating floods in any country rely heavily on the protected forest and conservation area system. Mining is encroaching into many fragile ecosystems and a healthy environment and livelihood of poor population is at stake.

The vegetation along with the underlying still and the rock masses in the mining areas are removed so that topography and landscape in the related area is destroyed. Mining, like timber extraction, has various adverse effects, as,

1. Deforestation.
2. Landslides.
3. Land degradation.
4. Water pollution due to discharge of mineralized mine water.
5. Lowering of ground water table, making it less available for industrial agricultural and drinking purposes.
6. Air pollution due to release of toxic gases.
7. Fire risks.
8. Drying up of water sources, as springs.
9. Migration of tribal people residing in that area due to loss of employment for warkers.
10. Sedimentation.
11. Destruction of agricultural land.
12. Threat to livelihood.
13. Human rights violation.
14. Damage to roads and bridges.

Mining and oil development may also pose risks to environment and forest ecosystem. Forests are the most biologically diverse terrestrial ecosystem. Tropical forests and particularly diverse and provide the greatest source of endemic plant species in the world. The direct impact of mining on forest ecosystem is the removal of vegetation. Indirect impacts include road building and pipeline development, which may result in habitat fragmentation and increased access to remote areas. While larger intact forest ecosystems may withstand the impact of mining and oil development, smaller fragments are likely to be particularly sensitive to clearing.

4.6.2 Mineral Resources

Minerals are naturally occurring inorganic, crystalline solids, obtained from various types of deposits in the earth. Some of these are necessary for human beings for their social, economic and cultural growth. A number of minerals, as iron, copper aluminium etc. are indispensable and are included in our daily requirement.

Mining and recovery of mineral resources has been in practice for a long time. National wealth basically comes from agriculture, manufacturing and mineral resources. Our complex modern society is built around the exploitation and use of mineral resources. Modern agriculture and the ability to feed an overpopulated world is dependent on mineral resources to construct the machines that enrich the soil with mineral fertilizers.

Mineral extraction is one of the many pressures on land use in a nation. As with most forms of development, mineral extraction and related activities require good development strategy stresses that economic growth needs to take place in a way that ensures effective protection of the environment and prudent use of natural resources. The overall aim is to meet society's needs for minerals at least social, economic and environmental. Some major principles for achieving a sustainable supply of minerals are :

- Minerals should be conserved, as far as possible, whilst securing an adequate supply to meet the needs of the economy and society.
- Environmental impacts caused by mineral operations and the transport of minerals should be kept to an acceptable minimum through good operational and management practices.
- The production of waste should be minimized and minerals should be used efficiently, taking appropriate account of the use of high quality materials and opportunities to recycle wastes.
- Restoration to sites should preserve or enhance the overall quality of the environment after working has ceased and where appropriate, make contribution to improved habitats and biodiversity.
- Areas of designated landscape, nature conservation or heritage value should be protected, as far as possible, from mineral development.
- The unnecessary sterilization of mineral resources by using the land permanently for other purposes should be avoided.

The origin and distribution of minerals is unequal and is related to the biosphere and the geologic cycle. If we discuss about our own country, we have sufficiently rich sources of minerals. India has a large number of economically useful minerals and they constitute one-quarter of the world's known mineral resources. Until 1993, India's mineral sector was state-owned. The Government has made a number of changes to its mining regulations in order to encourage exploration.

Different characteristics of mineral reserves of India are as follows :

1. India has a significant mineral industry that produces 87 minerals and metals, including fuel, metals, non-metals and minor minerals.
2. In 1999–2000, mineral production was worth US $10 billion, approximately 2% of the GDP.
3. India is an important producer of mica, barites, chromite, bauxite, iron ore, coal and lignite.
4. India's minerals and mining sector offers a number of opportunities, particularly for suppliers of the following :
   (*i*) Exploration, mining and ore-treatment equipment.
   (*ii*) Related technologies, such as technology for environmental management.
   (*iii*) Engineering, procurement and construction services.

There are large reserves of mica, bauxite, manganese, chromite, dolomite, gypsum, feldspar, limestone, gold, etc. Some minerals, such as mica and bauxite are even exported from India. Important areas of India providing some specific minerals are :

1. West Bengal, Jharkhand, Orissa, M.P. and A.P. –**Coal and lignite.**
2. Jharkhand, A.P., Meghalaya and Rajasthan-**Uranium, in the form of Pitchblende and urbanite ore.**
3. Jharkhand, West Bengal, Maharashtra, M.P. and Tamilnadu-**Aluminium in the form of bauxite ore.**

4. Jharkhand, Orissa, M.P., A.P., Tamilnadu, Karnataka, Maharashtra and Goa-**Iron in the form of haematite and magnetite ore.**
5. Rajasthan, Bihar, Jharkhand, Karnataka, M.P., West Bengal, A.P. an Uttranchal-**Copper in the form of copper pyrites.**
6. Bihar, A.P. and Rajasthan-**Mica.**
7. Tamilnad and Rajasthan-**Gypsum.**
8. Cuttack (Bihar) and Mayurbhanj (Orissa)-**Nickel ore.**
9. Kerala and West coastal beaches-**Ileminite.**
10. Sanapahar (meghalaya) and Pipra (M.P.)-**Silimanite.**
11. Ramagiri fields (A.P.), Kolar and Hutti (Karnataka)-**Gold.**
12. Ranna, Chatarpur & Satna (M.P.) and Banda (U.P.)-**Diamond.**
13. Assam and Gujrat – **Petroleum**
14. Assam, Tripura, Manipur, West Bengal, Punjab, H.P., Kutch and the Andamans-**Oils.**

4.6.3 Classification of Minerals

Different criteria may be chosen for classifying minerals. Some of the important classifications of minerals are :

I. **Classification Based upon their Structure :** On this basis, minerals may be classified into following groups.

1. **Native Elements :** These are minerals of single element. With the exception of gases, they make up about 20 minerals.
   These are further divided into metals, semi-metals and non-metals. *e.g.*, Au, Ag, Cu, Pt, Pd, Fe, As, Sb, Bi, Se, Te, S, C, etc.
2. **Sulphides:** They contain metallic cation and sulphur as anion. They include sulfarsendies, arsenides and tellurides *e.g.*, Galena (PBS), Pyrite (FeS_2), etc.
3. **Sulphosalts:** They contain As and Sb as cations and play a more important role than anions. *e.g.*, some important Ag minerals.
4. **Oxides and Hydroxides:** They contain anions of oxygen (O^{-2}) or hydroxyle ion (OH^-). *e.g.*, Hematite (Fe_2O_3), Brucute [$Mg(OH)_2$], etc.
5. **Halides:** They contain Cl^-, Br^-, F^- and I^- as anions. *e.g.*, Halite (NaCl), Fluorite (CaF_2), etc.
6. **Carbonates:** They contain CO_3^{-2} as anions. *e.g.*, Calcite ($CaCO_3$), Malachite [$Cu_2(CO_3)(OH)_2$], etc.
7. **Niltrates:** They contain NO^-_3 as anion. Only 7 minerals of this type are known.
8. **Borates:** They contain BO_3^{-3} as anion. About 100 minerals of this type are known and they can form polymers.
9. **Phosphates:** They contain PO_4^{-3} as anion. *e.g.*, Apatite [$CaF_5((PO_4)_3(F, Cl, OH)^-$], etc.
10. **Sulphates:** They contain SO_4^{-2} as anion. These can be hydrous or anhydrous in nature, *e.g.*, Gypsum ($CaSO_42H_2O$), Barite ($BaSO_4$), etc.
11. **Tungstates:** They contain WO_4^{-2} as anion. *e.g.*, Schelite, Wolframite, etc.
12. **Silicates:** They contain SiO_4^{-4} as anion. *e.g.*, Olivine [$(Fe, Mg)_2(SiO_4]$, Quartz (SiO_2), etc.
13. **Organics:** They are not true minerals. *e.g.*, Amber or Kidney Stones-Whewellite ($CaC_2O_4H_2O$), etc.

II. **Classification Based upon their Requirement :** On this basis minerals may be of two types :
1. **Critical Minerals,** required for the economy of a nation, *e.g.*, Fe, Al, Cu, Au, etc.
2. **Strategic Minerals,** required for the defence of a nation, *e.g.*, Mn, Co, Pt, Cr, etc.

III. **Classification Based upon their Properties :** On this basis too, minerals may be of two types:
1. **Metallic Minerals,** as aluminium, beryllium, chromium, cobalt, etc.
2. **Non-Metallic Minerals,** as concern felsporflo, nitrogen etc.

4.6.4 Uses of Minerals

Minerals have always been important because they provide a number of useful products. Since almost everything on earth is made from minerals, it may thus be stated that minerals are pretty important to life on earth and are required in one's daily diet to be strong and healthy.

Minerals play a vital role in our everyday life. Nearly every product we use contains minerals or is manufactured with machines or tools that are made of minerals. Minerals are used in a broad range of items from those we use everyday at hove, school and work. Minerals have widely been used by human beings since very early civilizations.

In general, minerals are used :
1. For generating atomic energy.
2. For the development and designing of industrial plants and machineries.
3. For weapons.
4. For transport and communication.
5. For construction of house.
6. In medicinal systems, particularly in Ayurveda.
7. In alloys.
8. In ornaments.

Iron and steel are most widely used minerals for different purposes. A brief discussion of different minerals is summarized in the following table:

Table 4.1. Occurance of mineral in India

	Mineral	*Reserves (in million tones)*	*Main reserves in India*	*Uses*
1	Bauxite	2,525.00	(*i*) Jharkhand, (*ii*) W.B. (*iii*) M.P. (*iv*) Maharshtra, (*v*) Tamilnadu	(*i*) Packaging food items. (*ii*) Transport. (*iii*) Utensils. (*iv*) Electronics. (*v*) Building material. (*vi*) Aircraft.
2	Chromite		(*i*) A.P. (*ii*) Bihar (*iii*) Karnataka (*iv*) Maharashtra (*v*) Manipur (*vi*) Orissa (*vii*) Tamilnadu	(*i*) High strength steel alloys. (*ii*) In texile industry. (*iii*) In tanning industry. (*iv*) In metallurgy. (*v*) In chemicals.

3	Iron (Haematite)	9606.00	(*i*) Jharkhand	(*i*) Heavy machinery.
			(*ii*) Orissa (*iii*) M.P. (*iv*) A.P. (*v*) Tamilnadu (*vi*) Karnataka (*vii*) Maharashtra (*viii*) Goa	(*ii*) Steel production. (*iii*) Transport. (*iv*) Building materials. (*v*) In industries.
4	Copper	431.04	(*i*) Rajasthan (*ii*) Bihar (*iii*) Jharkahnd (*iv*) Karnataka (*v*) M.P. (*vi*) W.B. (*vii*) A.P. (*viii*) Uttranchal	(*i*) Electric goods. (*ii*) Electronic goods. (*iii*) Construction of buildings. (*iv*) Vessels. (*v*) Alloys. (*vi*) Piping. (*vii*) Metal alloys.
5	Lead	189.55	(*i*) Rajasthan (*ii*) Gujrat (*iii*) W.B. (*iv*) A.P. (*v*) Meghalaya (*vi*) Orissa (*vii*) Sikkim (*viii*) Tamilnadu (*ix*) U.P	(*i*) Leaded gasoline. (*ii*) Car batteries. (*iii*) Paints. (*iv*) Alloys.
6	Manganese	233.33	(*i*) Karnataka (*ii*) Orissa (*iii*) M. P. (*iv*) Maharashtra (*v*) Goa	(*i*) High strength, heat resistant alloys. (*ii*) Steel. (*iii*) Disinfectants.
7	Gold	17.69 (tons)	(*i*) Kolar (*ii*) Karnataka (*iii*) A.P.	(*i*) Ornaments. (*ii*) In medicines. (*iii*) In electronic goods. (*iv*) In aerospace. (*v*) In dentistry. (*vi*) In electrical conductors. (*vii*) Face shields of fire fighters. (*viii*) In building windows.
8	Nickel	183.5	(*i*) Orissa (*ii*) Bihar	(*i*) Chemical industry. (*ii*) Steel alloys.

9	Limestone	76,466.00	(*i*) M.P. (*ii*) Tamilnadu (*iii*) A.P. (*iv*) Gujrat (*v*) Bihar (*vi*) Orissa (*vii*) Rajasthan (*viii*) Karnataka	(*i*) Concrete. (*ii*) Building stone. (*iii*) Agriculture. (*iv*) Cement industry.
10	Gypsum	239.00	(*i*) Rajasthan (*ii*) Tamilnadu (*iii*) Jammu and Kashmir (*iv*) U.P. (*v*) H.P. (*vi*) Gujrat	(*i*) Plaster wall board. (*ii*) Agriculture.
11	Rock phosphate	145.95	(*i*) M.P. (*ii*) Rajasthan (*iii*) U.P.	(*i*) Fertilizers. (*ii*) Chemicals.
12	Tungsten	39.65	Rajasthan	(*i*) Alloys. (*ii*) Chemicals.
13	Fluorspar	3.04	(*i*) Rajasthan (*ii*) Gujrat	(*i*) Flux. (*ii*) Refreigerants. (*iii*) Propellants. (*iv*) In acids.

4.6.5 Exploitation of Minerals

Minerals are valuable natural resources being finite and non-renewable. They constitute the vital raw material for many basic industries and are a major factor for development. The rate of consumption of minerals is increasing continuously. Human population growth and increased modern industry are depleting our available resources at increasing rate. The consumption of natural resources increased at a phenomenal rate during the past hundred years. The population and production increase cannot continue without increasing pollution and depletion of mineral resources. Management of mineral resources has, therefore, to be closely integrated with the overall strategy of development and exploitation of minerals is to be guided by long term national goals and perspectives. Some steps must be taken to cause the minimum wastage of this valuable resource. Some of the important measures useful in this direction are :

1. Recycling and recovering.
2. Adoption more efficient technologies.
3. Exploiting untapped deposits.
4. Deep sea-mining.

Besides the above methods, some legal steps have also been adopted to save these precious minerals. The regulation of mines and development of mineral resources in accordance with the national goals and priorities has been considered to be the responsibility of the Central and State Governments. The basic objectives of the mineral policy with respect to minerals are as follows .

1. To explore for identification of minerals wealth on the land and in off-shore areas
2. To develop mineral resources taking into account the national and strategic considerations and to ensure their adequate supply and best use keeping in view the present needs and future requirements.
3. To promote necessary linkages for smooth and uninterrupted development of the mineral industry to meet the needs of the country.
4. To promote research and development in minerals.
5. To ensure establishment of appropriate educational and training facilities for human resources development to meet the manpower requirements of the mining industry.
6. To minimize adverse effects of mineral development on the forest, environment and ecology through appropriate protective measures.
7. To ensure conduct of mining operations with due regard to safety and health of all concerned.

4.6.6 Environmental Effects of Extracting and Using Mineral Resources

Minerals are actually the ores containing different inorganic elements. In order to recover these elements, their respective ores are subjected to a series of operations, collectively known as extraction. The extraction of elements from these minerals includes some processes which have an adverse effect on the related environment around it. The effects so produced in this procedure are not uniform everywhere and depend upon different factors, such as

1. Quality of ore.
2. Hydrological conditions.
3. Climate.
4. Type of rocks.
5. Topography.

Minerals is an activity with severe environmental effects. The threat affects 40 % of forests and major ecosystems on the planet. Environmental issues can also include erosion, formation of sinkholes, dangerous sites such as deep holes in child friendly communities, destruction of the environment, loss of biodiversity and contamination of ground water by chemicals from the mining processes and products.

Further, ore extraction distrupts the topsoil and can displace local animals and plants and sometimes native human population. Runoff can contaminate nearby water sources with pollutants such as mercury and sodium cyanide used in gold mining. Waste materials and smelters can cause sulphurous dust clouds that result in acid rain.

1. Exploration stage.
2. Mining stage.
3. Processing stage.
4. Dereliction stage.

1. **Exploration Stage :** Exploration and testing is the first stage involved in extraction process involving surface mapping and drilling. This stage causes a much less effect on the environment.

2. **Mining and Processing Stage :** Exploration and testing is followed by mining and processing. Mining is done to extract minerals from deep deposits under ground. Mining may be of two types:
   (*i*) Sub-surface mining.
   (*ii*) Surface mining.

   Surface mining may be done by three methods:
   (*i*) Open-pit mining involving digging holes by machines to scrap out ores.
   (*ii*) Dredging involving use of chained buckets and draglines to drag minerals.
   (*iii*) Strip mining involving use of bulldoxers, power shovels and stripping wheels.

   Mining and processing have severe effects on environment as well as on society. It causes :
   (*i*) Land degradation.
   (*ii*) Devegetation of landscapes.
   (*iii*) Surface and ground water pollution due to release of certain harmful elements in trace amounts.
   (*iv*) Reduced growth of vegetation.
   (*v*) Air pollution due to emission of dust and gases.
   (*vi*) Deforestation.
   (*vii*) Tilting of buildings.
   (*viii*) Cracks in houses.
   (*ix*) Bending of rail tracks.
   (*x*) Gas leakage from cracked pipelines.
   (*xi*) Physical changes in the associated land, soil, water and air.
   (*xii*) Changes in nutrient cycling, biomass, species diversity and ecosystem stability.
   (*xiii*) Alteration in surface and ground water availability.
   (*xiv*) Rehabilitation of affected population.
   (*xv*) Stress on water suppliers, sewage and solid waste disposal systems.
   (*xvi*) Shifting of land from open forest and agricultural to urban patterns.
3. **Dereliction Stage :** This stage results from the uncontrolled mineral extraction and processing. Its effects are :
   (*i*) Waste of agricultural and industrial land.
   (*ii*) Unpleasant outlook.
   (*iii*) Health hazards including respiratory and skin diseases.
   (*iv*) Accidental hazards due to unfilled shafts.
   (*v*) Permanent damage to landscape.

4.7 FOOD RESOURCES

Food is an essential requirement of body. Food materials are the materials and articles intended to come into contact with foodstuffs, including packaging materials along with cutlery, dishes, processing machines, containers, etc. The term also includes materials and articles which are in contact with water intended for human consumption. Many food sources are seasonal (*e.g.*, fruits, Forde 1986), having a regular seasonal occurrence during a particular period of the year while others are irregular. They may be unpredictable from year to year (e.g. nectar. Mac Nally & Mc Goldrick 1997) or have cycles that include a variable number of years between years of relatively abundant supply. The role of food resources is very important as well as complex. Healthy food is required for the proper growth and development of body. Human diet cannot be restricted to a particular article of food, rather each and every constituent is required in suitable amount. Main sources of food are:

1. **Crops :** About 2,50,000 species of plants are known, but among them only 3,000 species are grown as agricultural crops. Out of these, only 300 are grown for food and other purposes. It is surprising to note that out of these 300 crops, about 100 crops may be used for food but only 20 of them constitute the major crops consumed by human being. These include,

(*i*) Wheat	(*ii*) Rice	(*iii*) Corn	(*iv*) Potatoes
(*v*) Barley	(*vi*) Sweet potatoes	(*vii*) Cassavas	(*viii*) Soyabeans
(*ix*) Oats	(*x*) Sorghum	(*xi*) Millet	(*xii*) Sugarcane
(*xiii*) Sugar beet	(*xiv*) Rye	(*xv*) Peanuts	(*xvi*) Field beans
(*xvii*) Chick peas	(*xviii*) Pigeon peas	(*xix*) Bananas	(*xx*) Coconuts

Health maintenance without balanced and proper nutrition is not possible. A minimum amount of dietary energy intake is essential for body weight maintenance and work performance. For good health people need not only a certain amount of calories but also food quality that provides them with the proper amount of proteins, carbohydrates, lipids, vitamins and minerals. Some of the important groups are discussed here in brief.

Carbohydrates, like starch account for most of our daily intake of calories. Common sources of carbohydrates include white rice, cassava, potatoes, grains and sugars.

Lipids are organic compounds rich in fats and oils. They supply us with about twice as many calories per gram as starch or proteins. Fats and oils are used by our body to construct cellular membranes.

Protein is required to supply our body with ten essential amino acids which our body cannot manufacture itself. An average individual requires about 40 grams of protein every day.

Mineral nutrients form another important food requirement for human beings for survival. Minerals, *i.e.*, inorganic elements such as iron and calcium are essential for the normal functioning of the body. The most common mineral deficiencies globally are those of calcium, iron and iodine.

Finally, human beings also require vitamins, specialized organic molecules that we cannot synthesize ourselves.

Fruits and vegetables are also essential in the human diet as they contain vitamins, minerals, dietary fibres and complex carbohydrates.

Of the 350,000 species of plants catalogued by scienlists, only about 100 crops are primarily used to feed the world. About 15 plants and 8 animal species supply 90% of our food. Wheat, rice, corn and potato are the primary crops that provide us with the bulk of the starch we consume. Other important crops include barley, sweet potato, cassava, grapes, soyabean, oats, sorghum, sugarcane, millet, banana, tomato, sugar beet, rye, orange, coconut, cottonseed, apple, yam, peanut, watermelon, cabbage, onion, bean, pea, sunflower seed and mango.

2. **Livestock:** Domestic animals also provide a good source of many nutritious requirements such as milk. Some animals are also used as food for meat. Milk and meat consumption is more in developed countries of North America, Europe and Japan.

4.7.1 World Food Problems

The amount of caloric energy needed for human survival varies with health, climate, body weight, fitress and level of metabolism. The basic minimum requirement for human beings is 2,700 calories for men and 2,000 calories for women. People living an active life in a cold climate may require as much as 7000 calories of energy for survival. The average daily caloric intake of people living in industrialized western countries is about 3500 calories.

A perfect diet containing the required quantity of every nutrient required by the body is known as balanced diet. A balanced diet contains an appropriate quantity of each of its constituents. Balanced diet is necessary for good health. But it has been investigated by 'The Food and Agriculture Organization' (FAO) that a very large number of people in the world do not get the so called balanced diet and hence suffer from a number of problems. The World Bank estimates that more than 630 million people get less than 80% of their basic needs of calories, nutrients or both. Many of the world's people do not get enough protein in the balanced quantities needed for survival. In more developed countries, people suffer from a diet too rich in refined sugar and too low in roughage while people in less developed countries have diets rich in carbohydrates, but lacking in proteins, fats, minerals and vitamins.

Increasing population and inequitable distribution of income are the two chief causes of the food insufficiency. Food insufficiency, *i.e.*, improper dietary intake leads to:

1. Undernourishment
2. Malnourishment.

1. **Undernourishment :** The intake of less than 90% of the minimum dietary requirement is known as undernourishment. It is based on a comparison of usual food consumption expressed in terms of dietary energy (kcal) with certain energy requirement norms. According to FAO, the average minimum daily calorific requirement for an individual is 2500 calories per day. People taking less than 90% of daily requirement are said to be undernourished. According to World Health Organization about 1.3 billion or roughly 1 out of 4 people are underfed and undernourished. On an average, people in the richest countries eat 30–40% more calories than needed, while people in the poorest countries eat 10% less than needed and may be considered undernourished. Undernourishment is a serious, potentially life-threatening situation in the elderly. In the condition of undernourishment, the living being does not have sufficient energy to lead an active and productive life. Consequently, it has a little or no ability to work. Further, due to insufficient energy and nutrients, their body becomes susceptible to a number of infectious diseases.

   One more severe condition of undernourishment is 'serious undernourishment' in which people take less than 80% of their daily calorific requirement. It leads to stunted growth, mental retardation and other types of disorders in the living beings.

2. **Malnourishment :** Deficiency of specific and essential components of food, such as proteins, vitamins, etc., is known as malnourishment. Malnourishment strictly does not mean insufficient intake of food. It occurs when a person's diet does not meet the quality required in terms of minimum daily requirements for proteins, vitamins, minerals and other nutrients, *i.e.*, it results from improper intake, as when a person is taking full diet to satisfy his appetite but not in proper proportion according to requirement, it may cause malnourishment. This condition is very common in richer countries. Otherwise, in poorer countries, it results due to economic reasons. Most people in less developed countries suffer from nutritional imbalance and are

malnourished because their diet has low protein but high starch. According to the World Bank and the United Nations, from 1 to 2 billion human beings are now malnourished, indicating a combination of insufficient food, low income and inadequate distribution of food. This is the largest number of hungry human population ever recorded in history. In China about 80 million are now malnourished and hungry.

The effects of malnourishment are long term and result in a number of diseases, summarized here along with their deficiency cause.

S.No	*Disease*	*Deficiency cause*
(*i*)	Marasmus	Proteins and calories
(*ii*)	Kwashiorkor	Proteins
(*iii*)	Anaemia	Iron
(*iv*)	Goiter and hypothyroidism	Iodine
(*v*)	Pellagra, scurvy, beriberi, anemia, ricket and xeropthalmia	Vitamins
(*vi*)	Chronic hunger	Insufficient food

Both, undernourishment and malnourishment lead to a reduction in metal and physical efficiency and lowering of resistance to diseases. In the developing world, lack of adequate food is a common cause of death. Infectious diseases are the immediate cause of death for most of the 11 million children under the age of five, who die each year in developing world. The four biggest killers of children are dminlor, acute repository, illness, malaria and measles. The risk of fatality due to of these disease is far greater for children who are hungry and malnourished.

4.7.2 Changes Caused by Agriculture

The continued production of an adequate food supply is directly dependent on ample fertile land, fresh water, energy along with the maintenance of biodiversity. More than 99% of the world's food supply comes from the land while less than 1% is from oceans and other aquatic habitats (Pimentel et al., 1994). As the world population continues to grow geometrically, great pressure is being placed on arable land, water, energy, and biological resources to provide an adequate supply of food while maintaining the integrity of our ecosystem. At present, fertile cropland is being lost at an alarming rate. For instance, nearly one-third of the world's cropland (1.5 billion hectares) has been abandoned during the past 40 years because erosion has made it unproductive (Pimentel et al., 1995). Most replacement of eroded agricultural land is now coming from marginal and forest land.

Agriculture is one of the most widely practiced occupations all over the world. Due to different factors and techniques employed in agriculture, life and environment have been affected to a great extent. These effects cause some adverse changes at the site of agriculture as well as in the places away from it. Some of the important direct and indirect impact of agriculture are:

1. Deforestation due to slash and burning of trees in forests to clear the land for cultivation. The pressure for agricultural land accounts for 60 to 80% of the world's deforestation.
2. Soil erosion due to running off of fertile layer of soil.
3. Devastation of crops by some diseases caused by the attack of pathogens.
4. Micronutrient imbalance in the crops due to excessive use of fertilizers containing nitrogen, phosphorus and potassium.

5. A serious health disease, called blue baby syndrome or methaemoglobinemia caused by the concentration of nitrates in water beyond limit by excessive use of nitrogenous fertilizers.
6. Eutrophication, *i.e.*, over nourishment of lakes receiving washings and running off water containing large proportion of nitrogen and phosphorus from crop fields.
7. Biological magnification caused by the accumulation of non-biodegradable pesticides in the food chain.
8. Water logging due to overirrigation of crops.

4.7.3 Changes Caused by Overgrazing

Livestock play an important role in the rural life of a country. Livestock require grazing land for their existence. But land can provide food to a definite number of livestock according to its carrying capacity. The maximum population that can be supported on a particular piece of land is called its carrying capacity. When more than the optimum number of livestock depend on this land, the land is said to be overgrazed and the condition as overgrazing. Overgrazing has a number of negative impacts on land such as :

1. Reduction in the diversity of plane spacie.
2. Reduction in vegetation.
3. Abundance of some particular type of crops not liked by livestock.
4. Soil erosion.
5. Land degradation.
6. Loss of useful species.

4.7.4 Effects of Modern Agriculture

Traditional methods of agriculture have been improved by making use of good quality seeds or hybrid valfle high quality and modern technologised equipments and a number of fertilizers, pesticides, etc. During mid-1800s began an era of great change in the field of agriculture, leading to agricultural revolution, which brought advancement in cultivation methods, breeding of improved crop varieties and use of fertilizers and crop rotations to maintain soil productivity. The use of pesticides also began in the mid 1800s, when it was discovered that dusting of grape plants with sulphur provided a cure for powdery mildew. Chemical control of agricultural pests expanded rapidly after these initial discoveries and by 1893 there were 42 patented insecticides were also offered by several manufactures. All this has resulted in tremendous increase in the production of crops. But, besides this positive aspect, these techniques have also some negative impacts, both on environment and health.

1. **Effect of Fertilizers :** In addition to sunlight, carbon dioxide and water, plants require some inorganic nutrients, in small amounts for their proper growth. These inorganic elements are N, P, K, Ca, Mg and S. Among these six, first three, *i.e.*, N, P and K are most important and stimulate growth and increase crop yield. These are called NPK fertilizers. Of these, nitrogen is the most readily lost because of its high solubility in the nitrate form. Leaching of nitrate from agricultural fields can elevate concentrations in underlying groundwater to levels unacceptable for drinking water quality. Phosphorus does not leach as readily as nitrate because it is more tightly bound to soil particles. However, it is carried with eroded soils into surface water bodies, where it may cause excessive growth of aquatic plants. Potassium, the

third major nutrient in fertilizers, does not cause water quality problems because it is not hazardous in drinking water and is not a limiting nutrient for growth of aquatic plants. It is tightly held by soil particles and so can be removed from fields by erosion, but generally not by leaching.

Advantages : Substances supplying these essential nutrients are generally artificial in their preparation and have several advantages over naturally occurring organic fertilizers as,

(*i*) They can be easily stored, handled, applied and transported.

(*ii*) They provide grazing land very soon after their application.

Disadvantages: Besides above advantages, fertilizers also suffer from some disadvantages, especially from hygienic point of view.

(*i*) An uneven application may lead to contamination resulting in less crop production.

(*ii*) Nitrogenous fertilizers generally have a tendency to go deep into the soil so that they get concentrated in the ground water. This becomes the cause of a serous health disease, viz. blue baby syndrome or methaemoglobinemia leading to the death of infants.

(*iii*) Phophatic fertilizers have resulted in the pollution of waterbodies causing eutrophication.

2. **Effects of pesticides :** Pests are the undesirable competitors, parasites or predators including a number of organisms, such as insects, nematodes, bacteria, virus, weeds and vertebrates. These pests are harmful for the crops. The growth of pests may be checked by the use of certain pesticides which actually kill these pests. Besides killing pests, pesticides are used:

(*i*) To increase crop production.

(*ii*) To reduce the diseases in plants caused by pests.

(*iii*) To improve the outlook and appearance of crops.

(*iv*) For preservation and maintenance of buildings, cloths and furniture.

A number of pesticides are used in agriculture. Some important ones are DDT (*p*, *p*-dichlorodiphenyltrichloroethane, and the compounds containing sulphur, arsenic, etc.

Advantages : Pesticides are particularly useful in increasing crop production and decreasing different diseases caused by pests.

Disadvantages : Undoubtedly use of pesticides has been an important source of the social, agricultural as well as of economic benefit, but it also causes a number of hazards.

(*i*) Besides their targets, pesticides also affect the side crops which are not intended to be affected.

(*ii*) They are not selective in nature. It means that they don't know which pest is to be killed and which is just to be injured.

(*iii*) They are toxic in nature.

(*iv*) They have a tendency to get accumulated in food web.

(*v*) Sometimes they kill useful predators, killing pests thus creating new pests.

(*vi*) They can cause acute poisoning and illness in human being by their high dose and accidental exposure.

(*vii*) They may even lead to cancer, birth defects, immunological problems and other chronic degenerative diseases.

3. **Effects of Water Logging:** Water logging is the condition of surface flooding and the formation of high water table. It generally arises due to over-irrigation of croplands by farmers for better growth of crops. Another major cause of water logging is improper drainage system.

Disadvantages: (*i*) The water logged soil has less oxygen available for respiration by plants.

(*ii*) It reduces the mechanical strength of soil.

(*iii*) It results in lower productivity of soil.

4. **Effects of salinity:** Condition of increased concentration of soluble salts in soil is known as salinity. One major cause of salinisation of soil is excessive irrigation. In absence of proper irrigation and drainage of flood water, dissolved salts in water accumulate on the soil surface. In the areas of low rainfall and high temperatures, water evaporates leaving the salts behind in the soil.

Disadvantages :

(*i*) Salinity is injurious to plant growth and survival due to formation of a crust of salts on the soil surface.

(*ii*) It also affects proper absorption of water by plants.

Besides this, environmental and human health effects of modern agriculture are wide ranging and well-documented and include:

1. Pesticides contaminating water and harming wildlife and human health.
2. Nitrates and phosphates from fertilizers, livestock wastes and silage effluents contaminating water and so contributing to algal blooms, deoxygenation, fish deaths and nuisance to leisure users.
3. Soil erosion disrupting water courses and runoff from eroded land causing flooding and damage to housing and natural resources.
4. Harm to consumers exposed to harmful residues and micro-organisms in foods.

4.8 ROLE OF AN INDIVIDUAL IN CONSERVATION OF NATURAL RESOURCES

Natural resources are very valuable for us and play a vital role in deciding the development and financial status of a nation. But as discussed above over-use or over-exploitation of these resources has led to drastic hazards, both to human life and environment. Hence, it is necessary to conserve these gifts provided to us by nature. Conservation may be done and is being done on large scale by different organizations, but it is necessary that each and every individual must wake up to ensure his/her participation in this act on his/her personal level.

Most conservation problems are caused by human behavior and solving these problems requires understanding and changing that behavior. Although psychology is the discipline most focused on the study of human behaviour bit its role in biodiversity conservation has been surprisingly minimal compared to the other science. Conservation psychology is a relatively new field that seeks to understand why people help or hurt the natural environment and how to promote conservation.

Growing population density lead to lands scarcity and widespread changes in land use. Human activities result in environmental problems such as deforestation, overgrazing and the depletion and contamination of land and water resources. To solve these problems and ensure that future generations may can enjoy the benefits of the earth's resources, improved management of these resources is needed. Planners, managers, policy makers and researchers alike need to understand the complexity of factors involved. They must collect and interpret the required data and work together with professionals from other disciplinary fields. Geo-information technology and particularly remote sensing, plays an important role in these tasks.

To solve environmental problems and manage resources, people need to have skill in many areas-science, policy, economics, communication, history and design.

Environmental education is a learning process concerned with the interrelationship within and between the various components of the natural and man-made world resources, producing growth in the individual leading to responsible stewardship of the earth.

In order to provide appropriate training and capacity building to effective individuals and institutions for conservation, it is necessary to :

1. Define the different types of individual conservationists and institutions.
2. For each of these, define the required competencies for effective individuals and attributes for effective institutions.
3. Identify the most appropriate ways in which those competencies and attributes can be realized.

The goal of environmental education is to develop a citizenry that is aware of and concerned about the total environment and its associated problems and which has the knowledge, attitude, motivation, commitment and skill to work individually and collectively toward solution of current problems and the prevention of new ones.

The important goals of conservation education programs are :

1. To increase awareness and understanding of interrelationship in natural systems and between people and the land.
2. To enable people to recognize the increasing complexity of natural resource management and to give them information that will allow them to make choice within social, cultural, political and scientific realities.
3. To foster the individual's responsibility to conserve and wisely use natural and cultural resources in forest, grassland and aquatic ecosystems.

The Resource Conservation Challenge (RCC) is a national effort to conserve natural resources and energy by managing materials more officially. We conserve energy and preserve natural resources by committing ourselves to :

1. Reduce waste
2. Reuse and recycle more products.
3. Buy more recycled and recyclable products.
4. Reduce toxic chemicals in waste.

The RCC is working hard to reduce waste and increase the reuse and recycling of materials. We've targeted the nation's largest waste streams and set priorities for :

1. Recycling municipal solid waste.
2. Reusing and recycling industrial materials.
3. Buy more recycled and recyclable products.
4. Promoting green initiatives, especially the safe design and recycling of electronics.

Some basic precautions which may be taken by an individual to conserve different natural resources are given here :

1. **Conservation of Forest Resources :**

   (*i*) One should oppose the cutting of trees.

(ii) Crops must be cut in such a manner as to avoid soil and tree damage.
(iii) The forest wealth must be kept safe while working in forests.
(iv) Plantation must be enhance.
(v) Working techniques must not affect the environment of forests.

Conservation education programs are strong elements of the Forest Service mission, vision and guiding principles. These programs offer the American public opportunities to better understand forest ecosystems and the active role they play in the future of these systems.

2. **Conservation of Water Resources :** Environmental problems caused by industrial waste and pesticide residues occur in river basins. Improperly processed industrial waste has caused water pollution, which has led to the destruction of water organisms and the deterioration of people's health. Household waste released by people lining along the rivers and the use of dynamite for fishing have worsened the situation. Some basic steps for conserving water are :

(i) Water taps must be closed after use.
(ii) Water must not go on flowing while washing, brushing, bathing or shaving.
(iii) Flushes used in toilets must be of small capacity.
(iv) Leakage must be checked in pipes.
(v) The washings of soapy water may be used for washing vehicles and for plants.
(vi) Water the plants in early mornings or in late nights when evaporation losses are minimum.
(vii) Drip irrigation and sprinkling irrigation techniques must be employed.
(viii) Rain water may be collected and may be used for different activities other than drinking and cooking.

Recent national reports in the United States have documented the importance and decline of marine resources. New efforts and approaches are needed to create and maintain healthy, sustainable and resilient marine ecosystems. Traditional approaches to science, restoration and management within the marine environment have unsuccessfully linked the needs of the environment, economy and local communities. These traditional approaches must be supplemented by new, innovative approaches that unity environmental, economic and social needs.

Transboundary Natural Resources Management (TBNRM) is a recent initiative whose objectives cover conservation and dflsajd development needs wide a holistic approach in border regions.

3. **Conservation of Food Resources :**

(i) Food must not be wasted.
(ii) Use of fertilizers and pesticides must be decreased.
(iii) Drip irrigation must be employed.
(iv) Seasonal vegetables, easy to avail, transport, store and preserve must be used.

4. **Conservation of Energy Resources :**

(i) Lights, fans and other electrical appliance must be turned off when not in use.
(ii) Use of natural sources of energy may be preferred over electric sources.
(iii) Solar cookers may be used for cooking instead of fuel or LPG.
(iv) While drawing the map for a building, proper natural light and heat provision must be kept in mind.

(*v*) Plantation must be enhance.

(*vi*) Engines of vehicles must be stopped when not in motion.

5. **Conservation of Land Resources :**

(*i*) Trees must not be uprooted for the construction of buildings.

(*ii*) Plantation must be done.

(*iii*) Watering of plants must not be done by a strong flow of water as it may wash away the soil with it.

(*iv*) Sprinkling irrigation must be adopted.

(*v*) Green manure must be used.

(*vi*) Fields must not be over-irrigated.

4.9 EQUITABLE USE OF RESOURCES FOR SUSTAINABLE LIFESTYLES

Nature has rewarded numerous gifts to mankind in the form of different resources. Proper and appropriate utilization of these resources requires the application of suitable technologies.

In the early ages, man had limited knowledge and he simply used to consume plants and animals around him. But with time, he started cultivation, using wind and water and as time passed industrialization took place. The consumption of resources in the industries has increased rapidly and has added a lot to the financial status and economy of mankind. The consumption of resources is to different extents in different countries. It is important to note that the more developed countries consume about 80% of the total natural resources although their population is only about 23% of the total world's population. On the other hand the developing countries consume only about 12% of the natural resources whereas their population is about 78%.

Sustainable lifestyles and ways of working are central to overcoming poverty and conserving and protecting the natural resources base for all forms of life. Sustainable methods of production are needed in agriculture, forestry, fishing, and manufacturing. Use of resources needs to be minimized and pollution and waste reduced.

Indeed there is a need to reduce the impact of lifestyle consumption habits on society and resources to ensure equitable availability of resources for all around the world. Education and training for sustainable production and consumption play an important role in achieving this by creating a more critical and responsible attitude towards consumerism in our everyday life.

Sustainable livelihoods are those that meet the financial needs of average citizens while supporting cleaner, healthier environment and protecting the natural resource base for future. Sustainable livelihood encompasses a wide range of innovative activities such as ecotourism, organic farming, sustainable business and community-owned and managed natural resources.

To have a sustainable life style, there must be a balance in the consumption of natural resources between developed and developing countries. But due to only one sided advancement of developed countries, the sustainability of earth's life system is continuously deteriorating and is under threat of destruction. Actually, this condition is a movement towards unsustainability. There are two basic causes of unsustainability :

1. Overpopulation in poor countries resulting in underconsumption of resources.
2. Over consumption of resources by rich countries.

This problem of unsustainability may be solved by an equitable distribution of resources and wealth worldwide. In the same direction a global consensus has to be reached for more balanced distribution of basic resources like drinking water, food, etc. so that poor people in-developing countries may at least meet their basic needs to lead a sustainable life. Sustainable development and global equity also require a redirection of our lifestyle in the industrialized countries. Even today, four planets would be required if all people were to enjoy the lifestyle of the industrialized nations.

Now, if we concentrate on resources, non-renewable resources are limited in their quantity and quality. Even the renewable resources may become unsustainable if their use exceeds their regeneration. Thus, the management of these resources should be based on the concept of maximum sustainable yield, according to which biological resources should not be exploited above their biological reproduction rate without management.

Due to this increasing global concern of man's mismanagement of natural resources, United Nations Conference on Environment and Development in Rio de Janeiro in 1992 established a new UN body, the Commission on Sustainable Development, which is working with its full potential in this direction.

4.10 TECHNOLOGICAL GROWTH AND DEPLETION OF OIL RESERVES

In the 21st century the world must solve two great problems. These problems are rarely discussed by the public and have received little media attention. Neither are they discussed by those in powers, at least not publicly. The two problems are:

- Overpopulation in the developing world.
- Over consumption in the developed world.

Of course it's all about resources. Either too many people consume a few resources each totaling a lot of resources, or a few people each consume too many resources again totaling a lot of resources. What is important to note is the absolute consumption, not the relative consumption.

It is difficult to differ between the developing world and the developed world, but the canonical values usually used are that:

- The developing world has 80% of the world's population and consumes 20% of its resources.
- The developed world has 20% of the world's population and consumes 80% of its resources.

Today (spring 2002) there are about 6,200 million persons in the world. In 1998 energy was consumed at a rate of about 12.69 (US) trillion Watts (12.69TW) [USGSE98].

The Problem

The main reasons are;

- The ecological systems from which the developing world derives their food, water, and energy can not be sustained with an increasing demand from an increasing population base. If the demand keeps increasing the systems will eventually collapse. As the systems collapse the population which is a part of these systems will collapse, die off, or try to escape to places where resources are more abundant.

- The resource base from which the technically more adept developed world derives their food, water, restrictions. This will change the geopolitics and the economic system of the developed world. Eventually this will lead to a paradigm shift.

The Focus

The problems mentioned above have similarities and they are not separated as the developed world interacts with the developing world in many ways. The problems in the developing world are water shortages, droughts, floods, famines, forest losses, soil erosion and loss of biodiversity. These all decrease the carrying capacity of the life support system. The number of people and their behavior determine the rate of loss of this support system. Usually their behavior is motivated by personal survival instincts. *e.g.*, energy is required to cook and make the available low-quality foods edible, but as most of the trees have been cut down already, dung is collected and burned which in turn decreases soil fertility and increases erosion. Many of these people already live using a minimum of resources. Their problems are determined by their huge number which increase rapidly since most choose to have many children.

The problems in the developed world-and these problems are not so visible-are pollution, damage of eco systems, loss of biodiversity, loss of forests, soil erosion and depletion of the mineral and energy resource base on which their society is based. The loss of resources is determined by the behavior and rate of consumption of these people. Their behavior is motivated by greed-to sell and buy and increasing number of consumer goods, but also by necessity, since it is easier to successfully fit into society and swim with the stream than to drop out or swim against the stream. Their problem is their overconsumption and dependence on a great and cheap resource base to fuel their economy and the setup of their society.

The problems are grand in scope and it is unfortunately impossible to be an expert on all aspects of the problems. Experts tend to focus too much on their own area and might even be completely ignorant on other areas which is worse, but with this caveat in mind, I will try to give a general amount of power which indicate that they also have the greater power in the geopolitical sense. Furthermore I am lucky to live in the developed world, which makes a focus on this part of the problem personally relevant.

4.10.1 Oil Depletion: The Primary Problem of the Developed World

In dealing with several problems it is worth to spend time singling out the most important problem of the set such as not to waste too much time on relatively trivial matters. I believe that the *primary problem* is that of a sustained energy supply. Energy unlocks all other resources. Energy is required to transport raw minerals, to refine these, to manufacture goods, and to transport the goods to their destination. Without energy none of these functions which are essential to society are possible.

Today the primary energy source is oil. **Oil accounts for 40% of all energy use** [USGSE98], therefore continued access to this resource or an equivalent or an improved replacement is essential to continue the world as we know it.

A reader with some foresight may suspect that I believe that continued access to the oil resource is not possible. He may also suspect that I do not believe that any immediate replacements exist. He would be absolutely correct. Now I will try to explain why.

4.10.2 The Scale of the Problem

To solve a problem one must understand the magnitude of the problem. To reiterate previous numbers energy is consumed by humans at a rate of about 13TW (ITW equals one (US) trillion Watts). A very large fraction (around 40%) comes from oil. Oil is therefore the primary energy source, and it is the primary energy source which should concern us.

The world consumes 77 million barrels (one barrel is 42 (US) gallons or 159 (SI/metric) liters) of petroleum daily, which makes 26 billion barrels annually. The biggest extractors are Saudi Arabia, Russia, the United States, and Mexico. The biggest exporters are Saudi Arabia, Russia, and Norway. The biggest importers are the United States, Japan, Korea and Germany [IEAKEY2003].

A nuclear power plant produces about 0.5–1.0GW. It does not run continuously and is offline some 20–40% of the time. A rough calculation shows that a replacement of the energy of oil by nuclear energy will require the construction of at least 5,000 nuclear power plants. A modern off-shore wind turbine produces about 2MW depending on the wind speed. Hydro power plant depends on the site, so I can not give an estimate. Solar power using PV cells depends on the sun facing area of the cells. I do not know the output of coal or gas fired plants. If anyone has some good numbers, please tell me.

These numbers do not tell the entire story. A nuclear power plant produces electricity, and one can not use electricity to make plastics, fertilizer, and a whole bunch of other industrial products. Additionally the world's transport system is based on the internal combustion engine which in terms of output/mass is much more effective than any other engine (steam, Stirling, electric, etc.,) only gas turbines are more effective, but they are not as robust, and they also depend on fossil fuels. Electricity is an inconvenient source of energy for many purposes. It is only transportable through batteries, by cable, or by converting water into hydrogen. Both conversion methods loose energy in the process, especially the former. In conclusion oil is a source of energy as well as a mineral source which is difficult or impossible to replace.

Do not forget Liebig's law. Even if the energy problem is solved the world will still be facing water shortage, topsoil loss, and loss of biodiversity. One can only hope that possible replacement sources will be used wisely.

4.10.3 Oil Follows the Extraction Pattern of all Other Resources

Generally the easy to reach and rich resources will be found and used first. These resources can be exploited using simple technology and often a hole in the ground in the right place which is not so hard to find will do the trick. Later more complex technology is required. Oil fields will be smaller, require more effort, and ultimately yield less oil than the first big fields. Later still, advanced technology like 3D or 4D seismic searches, directional drilling, steam injection, and drilling in difficult terrain like off shore or arctic conditions is required. At some point the effort, namely the energy which goes into the process *i.e.*, of manufacturing drilling rigs, actually finding the oil, keeping the crews supplies, and getting the oil to the surface will surpass the energy yield at which point further drilling makes no sense. Presently the limits are determined by economical arguments *i.e.*, the money-price of oil since the energy yields are still much greater than the energy efforts. If the development of an oil field will cost less than the estimated price of the oil which is pumped up the field will be developed.

There is some uncertainty attached to the estimation of oil resources. Although it is reasonable clear how much oil has been extracted from the ground (cumulative production) and what the current rate of extraction is, it is debated how much oil is actually left (resources) and how much of it will be extracted (reserves).

4.10.4 How Oil Reserves are Estimated?

It turns out that there are actually two quite opposing views, because people confuse reserves with resources and tend to focus too much on one or the other. Reserves include the amount of oil which can and will be extracted with a given probability. The P90 reserve i.e. the amount of oil which can be extracted with a 90% probability is usually referred to as *proven* reserves. A P50 reserve is called *proven* and probable.

The two different views are:

1. Since the amount of oil is in principle unknown we can never know how big the resources are. What is relevant is whether we can find more oil or not. A rise of oil prices rise will motivate people to develop new methods to find more oil, to extract more oil from known wells, or to make development of shut-in wells economically possible.
2. Although the total amount of oil is unknown it is still finite. Once all resources have been turned into reserves no more oil can be discovered. The total amount of reserves can be estimated by noticing that the biggest fields are found and developed first after which increasingly smaller fields are found and brought into production. Eventually the new fields will be small and hard to find. Plotting the so-called creaming curves *i.e.*, the cumulative discoveries against the cumulative number of wildcats (exploration wells) the asymptotic value will indicate the ultimate amount of reserves.

Governments and oil companies some of which have economic turnovers compared to the GNP of entire countries might not benefit from revealing their true reserves, since many political and economic decisions depend on these numbers. Thus data are divided into freely available *official* data and confidential "technical" data which determine the development strategies of the oil companies. The freedom in reporting *official* data leads to so-called "reserve growth" which is not true growth, since the amount of oil in the ground does not increase in the reported *official* number.

4.10.5 Reserve Growth

Reserves are inherently unknown, but they can be estimated within a range assigned to a probability. A P90 estimate denominates the amount of oil which can be extracted economically with a probability of 90%. Naturally a P10 estimate will be higher than a P90 estimate, and oil companies are free to report whatever number suits their purposes. Usually they will not even give the probability but simply give the *official* reserves.

Economists may then add all these official numbers and arrive at the total world reserve which tend to grow suggesting that more and more oil is discovered/available. This growth is not surprising. Initial estimates will be conservative since a company would develop the field only if they were quite sure that the investment would be returned. As the initial estimate is conservative, later times will demonstrate that the field most likely contains more oil. US companies are required by SEC to report "proven reserves with a reasonable certainty". As the fields are likely to be bigger than what is reported most likely the reserves will "grow" in subsequent reports. Updating of the *official* number will tend to increase the stock value of the company.

The opposing view is that **the initial estimate was wrong** and the reserves (amount of oil) has not grown. Instead of marking the increase as newly discovered oil the addition is **backdated** and added to the original estimate of the size of the field at the time when it was first discovered. This gives a discovery curve which has been corrected for wrong initial estimates.

Today about 6Gb are discovered and 26Gb are consumed each year. Since **oil has to be discovered, before it can be extracted** the known reserves are being depleted. This is clearly not sustainable... we estimate that world oil and gas production from existing fields is declining at an average rate of about 4 to 6 percent a year. To meet projected demand in 2015, the industry will have to add about 100 million oil-equivalent barrels a day of new production. That's equal to about 80 percent of today's production level. In other words, by 2015, we will need to find, develop and produce a volume of new oil and gas that is equal to eight out of every 10 barrels being produced today. In addition, the cost associated with providing this additional oil and gas is expected to be considerably more than what industry is now spending.

Equally daunting is the fact that many of the most promising prospects are far from major markets—some in regions that lack even basic infrastructure. Others are in extreme climates, such as the Arctic, that present extraordinary technical challenges.

4.10.6 Why the Extraction Rate is More Important than the Total Amount

Policy makers have previously been concerned with R/P rations to strategically account for resource depletion. R/P means the total amount of reserves divided by the current production rate. This number does not take into account that the reserves may grow, and in particular it does not consider that the extraction rate will change later on. The R/P ratio given the false impression that the current rate of extraction can continue for a time of R/P until we one day abruptly run out. In a more realistic scenario one would expect that the extraction rate would decrease and finally slow to a trickle after which it might not even be feasible using either money-economy or energy-economy to extract the last drop. In that case the resource might in principle last forever, but that is irrelevant to society. What we are interested in is the extraction rate, at present and in the future. The extraction rate determines the amount of oil which will reach the market in the near future. After it has passed through the refineries and the distribution system the free market will try to determine the market value. The market value is quite susceptible to the supply rate. So-called swing producers use this to control/increase the prices. Oil importers can counteract/decrease the prices by selling oil from their *strategic reserves* to the market or by decreasing their demand, perhaps involuntarily.

4.10.7 The Peak of the Extraction Rate

Adding the extraction rates for all the wells in an oil field, given the total extraction rate for the field as a function of time. According to the central limit theorem the total extraction rate of several such fields is distributed as a Gaussian (bell shaped) curve, if they all have the same individual distribution. The integrated area under this curve is the total amount of oil which will ever be extracted. It is evident from the curve that the extraction rate will decrease after half of the oil has been extracted. This point is called the **peak year.** The peak year is dependent on the estimate of the total reserves.

It is not known when the extraction rate will peak. There has been a few local peaks in history, so the peak year will not be known until several years after when it can be confirmed that the extraction rate will never again reach its previous maximum.

The peak year depends on the total amount of oil (unknown), the future extraction rate (unknown) which is correlated to future demand (unknown). Prediction is a highly uncertain business. People have been wrong before and they will be wrong again. However, what is important is not the exact peak year rather it is a range of years. Everybody who subscribes to the Hubbert school calculate a peak year within the next two decades. The most popular year at the moment of writing seems to be around 2007. This is certainly within the life time of most people alive today.

It is obvious that oil has to be found before it can be pumped from the ground. A plat of the (backdated) discovery curve shows the extraction curve lagging by about 40 years. The discovery curve peaked in the 1960s.

One can get a good estimate of the total reserves by plotting the integrated discovery rate. This graph approaches an asymptotic value as the newly discovered fields are becoming increasingly smaller and more rare. Such a plot is called a creaming curve.

4.10.8 Replacements as a Possible Solution

The world consumes energy at a rate of about 13TW and petroleum accounts for 40%. Predictions estimates a world drop in supply by 3% a with less in some regions and more in other regions *e.g.*, the North Sea. Therefore the world must find a substitute and construct and bring it online quickly enough to alleviate the effects of a diminishing oil supply after the peak year.

Using the above figures about 150,000MW has to be brought online each year for the next several decades in order to continue to meet demand which will increase if possible. Compare this number to a large nuclear power plant (1000MW) or a modern off-shore wind turbine (2 MW).

Obviously the current energy infrastructure is not designed to handle alternative forms of energy so this has to be replaced. Neither are the current users like aero planes, ships, cars, etc. These will have to be replaced as well.

Some replacements under consideration are listed below:

- Solar power
- Hydrogen
- Natural gas
- Coal
- Bio-fuels *e.g.*, ethanol
- Controlled fusion.
- Wind power
- Hydro power
- Tar sands
- Coal-bed methane
- Nuclear fission

QUESTIONS

1. Define natural resources. Discuss the importance of forests. What are the adverse effects of deforestation?
2. Explain consequences of flood and drought. Suggest measures to control floods.
3. Explain the benefits of dam and the problems arising from their construction.
4. Explain the uses and over-exploitation of mineral resources.
5. What are the various causes of land degradation?
6. What are the main causes of over-exploitation of water resources?
7. What are various reasons for depletion of oil reserves?
8. What is role of an individual in conservation of natural resources?

5

GLOBAL ENERGY CRISES AND RENEWABLE ENERGY RESOURCES

5.1 INTRODUCTION

Energy is defined as the ability or the capacity to do work. We use energy to do work and make all movements. When we eat, our bodies transform the food into energy to do work. When we run or walk or do some work, we 'burn' energy in our bodies. Cars, planes, trolleys, boats, and machinery also transform energy into work. Work means moving or lifting something, warming or lighting something. There are many sources of energy that help to run the various machines invented by man.

5.1.1 Human Energy Demands: Historical Perspectives

The discovery of fire by man led to the possibility of burning wood for cooking and heating thereby using energy. For several thousand years human energy demands were met only by renewable energy sources-sun, biomass (*wood, leaves, twigs*), hydel (*water*) and wind power.

As early as 4000-3500 BC, the first sailing ships and windmills were developed harnessing wind energy. With the use of hydropower through water mills or irrigation systems, things began to move faster. Fuelwood and dung cakes are even today a major source of energy in rural India. Solar energy is used for drying and heating.

With the advent of the Industrial Revolution, the use of energy in the form of fossil fuels began growing as more and more industries were set up. This occurred in stages, from the exploitation of coal deposits to the exploitation of oil and natural gas fields. It has been only half a century since nuclear power began being used as an energy source. In the past century, it became evident that the consumption of non-renewable sources of energy had caused more environmental damage than any other human activity. Electricity generated from fossil fuels such as coal and crude oil has led to high concentrations of harmful gases in the atmosphere. This has in turn led to problems such as ozone depletion and global warming. Vehicular pollution is also a grave problem.

There has been an enormous increase in the demand for energy since the middle of the last century as a result of industrial development and population growth. World population grew 3.2

times between 1850 and 1970, per capita use of industrial energy increased about twenty fold, and total world use of industrial and traditional energy forms combined increased more than twelve fold.

5.1.2 Non Conventional Sources of Energy

Due to the problems associated with the use of fossil fuels, alternative sources of energy have become important and relevant in today's world. These sources, such as the sun and wind, can never be exhausted and are therefore called renewable. Also known as non-conventional sources of energy, they cause less emission and are available locally. Their use can significantly reduce chemical, radioactive, and thermal pollution. They are viable sources of clean and limitless energy. Most of the renewable sources of energy are fairly non-polluting and considered clean. However, biomass is a major polluter indoors.

Renewable energy sources include the sun, wind, water, agricultural residue, fuelwood, and animal dung. Fossil fuels are non-renewable sources. Energy generated from the sun is known as solar energy. Hydel is the energy derived from water. Biomass-firewood, animal dung, and biodegradable waste from cities and crop residues-is a source of energy when it is burnt. Geothermal energy is derived from hot dry rocks, magma, hot water springs, natural geysers, etc. Ocean thermal is energy derived from waves and from tidal waves.

Through the method of co-generation a cleaner and less polluting form of energy is being generated. Fuel cells are also being used as cleaner energy source. In India a number of initiatives have been taken.

5.2 ENERGY EXPLOITED

5.2.1 Per Capita Energy Consumption

Energy exploited is energy extracted out of the resources and utilized. The standard of living of a country can be related to per capita energy consumption. If we take the annual per capita income of various countries and plot them against per head energy consumption, it will appear that the per capita energy consumption is a measure of the prosperity of the nation. The per capita income of U.S.A. is about 50 times more than per capita income of India, and so also is the per capita energy consumption. U.S.A. with 7% of world's population consumes 32% of the total energy consumed in the world, whereas India, a developing country with 20% of the world's population consumes only 1% of the total energy consumed in the world. Therefore, one might conclude that to be materially prosperous; a human being needs to consume more and more energy than his own.

5.2.2 Continued Growth of Energy Demand

If the present trend of energy exploitation continues than the supply of oil will fail to meet increasing demand before the year 2010, even if the oil-prices rise 50 percent above the current levels in real terms. Additional constraints on oil production will hasten this shortage, thereby reducing the time available for action on alternatives. Demand for energy will continue to grow even if governments adopt policies to conserve energy. Energy resources other than oil, which will be progressively reserved for uses that oil can satisfy, must increasingly satisfy this growth.

The continued growth of energy demand requires that resources be developed with the utmost vigour. The change from the economy dominated by oil must start now. The alternatives require 5 to

15 years to develop, and the need for replacement fuels will increase rapidly. Electricity from nuclear power is capable of making an important contribution to the global energy supply although world-wide acceptance of it, on a sufficiently large scale is yet to be established. Fusion power will not be significant before the next 20 years.

5.2.3 Increasing Importance of Renewable Sources

Coal reserves are abundant, but taking advantage of them requires an active programme of development by both producers and consumers. Natural gas reserves are large enough to meet projected demand, provided the incentives are sufficient to encourage the development of extensive and costly inter-continental gas transportation systems. Although the resources-base of other fossil fuels such as oils sands, heavy oil and oil shale is very large, they are likely to supply only small amounts of energy before the year 2020. Other than hydroelectric power, renewable resources of energy *e.g.*, solar, wind, wave are unlikely to contribute significant quantities of additional energy during the century at the global level, although they could be of importance in particular areas. They are likely to become increasingly important in the 21st century. Energy efficiency improvements can further reduce energy demand and narrow the prospective gaps between energy demand and supply.

5.3 SOURCES OF ENERGY

5.3.1 Types of Energy

A source of energy is one that can provide adequate amount of energy in a usable form over a long period of time. These sources can be of two types:

1. *Non-Conventional:* These sources of energy can be used again and again in an endless manner. These can be generated continuously in nature and are inexhaustible *e.g.*, solar energy, wind energy, tidal energy, biomass energy, geo-thermal energy etc. These are also known as renewable resources of energy.
2. *Conventional:* These are accumulated in nature over a long span of time and can not be quickly replenished when exhausted. These are known as Non-renewable resources *e.g.*, coal, petroleum, natural gas and nuclear fuels like uranium and thorium.

5.3.2 Difference Between Conventional and Non-Conventional

The demarcation between Conventional and Non-conventional is not rigid. The sources which are non-conventional today will become conventional after a few years. For our reference, the energy resources which are considered for large scale use after 1973 oil crisis are called as alternate or non-conventional.

5.3.3 What Prompted the Development of Non-Conventional Energy Sources

The energy crisis which began in 1973 caused petroleum supplies to decrease and prices to rise very high. This crisis forced developing countries to reduce their important development programs, so that they could purchase to keep their economies operating. It created the urgent necessity to find and develop alternate (non-conventional) energy sources.

5.4 NON-CONVENTIONAL SOURCES OF ENERGY

The world is facing a severe energy crisis which has derailed the global economy and disrupted the environmental support system.

5.4.1 Seeking Energy Systems Not Based on Fossil Fuels

It is beyond the capacity of the global biosphere to absorb the emissions of a fossil fuel based energy system. Concentration of Carbon-dioxide-a green house gas is gradually increasing in the atmosphere and is likely to result in abrupt climatic disruptions by increasing the global temperature. A reduction in the carbon dioxide concentration can be achieved by limited use of fossil fuels. Reduced availability of fossil fuels, especially the petroleum, and the limited capacity of the world to cope with the pollution caused by fossil fuels are the two major considerations that have forced the world to seek an energy system not based on fuels. In the decades ahead, fossil fuel based energy system will be replaced by a sustainable energy system based on renewable, clean and non-polluting energy resources, in which probably:

1. There will be no nuclear power
2. Power will be derived:

   a. From solar energy resources which are renewed daily by incoming sun light.
   b. From wind energy resources.
   c. From bio energy resources.
   d. From geothermal energy resources.
   e. From ocean energy resources etc.

The challenge ahead is to meet the growing energy requirements by developing new technologies that use energy efficiently and harness the renewable energy resources economically to ensure equal access to it by future generations.

5.5 ENERGY DEMAND

5.5.1 Classification of Energy Requirements

The demand for energy has been rising rapidly with growing population and industrialization. We need energy in various forms for various activities. The energy requirements can be broadly classifies into the following categories :

1. Agriculture
2. Commercial and Industrial
3. Transportation
4. Domestic and Municipal
5. Defence, medical etc.

The usable energy is either purchased from energy supply company or obtained as free commodity from nature (*e.g.*, solar, wind etc.).

5.5.2 Reasons for Energy Demand

During past several decades, the energy demand of the world has continued to increase at an annual growth rate of 3 to 4%. The reasons for this energy demand are as follows :

1. With increasing standard of living, per capita energy consumption has increased.
2. Increase in population.
3. Increase in industrialization.
4. Increasing transportation.
5. Invention of big machines for energy conversion such as electric motors, gas turbines, I.C. engines has increased the demand for energy.
6. Development of energy supply system is also the cause of more energy demand.

5.5.3 Level of Energy Demand

Energy needs of a man vary with life style, climatic conditions, seasons, industrial progress etc.

Primitive man needed less energy. Rural man requires less energy than the urban man. Urban man is dependent on electrical energy, petrol, L.P.G. *etc.*

Today's industry needs coal, steam, electrical energy, diesel, chemicals, lubricating oils etc. Raw materials like steel, aluminium etc. are produced by energy-intensive methods. Water is pumped by using motor-pumps which consume electrical energy. Transportation by road, rail, ocean and air requires high energy input.

Higher per capita energy consumption of a country indicates industrial progress and prosperity. Per capita electrical consumption of india is 2,400 kwh against 92,000 kwh of USA.

5.5.4 World's Energy Demand

The world's energy demand has increased significantly after 1950. The projections uptc the year 2020 are as follows :

Growing energy demand

Year	*World energy demand in exajoules/Year*
1972	270
1985	390
2000	590
2020	840

Note : 1 exajoule = 10^{18} joule.

5.6 SOLAR ENERGY

5.6.1 Radiant Energy of Sun and Its Distribution

Solar energy is a clean, cheap and abundantly available renewable energy. The total radiant energy of the sun is vast, out of which only small fraction gets incident on the outside of the earth's atmosphere. 30% of it is directly reflected back to outer space in the short wave radiation, 47% is directly converted to heat that is lost to the space in the long wave radiation and major portion of the rest 23% causes evaporation, precipitation and circulation of water in the hydrologic cycle of the earth.

Two-thirds of the solar energy falling on the whole earth land surface, fall on the northern hemisphere and one-third on the southern hemisphere.

5.6.2 Historical Perspectives

The first effort in exploiting solar energy for agricultural purposes is reported to be 5000 BC, and first application to concentrate solar rays by means of reflectors is reported in the way back to 200 BC.

5.6.3 Why Its Utilization is Still Neglected?

The commercial utilization of solar energy remains still neglected because of low and irregular solar power density, high installation and generation cost by Solar Photo Voltaic (SPV) panels and Solar Thermal Power Plants, although the Solar Photo voltaic Technology enables the conversion of sunlight into electricity in an environmentally clean manner.

5.6.4 SPV Systems

The SPV systems have emerged as powerful sources for applications including lighting, water pumping, telecommunications and power plants. Reports received from State Implementing Agencies indicate that recently about 13000 home lighting systems, 32000 solar lanterns and 1200 street lighting systems have been installed.

5.6.5 SPV Power Projects in India

Recently the Ministry of Non-Conventional Energy Sources canvassed the scheme to supplement grid power by installing SPV power projects in the range of 25 – 100 kw for rooftop systems on public buildings to demonstrate peak load saving applications in major urban areas and for transmission and distribution support systems in remote rural areas. Under this scheme of MNES, a subsidy of two-thirds of the total project cost has been provided to the State Electricity Boards (SEB).

For encouraging SPV systems, the World Bank has initiated a programme of finance allocation to the developing countries. Some US $ 15 million is likely to be allocated for Indian projects for SPV market expansion for meeting the energy needs.

5.6.6 Where are SPV Systems Successful?

In SPV systems several photovoltaic cells are placed in series or parallel in solar module. Solar cells convert solar radiant energy directly to low voltage DC electrical energy. SPV systems are very much successful for low power portable and medium power remote applications upto 350 kW.

5.6.7 Conversion of Solar Energy to Usable Energy for Utilization

Besides the SPV system, there may be six or seven solar 'energy chains' denoting a series of energy conversions from the primary energy source to final usable energy for utilization as follows :

1. Solar Low Temperature Passive Heating System
2. Solar High Temperature Heating System
3. Solar-Diesel Hybrid System
4. Solar Satellite Microwave System
5. Solar Electro-Chemical Energy Plant
6. Solar Central Receiver Thermal Power Plants
7. Cogeneration Plants.

5.6.8 Low Temperature Passive Heating System and Utilization

In the *Low Temperature Passive Heating System,* the solar radiation is collected by flat plate collectors using medium temperature hot water. This system is used for heating houses, commercial, industrial, agricultural purposes.

5.6.9 High Temperature Heating System and Utilization

In the *Solar High Temperature Heating System,* two methods have become commercially successful, one is use of central receiver containing water and steam, the other is use of distributed trough collectors with the transportation of hot water to central tank. This system is used for process industry, feed water heating in power plant etc.

5.6.10 Solar Diesel Hybrid System for Remote Plant Application

Solar Diesel Hybrid System is operated in conjunction with a diesel generating set and a storage battery energy storage system. This system is economical for remote, stand-alone plants rated from 5 KW to 350 KW.

5.6.11 Solar Satellite Microwave System

Solar Satellite Microwave Power Plants are in planning stage. In this system solar PV panels can be located on satellite orbiting in a fixed geosynchronous orbit. Solar energy can be collected by PV panels for conversion into electrical energy, which, in turn, can be converted to microwave energy in the orbiting satellite. The microwaves to be transmitted through the space by transmitters are beamed to the ground station. This microwave energy is inverted to power frequency electrical energy which is fed into the electrical network.

5.6.12 Solar Electrochemical Power Plant and Its Utilization

As the Solar satellite microwave project is only at the thought provoking stage, the *Solar Electro-Chemical Power Plant* is also only the futuristic application. In the future when fossil fuel cost is going to be exhaustively high, this idea can be transformed into a feasible one. In this system solar energy can be used for electrolysis of water to obtain hydrogen and is used as the fuel in Fuel Cell Plant, similarly Oxygen also can be used for fuel cell.

5.6.13 Solar Central Receiver Thermal Electric Power Plants with Heliostats.

In *Solar Central Receiver Thermal Electric Power Plants* a large central receiver is mounted on a tall tower which is surrounded by thousands of 'heliostats' on ground level by which the solar rays are reflected up to the central receiver. This central receiver produces high temperature steam which can be used for steam electric power plant, gas turbine/combined cycle power plant.

5.6.14 Cogeneration Power Plant for Desert Lands

Cogeneration power plant in the range of 10-200 MW can successfully be installed in arid, desert lands, dry lands away from coal, hydro and gas reserves.

5.6.15 Utilization of Solar Energy in India

The energy related CO_2 emissions through power plants are the single largest source factor for the global ground renewable during the current plan period, of which solar energy accounts for 150 MW.

In our country's industrial sector solar, thermal energy is utilized for pre-heating boiler feed water and for supplying direct process heat, resulting the boiler fuel saving. In domestic sector, the introduction of solar water heating systems replacing electrical geysers results in electrical energy saving.

5.6.16 Promotional Efforts in India by MNES and IREDA

MNES has provided various incentives in terms of subsidy and other fiscal benefits for the promotion of solar water heating systems, although gradually these subsidies have been phased out. Instead of these, soft loans through the Indian Renewable Energy Development Agency Ltd. and other stipulated banks have been made available. After providing subsidies on solar hot water systems and cookers for a decade, MNES decided to discontinue the subsidy in 1993 since a degree of technical and commercial viability had reached by them and a revolving fund to provide soft loans to industrial and institutional users was started in IREDA in 1993-94 by MNES. This has been replaced with an interest subsidy scheme in 1996–97. A total of 127 projects involving 51,300 sq. metres of a collector area have been sanctioned so far under these schemes of which about 20,800 sq. metres have been installed till the middle of 1997. These systems are estimated to be capable of delivering over 26 lakh litres of hot water per day and saving about 43.6 million kilowatt hours of energy per year.

5.6.17 Examples of Ujjain and Mehboobnagar

A *solar water heating system* of 240,000 litres of water per day at the M.P. Oil Fedration's oil mill at Uijain and 10,000 litres per day system at Balaji Foods and Feeds Ltd., factory in Mehboob Nagar in Andhra Pradesh are among the largest systems not only in India but in Asia. In most cases, the pay back through savings in conventional fuels can be demonstrated as 3 or 4 years in the case of electricity substitution, and 5 to 6 years in the case of other fuels.

5.6.18 Use of Dryers in Tea-plantation

Solar air heating system and *dryers* are now being increasingly used in the plantation industry. MNES has supported the development of drying systems for use in tea plantations.

Timber drying is also fairly well established and there are over 70 solar timber kilns in the country.

5.6.19 Solar Lantern Vers. Kerosene Lantern

Solar lantern proves to very useful in a variety of situations where portable lighting units are required. MNES has been implementing a major programme for the propagation of solar lanterns as a substitute for kerosene lantern. A subsidy of Rs. 1500 per lantern is being offered to users. At an average saving of 100 litres kerosene per lantern, and considering the existing subsidy of Rs. 4/-per litre from government, the subsidy of the solar lantern is recovered fully in less than 4 years with the lantern usable for another decade. In addition, there are advantages in terms of better light, safe operation and no pollution.

5.7 BIOMASS ENERGY

5.7.1 What is Biomass?

Organic matter produced from biological organisms like plant, by photosynthesis using solar energy is called Biomass. Biomass may be classified into 3 categories - Fuel wood, organic waste and photosynthesis fuel.

5.7.2 Sources of Biomass

Fuel wood is obtained from trees. Organic wastes include farm waste and urban garbage. Photosynthesis fuel is obtained from the plants like water hyacinth, algae and grasses. Water flowers and green leaves.

Biomass resources may be obtained from cultivated fields, crops, forests and harvests periodically. Biomass can be derived from the municipal waste, agricultural waste, animal dung etc. All these biomass materials have a high potential for meeting man's energy needs. Biomass energy is produced by green plants by photosynthesis.

Biomass cycle maintains the environmental balance between oxygen, carbon-di-oxide and rain leading to the Environment-friendly technology.

5.7.3 Contribution of Biomass Energy

Presently biomass energy contribution is between 4 and 18% of total primary energy consumption of various developing and developed countries respectively, which is increasing rapidly and may reach even 30% by the end of the present decade.

5.7.4 Promotional Efforts in India

In India ministry of non-conventional energy sources has initiated a biomass R and D programme in early eighties of the last century for establishment of BRC (Biomass Research Centre). Various areas indentified for R and D were tree and plantation improvement, nitrogen-fixing micro-organisms, aquatic biomass, petro crops, liquid fuels, solid fuels.

5.7.5 Objective of Biomass Production

Biomass production programme is supposed to have the objective of land reclamation (making land suitable for cultivation), augmentation (to increase) of fuel wood and generation of electricity through energy plantations. Manufacture of biogas from marine biomass can be taken either for supply as cooking gas or for power generation.

Liquid fuel production from biomass has not been given due importance in our country unlike USA. Also the biomass briquetting technology needs further consideration. (Note : Briquet means small block of compressed coal-dust used as fuel).

5.7.6 Power Generation in Sugar Industry

In sugar industry the huge amount of bagasse is generated which can be used as a biomass waste for bagasse based co-generation. As a rough estimate a potential of 35000 MW of power generation exists in the sugar industry of our country.

5.7.7 Bagasse-Based Co-Generation Plants

Although there is a major constraint of high capital investment setting up of bagasse-based cogeneration plants, the Ministry of Non-Conventional Energy Sources of India has been providing a subsidy on the equipment cost for projects and soft loans for modernization of projects using relatively established existing technologies. The Indian Renewable Energy Development Agency Ltd. Was sanctioned some 10 million US dollar from the Asian Development Bank, of which 20 million US dollar was ear marked (to assign something for special purpose) for financing cogeneration projects.

5.7.8 Wood Gasification System

More than one thousand wood gasification systems with generating capacity of 14 MW have been installed so far. Technological improvement by way of gasification of bagasse and the use of producer gas in a combined cycle mode can improve the overall efficiency of the Biomass gasified. Recently some 30 number of biomass gasifier systems with the capacity of 2500MW have been installed and 50 number of the systems with 5600 MW capacity have been sanctioned throughout the country, topping Andhra Pradesh with 10 systems with capacity of 1000 MW installed and some 30 number of systems with 3050MW capacity sanctioned in the recent past.

5.7.9 Examples of Biomass Combustion Based Power Projects

Biomass energy technology employs the plants of few watts to a few hundred MW. A domestic chulha burning wood or charcoal is rated less than 2kW, a large urban waste incineration (complete burning to ashes) power plant is usually rated to be 150MW. Recently three biomass combustion-based power projects have been implemented one is 6MW prosopis-based-project in Andhra Pradesh, the other is 5 MW rice-husk based project in Madhya Pradesh and the last one in cane trash and bagasse based 12MW private sector project in Tamil Nadu. All these projects have been subsidized by Ministry of Non-conventional Energy Sources.

Already some 32 biomass combustion-based power project with a total capacity of roughly 145MW have been installed in ten states for supplying power to the grid. Roughly 700 million units are being fed to the grids energy year saving 0.4 metric ton of coal and reducing 1 metric ton of CO_2 emissions into the atmosphere.

From this fact it is also clear how environment friendly technology is this Biomass Energy. Recently one National Bioenergy Board has been set up for evolving policy guideline and direction to launch for the proper treatment of urban and industrial wastes and the resultant recovery of energy from the urban and industrial wastes sources involving the technology of biomethanation. So far 3 million family type biogas plants based on cattle dung have been set up with the estimated potential of 12 million. Uttar Pradesh being on the top.

5.7.10 Biomass Conversion Technology

Biomass conversion technology can be classified generally under four major categories :

1. Anaerobic digestion
2. Pyrolysis
3. Combustion
4. Gasification.

5.7.11 Anaerobic Digestion

Anaerobic digestion is extensively used in our country for high-moisture organic materials like dung, also for treatment of industrial and municipal wastes and distillery effluent. The bacterial active process has a long cycle time like 4–6 weeks at a time and long rejection time with high moisture content resulting in relatively large plant size.

5.7.12 Pyrolysis of Biomass in Charcoal Making

Pyrolysis is used for production of various gaseous, liquid and solid fuels from biomass. Charcoal making kiln (oven) is an example of pyrolysis of biomass. Liquid fuel also can be produced by means of Pyrolysis.

5.7.13 Combustion of the Biomass for Electricity Generation

Combustion of the biomass is practiced throughout the world for producing heat resulting in electricity generation. By biomass combustion, alternate fuels can not be produced.

5.7.14 Biomass Gasifer and Its Application

Biomass gasifier is employed for a variety of thermal applications with the capacity rating up to 150Kg biomass/hour. The lower capacity gasifiers can be used for small-scale industries like silk-reeling, dyeing and drying and community cooking etc. The higher capacity gasifiers can be used in industrial applications viz. manufacturing of CO_2 or $MgCl_2$, process heat requirement for the industries like cement, sugar.

5.7.15 Biomass Gasification Technology and Its Failure in India

Biomass gasification technology has a high relevance to India where the rural energy scene is characterized by low efficiency of biomass utilization and low availability of electricity for irrigation pumping. Most of the biomass gasifier plants are wood chip based, which require cutting of wood to proper size. Skill the biomass based gasifier technology has not been popularized upto the mark in India, because of its operational problem with higher operational cost. Sometimes even the cost of generating electricity from the gasifier becomes higher than purchasing cost from electricity boards.

The electricity tariff rate for industries is about Rs. 3/ - per kWh; whereas the operation and maintenance cost for gasifier-based power generation only is about Rs. 2.50 per k Wh.

5.7.16 Application of Biomass Gasification

Biomass gasification converts solid biomass like wood or its waste, agricultural waste into a combustible gas mixture of CO and H_2, normally called 'producer gas' by partial combustion in absence of sufficient air supply.

A very important advantage of biomass gasification is that it can be employed at both small scale decentralized level as well as large scale centralized level.

There are various areas of application of biomass gasification technology like simultaneous charcoal and power production, irrigation pumping, village electrification, captive power generation and gratified power from energy plantation on waste lands.

Following are some of the typical important biomass gasification technology :

1. 3–4 kgs of biomass save a litre of diesel.
2. 65% to 85% replacement of diesel in C.I. engine.
3. 5 HP pump set operating for 1000 hours requires 4000 tones of biomass.

Biomass gasifier converts solid woody biomass fuel into a combustible producer gas which can be used to replace diesel even upto 75% in standard diesel engine for shaft power generation comparable to conventional electricity and substantially power cost than for diesel based generation.

5.7.17 Gasifier Used for Tea-Drying

The gasifier can be used for tea-drying. In this tea-drying project there are various advantages, viz., producer gas is used for generating clean hot air for tea dryer by burning in a burner and mixing the clean flue gas directly with suitable quantity of cold air. The product tea was found to have no ill effects, rather better tea quality.

5.7.18 Rice Husk Gasification

Rice husk gasification is still under developmental stage even after two decades research in India. Using 4.5 Kg of rice husk, one can save one litre of diesel. 250 kW DG set at 80% load can save 42 litres of diesel which means the fuel cost saving Rs. 13.25 lakh as of today's price.

5.7.19 Biogas from Biomass

Biogas is a cheap secondary renewable gaseous fuel energy obtained from biomass by the process of anaerobic digestion of the urban wastes and refuse, rural or agricultural waste, cow dung, animal waste etc.

5.7.20 Biogas and Biogas Plants

The biogas can be used for household purposes like burner in kitchen fuel for kilns, furnaces, internal combustion engines for driving pump sets, producing mechanical energy on electrical energy from I.C. engines. Biogas plants have urban applications for converting urban refuse, also rural applications for converting cow dung, agricultural waste etc.

5.7.21 Biogas Digesters

Biogas Digesters are widely used in India and China. In China recently there was a programme with 6 million farm household with biogas digesters having an annual production of about 1.7 billion m^3 corresponding to roughly 16500 barrels of petroleum equivalent per day. China has built also over 650 large and medium sized utilities using organic waste from animal and poultry farms, food factories. The total capacity of these biogas utilities is 220000 m^3 servicing 84000 households.

5.7.22 Objectives of NPBD

There are various objectives of the National Project on Biogas Development (NPBD) launched by Ministry of Agriculture subsequently transferred to the department of Non-conventional Energy Sources of India. These are to provide energy in clean, unpolluted form, to make available enriched fertilizer as a by-product for supplementing and optimizing the use of chemical fertilizers, to eliminate the smoke-filled cooling environment, and to bring about an improvement in rural sanitation. Biogas basically is methane gas produced through the anaerobic-digestion of cow dung and other organic wastes. Besides methane gas, it contains CO_2 traces of nitrogen, sulphur and also moisture.

5.7.23 Advantages of Biogas Plants

Biogas is produced primarily as a microbial process by which the carbohydrates in the organic matter break down in the absence of oxygen. The factors affecting production of biogas includes temperature, acidity and C/N ration.

The advantage of biogas plants is that they can digest almost any constant mixture of city waste, manure and plant residues due to the complex bacterial process involved. The CO_2 co-produced with CH_4 can be removed easily from the gas stream.

5.7.24 Biogas Estimations for India

As per as an estimate the biogas of India can generate nearly 215 billion kWh energy annually equivalent to 260 million tones of manure with a nitrogen content 3.9 million tones. The available dung of India can support 20 million family biogas plants having 1.8 m^3 average capacity and 0.65 million community plants with the average capacity of $145m^3$.

5.7.25 Community Biogas Plants

The concept of community biogas plant is basically sound with the large scope to be operated as a decentralized cooking gas systems. There are several suggestions as follows that can be used for improving the productivity and efficiency of such systems:

1. Generate power by means of a dual fuel engine and sell the power to the State Electricity Boards.
2. Use the exhaust of the power generating system to operate the plant at a higher temperature, thus increasing gas production.
3. Use the advanced methods like bacterial support structures, recycling of spent slurry which have the potential to reduce the system cost substantially.
4. Addition of material like water hyacinth with dung for biogas generation.

5.7.26 Operational Problem for Biogas Plant

There is some operational problem for biogas plant like huge requirement of dung for initial charging minimum 3000Kg of dung is needed for initial charging of a $2m^3$ plant. However the community biogas plants are envisaged as potential nuclei for development of decentralized energy systems to catalyze rural development programme which helps in transforming village life by providing not only efficient cooking fuel, but also electricity for lighting, cottage industries, irrigation and drinking water supply, and sanitary toilets besides fertilizer for improved agriculture.

5.8 WIND ENERGY

5.8.1 Wind Energy : A Safe and Clean Energy Source

Kinetic energy of wind is measured proportion to its velocity and is being used for centuries in water pumping, wind mills, irrigating fields and navigation. Wind power is one of the most attractive solutions of the searching over of safe and clean renewable energy sources since the oil crisis of 1973 world over together with environmental consciousness.

5.8.2 Worldwide Potential of Wind Power

According to an estimate there is a world-wide annual potential wind power of 10 k Wh or 3.6 GJ. Small windmills are innumerable. Large wind power units of over 100kW capacity are underway since 1930 in former Soviet Union, USA and UK. However, by late 1980s commercial production of wind turbine generator has started. Several electrical windmills are in commercial production in several countries including Australia, Canada, Denmark, Netherlands, Swedan, USA, UK, France, Germany. As per recent data, installed capacities of wind farms are USA-2100 MW, Denmark-725 MW, Germany-410MW, Uk-210 MW, the Netherlands – 2502 MW.

5.8.3 Large Scale Utilization and Problems

With overall technical efficiency of the electric windmills being 40%, the major problem in the large scale utilization of wind power is the cost of suitable plants. In 1990s conversion of wind energy to electrical energy has become economical in the areas of favourable wind *e.g.*, the states of Gujarat, Tamil Nadu, Mharashtra. Orissa of our country with the help of various initiatives taken by Ministry of Non-conventional Energy Sources.

5.8.4 Wind Farms in India

In Gujarat, the districts where there is ultimate potential of wind farms are Kutch, Jamnagar, Rajkot, Junagad. Bhavnagar, with the ultimate potential electricity generation of 4.1 million GWh cost of electricity being Rs. 2.6 per kWh, with the average wind speed of 5.5m/s. The districts of Tamil Nadu are Ramanaths, Coimbatore, Tirunelveli, Kanyakumari with the average wind speed of 5.6 m/s where the ultimate potential wind farms are located with potential electricity generation of 1.03 million GWh, cost of electricity generation being Rs. 2.45 per kWh.

Ratnagiri is the district of Maharashtra where wind farms are located with average wind speed of 5.2m/s with 0.04 million GWh, the cost of electricity being Rs. 2.92 per kWh.

In Orissa state Puri and Baleshwar are the districts where wind farms are located with average wind speed of 5.1 m/s, with 0.03 million GWh, the cost of electricity being Rs. 3.1 per kWh. The total installed wind power capacity as per recent statistics is 1000 MW in India placing her fourth in wind power installation throughout the world. Several state governments have announced policies relating to the purchase price of power generated by wind, wheeling and banking fees. Moreover, the Private Sector can also take soft loans from the Indian Renewable Energy Development Agency. International Development Agency has extended to the private sector a credit worth of 30 million US dollar in the form of grants nom the Global Environment Facility, because of the environmental benefits accrued through he installation of wind farms.

5.8.5 Wind Speed Data

India is blessed with natural meteorological and topographical settings which are conducive to a level of wind velocities suitable for power generation. Relatively high wind areas of India are located mostly on the west coast. The energy content in wind varies with the factors like latitude, altitude, season and geographical location with respect to the monsoon wind. The availability of the wind speed data is the basic requirement for determining the feasibility of wind power generation at any site.

5.8.6 Installed Capacity of Wind Energy in India

India made its beginning in wind energy in mid-eighties and within fifteen years she gained a respectable position globally. State wise distribution of installed capacity of projects till recent past indicated Tamil Nadu topped with total 670MW. In addition, a capacity of about 2000 MW is in pipeline for harnessing wind power to generate electricity. With the present pace India seems poised to secure a position second only to USA in wind power generation.

5.9 GEOTHERMAL ENERGY

5.9.1 Applications of Geothermal Energy

Geothermal energy is derived from high temperature geothermal fluids. This can be utilized for power generation and thermal applications like green house cultivation, hot water irrigation in cold countries, space heating and cooking, process heat, electrical power generation.

5.9.2 Geothermal Fields

Exploitation of geothermal heat in the form of hot water, steam, electricity and hot gases has been made successful in a very few number of regions of the world.

Generally geothermal energy is available inside the earth at a depth more than about 80 Km. In a few locations called geothermal field in the world the geothermal energy is available at depths of 300m to 3 km. These fields are mainly the countries like USA, former USSR, Chile, Philippines, Mexico, New Zealand, Japan, Italy, Indonesia, Hungary.

5.9.3 Major Geothermal Power Plants

There are three major geothermal power plants, one in Tuscany of Italy with 406MW. One in Wairakai of New Zealand and the third in California of USA with 600 MW installed capacity. Presently about 20 countries have known geothermal resources with total installed capacity being 10000MW.

5.9.4 Indian Geothermal Sites

India has about 150 known geothermal sites having geothermal fluid of moderate and low temperature. The worth mentioning hydro-geothermal resources locations are Puga Hydro-Geothermal Field, Jammu and Kashmir, West-Coast Hydro-Geothermal field, Maharashtra and Tattapani-Hydro-Geothermal Field, Madhya Pradesh.

5.9.5 Geothermal Energy Reserve in Earth

Average geothermal temperature gradient of 30^0C per km depth can be exploited for releasing geothermal energy of 0.06 W/m^2. Total estimated geothermal energy reserve in the earth is 4×10^{12} Exajoule, out of which 2–10 Exajoule/year can be tapped for production.

5.9.6 Hydro and Petro Geothermal Energy Deposits

There are two types of geothermal energy deposit, namely Hydro-geothermal energy deposit and Petro geothermal energy deposit. The first type is of the deposit of hot water and steam at relatively lower depth like 3 km. With the help of production well, hot water, stream or hot brine can be extracted from such deposit. The second type is of hot dry rock at a temperature around 200^0C and a depth of 2 km. Water is injected into the hot dry rock fracture to form a deposit of hot water and steam in the fracture within the rock and these are extracted from the geothermal deposit of the hot dry rock.

5.10 TIDAL ENERGY

5.10.1 World's Famous Sites for Tidal Power

The idea of using tidal power is a very old one. A basin is filled during flow tide and closed while the tide recedes. The difference in the water head during ebb tide is used for rotation of a turbine in the generator. Availability of favourable sites for tidal power is limited to only 50 throughout the world. There is a 800 MW project on seven river on the west cost of England. At Rance of France there is a 240 MW installation.

5.10.2 Sites in India

India has a long coast line with the identification of the location of tidal power like Gulf of Kutch, Cambay and Sunderbans with average tides of 6m height and with a cumulative power potential of nearly 10000MW. The Kutch Tidal Project has been envisaged by the Central Electricity Authority for generation of 900 MW, implementation having been entrusted to National Hydroelectric Power Corporation (NHPC).

5.10.3 Reason of Oceanic Tides

The Earth's gravitational pull generates oscillating oceanic tides which keep billions of litres of water in movement as they build up and the recede twice a day. Accordingly to an estimate made by National Oceanographic and Atmospheric Administration, USA, the potential energy of tides is approximately 3×10^{11} MW out of which 1×10^6 MW exploitable energy is available on the sea coasts.

5.11 OCEAN ENERGY

5.11.1 Electric Power Generation from Temperature Difference

Ocean energy is hidden in thermal gradient, waves and tides of the seas and oceans which cover about three fourth of the earth. In tropical ocean like India Ocean there becomes a temperature difference of nearly 20^0C between the warm surface water and deep cold water at the depth of 1–2 Km. This temperature difference is sufficient for electric power generation. World's total estimated ocean energy reserves are about 130×10^6 MW. With the Indian Ocean having a thermal energy conversion potential of 50000MW. There is an enormous amount of opportunity to exploit this potentiality.

5.11.2 OTEC

This is nothing but OTEC (Ocean Thermal Energy Conversion) process which essentially consists of evaporating a working fluid e.g. ammonia or frown using a heat exchanger with the turbogenerator for the generation of electricity.

5.11.3 World's Largest OTEC Plant

The world's largest OTEC plant of 600MW is being installed at Kulasekarapatnam in Tamil Nadu State of India, by the joint effort of the Sea Solar Power of USA and Tamil Nadu Electricity Board of India.

5.11.4 Sea Wave Power

Sea Wave Power can be generated by the sea waves, piled up by the wind being risen up and down the shorelines resulting the oscillating water column. The most important grid-connected wave power generating capacity of 350KW Tapered Channel Plant is located at Toftestallen of Norway, being operated since 1986. A single Megawatt plant is near the completion in Australia.

5.11.5 Wave Energy Exploration

In India wave exploration potentiality is about 20000MW. However, only the demonstration unit of 150KW has been installed in 1993 at Vizhinjam, Kovalam beach near Tiruvananthapuram by Kerala State Electricity Board in technical collaboration with IIT, Chennai.

5.11.6 Ocean Energy Conversion Technologies

There may be the following ocean energy conversion technology as follows :

(*a*) **Ocean Tidal Energy :** *Ocean Tidal Energy* refers to the hydro energy in ocean tides which occurs due to gravitational attractive force from sun and moon. The difference head of water between high tide and low tide is used for rotating hydro turbine-generator unit installed within dams for generating electricity.

(*b*) **Ocean Wave Energy :** *Ocean Wave Energy* is nothing but which is found by exploiting the available energy from the waves of water from ocean to the shore, occur due of the rotation of earth and the winds over ocean surface. Wave energy can be made available by installing ocean wave machines.

(*c*) **Ocean Thermal Energy :** *Ocean Thermal Energy* is that acquired by the ocean water from solar radiation. The difference of temperature between upper level of ocean being at about 250^0C and the bottom of the sea being at about 100^0C is used for this energy.

(*d*) **Ocean Chemical Energy :** *Ocean Chemical Energy* of the chemicals can be usefully converted to secondary energy from by means of photochemical process or photo biological process.

(*e*) **Others :** Biomass Energy of the organic matter from ocean like aquatic vegetation, algae and animals can be converted into methane rich biogas by means of anaerobic digestion process.

5.12 WAVE ENERGY

5.12.1 How the Waves are Generated?

Waves are created by gravitational action of the sun and the moon, as also by the force of winds. Waves are generated in oceans and lakes. Waves are sometimes strong enough to overturn large ships. Wave energy fluxes in the open sea or against coasts may vary from a few watts to kilowatts per metre. Wave motion consists of both vertical and horizontal movement of water. Individual particles of water undergo almost a circular motion, moving up as the crest approaches, forward at the crest, down as it recedes and backward in the trough.

5.12.2 Difference Between Tide and Wave

Unlike the tides which exhibit a regular but long periodic motion, waves keep the ocean water in continual motion. The vertical rise and fall of the successive waves motion is perpetual, ocean waves can be harnessed to give energy vaselesly.

5.12.3 Wave Machine

Ocean waves energy conversion has gained attention during recent years. Wave machine converts ocean energy to mechanical energy. The plant converts the energy to electrical energy.

5.13 THERMOELECTRIC CONVERSION SYSTEM

5.13.1 Thermoelectric Effects

The quest for a reliable, silent, energy converter with no moving parts that transforms heat to electrical power has led engineers to reconsider a set of phenomena called the Thermoelectric effects. These effects, known for over a hundred years, have permitted the development of small, self contained electrical power sources.

Seebeck (thermoelectric) Effect : The German Scientist Seebeck (in 1822) discovered that if two dissimilar materials are joined to form a loop and the two junctions maintained at different temperatures, an e. m. f. will be set up around the loop. This effect has long been used in thermocouples to measure temperatures.

This phenomenon offers one method of producing electrical energy directly from the heat of combustion, but its thermal efficiency is very low, of the order of 1 to 3 per cent.

5.14 THERMIONIC CONVERSION SYSTEM

5.14.1 Thermionic Converter

A thermionic converter can be analyzed from at least three different points of view :

1. In terms of thermodynamics, it may be viewed as a heat-engine that uses an electron gas as a working substance.
2. In terms of electronics, it may be vied as a diode that transforms heat to electricity by the law of thermionic emission.

3. In terms of thermoelectricity, it may be viewed as a thermocouple in which an evacuated space or a plasma has been substituted for one of the conductors.

5.14.2 Thermionic Emission

Regardless of the point of view adopted in analysis, a thermionic converter works because of the phenomenon of '**thermionic emission**'. Thermionic emission implies emission of electrons from the metal when it is heated.

5.15 PHOTOVOLTAIC CELL

5.15.1 Photovoltaic Effect and Solar Cell

Solar energy can be directly converted to electrical energy by means of photovoltaic effect which is defined as the generation of an electromotive force as a result of the absorption of ionizing radiation. Energy conversion devices which are employed to convert sunlight into electricity by the use of the photovoltaic effect are called solar cells. A single converter cell is called a solar cell or a photovoltaic cell.

5.15.2 Solar Array or Solar Module

To increase the electrical power output a number of such cells are combined and the combination is called a *solar array (or solar module).*

In a photovoltaic cell sensitive element is a semiconductor (not metal) which generates voltage in proportion to the light or any radiant energy incident on it. The most commonly used photovoltaic cells are barrier layer type like iron-selenium cells or Cu-CuO_2 cells.

5.16 FUEL CELLS

5.16.1 Fuel Cell and its Advantage

A fuel cell is an electrochemical device in which the chemical energy of a conventional fuel is converted directly and efficiently into low voltage, direct-current electric energy. One of the chief advantages of such a device is that because the conversion, can be carried out isothermally, the Carnot limitation on efficiency does not apply.

5.16.2 Fuel cell as a Primary Battery

A fuel cell is often described as a primary battery in which the fuel and oxidizer are stored external to the battery and fed to it as needed.

The fuel gas diffuses through the anode and is oxidized, thus releasing electrons to the external circuit, the oxidizer diffuses through the cathode and is reduced by the electrons that have come from the anode by way of the external circuit.

The fuel cell is a device that keeps the fuel molecules from mixing with the oxidizer molecules, permitting, however, the transfer of electrons by a metallic path that may contain a load.

5.16.3 Fuels for Fuel Cell

Of the available fuels, hydrogen has so far given the most promising results, although cells consuming coal, oil or natural gas would be economically much more useful for large scale applications.

5.17 AVAILABILITY

5.17.1 Availability of Solar Energy

Sun radiates its energy in all directions of the universe. Sun light received during day hours and clear sky has power density between 0.5 kw/m^2 and 1 kw /m^2. Solar energy varies with season, geographical location, day/night etc. other relevant details are as follows.

Energy radiated by the sun into space per year = 1.17×10^{31} kJ Solar energy that reaches the earth = 5.4×10^{21} kJ/Year.

5.17.2 Availability of Wind Energy

Wind energy is utilized by mean of a wind mill. These are mainly located near the sea coast, island or a mountain cliff where strong winds are present.

Its potential in India is estimated to be 2.5×10^4 MW. Unfortunately, the wind is neither strong nor steady in India. Gujarat, Rajasthan and South Indian states have better potential during March and September as the wind power can be developed economically only above the wind speed of 10km per hour.

Wind energy can only be economical in areas of good wind availability. Wind energy differs with region and season. Although winds speeds generally increase with height, varying speeds are found over different kinds of terrain.

5.17.3 Availability of Geothermal Energy

Geothermal power station have been installed at a numbers of places around the world, where geothermal steam is available.

In India, Northwestern Himalayas and the western coast are considered geothermal areas. The Geological survey of India has already indentified more than 350 hot spring sites, which can be explored as areas to tap geothermal energy. Satellite Like IRS-1 have played an important role, through infrared photograph of the ground, in Locating geothermal areas.

There are many hot water springs in India. In north they occur in Ladakh and H.P. In western parts they are found in the Cambay region of Gujrat and Maharashtra. They are also found in the Singbhum region of Jharkhand while there are some in Assam. The water from a hot spring at Garampani near Jawai in Assam is so hot that rice kept in cloth bag gets boiled in no time.

The engineers have commissioned an experimental 1 kW generated running on geothermal energy in the Puga area. This is the first production of electricity from a hot water spring in India.

5.17.4 Availability of Biomass

1. Biomass in the form of agro residue and industrial waste is available in all geographical locations in India and holds considerable promice as a source of electricity.

2. Agriculture and its by products yields more than 350 million tones of waste every year, a large portion of which is either waste or used inefficiently.

5.17.5 Availability of Ocean Tides

1. Ocean tides are produced by gravitational forces of sun and moon. There are only a few sites in the world where tidal energy can be suitably harnessed. The bay of Funday Canada having 17–18m high tides has a potential of 5000 MW of power generation.
2. The tidal mill at La Rance, France is one of the First modern tidal power mill.
3. In India Gulf of Cambay, Gulf of Kutch and the sunder ban delta are the tidal power sites.

5.18 ENERGY UNITS AND CONVERSIONS

$$1 \text{ MBDOE} = 1 \text{ million barrels per day of oil equivalent}$$
$$= 2.2 \times 10^{18} \text{ joules per year}$$
$$1 \text{ QUAD} = 1 \text{ quadrillion Btus} = 10^{15} \text{ Btus}$$
$$1 \text{ MTCE} = 1 \text{ million short tons of coal equivalent}$$
$$1 \text{ Btu} = 0.000293 \text{ kWh} = 10^{55} \text{ Joules}$$
$$1 \text{ Kcal} = 3.97 \text{ Btu} = 4184 \text{ Joules}$$
$$105 \text{ BTU} = 1 \text{ Therm} = 29.3 \text{ kWh}$$
$$1 \text{ kWh} = 3.6 \times 10^{6} \text{ Joules.}$$

1 BTU (British Thermal Unit) is the amount of heat necessary to raise one 1 pound of water by 1^0F.

$$1 \text{ exajoule} = 10^{18} \text{ Joule}$$
$$1 \text{ giga joule} = 10^{9} \text{ Joule.}$$

Power Conversion:

1 horsepower (*hp*) = 745.7 watts

Different types of energy are measured by different physical units : barrels or gallons for petroleum; cubic feet for natural gas; tons for coal; kilowatt-hours for electricity. To compare different fuels, we need to convert the measurements to the same units.

BTU Content of Common Energy Units:

1 barrel (42 gallons) of crude oil = 5,800,000 Btu
1 gallon of gasoline = 1,24,000 Btu
1 gallon of diesel fuel = 139,000 Btu
1 cubic foot of natural gas = 1,026 Btu
1 gallon of propane = 91,000 Btu
1 kilowatt hour of electricity = 3,412 Btu.

5.19 FULL FORMS OF IMPORTANT TERMS

INEP : Integrated National Energy Planning
ESDM : Energy Supply and Demand Management
IREDA : Indian Renewable Energy Development Agency Ltd.

MTOE : Million Tones of Oil Equivalent
NEEP : National Energy Efficiency Programme
SEB : State Electricity Board
DMS : Demand Side Management
MHD : Magneto Hydrodynamics
MNES : Ministry of Nonconventional Energy Sources
NASA : National Aeronautics and Space Administration
LST : Local Solar Time
CPC : Compound Parabolic Concentrator
BTU : British Thermal Unit
1 MBOE : One Million Barrel of Oil Equivalent
QUAD : Quadrillion Btus
MTCE : Million short Ton of Coal Equivalent
WECS : Wind Energy Conversion System
KVIC : Khadi and Village Industrial Commission
PRAD : Planning, Research and Action Division
ASTRA : Application of Science and Technology to Rural Areas
HDR : Hot Dry Rocks
OTEC : Ocean Thermal Energy Conversion

Advantages of Renewable Energy

Following are the main advantages of renewable energy:

1. Renewable energy is an indigenous resource available in considerable quantities to all developing nations. It is capable of having a significant local, regional or national economic impact.
2. Several renewable options are financially and economically competitive for remote locations, where the costs of transmitting electrical power or transporting conventional fuels are high.
3. There is lower interest on borrowed capital because of shorter lead times and reduced transmission and distribution coasts for dispersed rural locations.
4. Rapid scientific and technological advantages are expected to expand the economic range of renewable energy application over the next 8–10 years, making it imperative for international decision makers and planners to keep abreast of these developments.

5.20 OBSTACLES IN THE IMPLEMENTATION OF RENEWABLE ENERGY

Obstacles to the implementation of renewable energy are as follows:

1. Inadequate documentation and evaluation of past experience and lack of clear priorities for future work.
2. Weak or nonexistent institutions and policies to finance and commercialize renewable energy systems.
3. Technology and economic uncertainties in many renewable energy systems, high economic

and financial coasts for some systems in comparison with conventional supply options and energy efficiency measures.

4. Skeptical (unwilling to believe something) attitudes towards renewable energy systems on the part of the energy planners and a lack of qualified personnel to design, manufacture, market operate and maintain such systems.
5. Inadequate donor coordination in renewable energy assistance activities, with little or no information exchange on successful and unsuccessful projects.

5.21 ENERGY CRISES

It is possible that the world is heading towards a global energy crisis due to a decline in the availability of cheap oil and recommendations to a decreasing dependency on fossil fuel. This has led to increasing interest in alternate power/fuel research such as fuel cell technology, hydrogen fuel, bio-methanol, bio-diesel, Karrick process, solar energy, geothermal energy, tidal energy and wind energy, and fusion power. To date, only hydroelectricity and nuclear power have been significant alternatives to fossil fuel (see Future energy development), with big ecological problems (residues and water spending). Hydrogen gas is currently produced at a net energy loss from natural gas, which is also experiencing declining production in North America and elsewhere. When not produced from natural gas, hydrogen still needs another source of energy to create it, also at a loss during the process. This has led to hydrogen being regarded as a 'carrier' of energy rather than a 'source'.

There have been alarming predictions by groups such as the Club of Rome that the world would run out of oil (and out of food) in the late 20th century. Although technology has made oil extraction more efficient, the world has to struggle to provide oil by using increasingly costly and less productive methods such as deep sea drilling, and developing environmentally sensitive areas such as the Arctic National Wildlife Refuge. The world's population continues to grow at a quarter of a million people per day, increasing the consumption of energy. The per capita energy consumption of China, India and other developing nations continues to increase as the people living in these countries adopt more energy intensive lifestyles. At present a small part of the world's population consumes a large part of its resources, with the United States and its population of 300 million people consuming far more oil than China with its population of 1.3 billion people.

Between 1950 and 1984, as the Green Revolution transformed agriculture around the globe, world grain production increased by 250%. The energy for the Green Revolution was provided by fossil fuels in the form of fertilizers (natural gas), pesticides (oil), and hydrocarbon fueled irrigation. The 20th century population explosion is strongly correlated with the discovery and extraction of hydrocarbons. The peaking of world hydrocarbon production may test Malthus critics.

Efficiency mechanisms such as Megawatt power can encourage significantly more effective use of current generating capacity. It is a term used to describe the trading of increased efficiency, using consumption efficiency to increase available market supply rather than by increasing plant generation capacity. As such, it is a demand-side as opposed to a supply-side measure.

5.22 ENERGY SECTOR DEVELOPMENT AND PLANNED ECONOMY IN INDIA

India ranks sixth in the world in terms of energy demand accounting for 3.5% of world commercial energy demand in 2001. With a gross domestic product (GDP) growth of 8 % set for the Tenth Five-Year Plan, the energy demand is expected to grow at 5.2%. Although, the commercial energy consumption has grown rapidly over the last two decades, a large part of India's population does not have access to it. At 479 kg of oil equivalent (KGOE), the per capita total energy consumption is also low even compared to some of the developing countries. In terms of commercial energy consumption, the figure is only 0.32 TOE only as compared to 8.80 TOE for the USA.

India is fortunate to be endowed with both exhaustible (particularly coal) and renewable energy resources. Despite the resource potential and the significant rate of growth energy supply over the last few decades, India faces serious energy shortages. This has led to reliance on increasing imports for meeting the demand of oil and coal. As per the current projections, India's dependence on oil imports is expected to increase. The demand of natural gas also outpaces supply and efforts are being made to import natural gas in the form of liquefied natural gas (LNG) and piped gas. The power sector has also been experiencing server shortages.

The fossil fuel, that is, coal, oil and gas will continue to occupy a major position in the energy mix of the country which is broadly met by coal (52%), oil and gas (44%), hydro power (2%), nuclear (1.7%) and the balance from non-conventional sources. While the country is comfortable with coal reserves, there is around 0.5% of world's proven crude oil reserves in India as against 2.8% of world's consumption. The gap between the domestic availability and the demand indicates vulnerability to energy imports. India currently imports about 67% of its crude oil from this region. In the year 2002–2003. India imported about 82 million tones (Mt) of crude oil valued at about Rs. 76,200 crore.

In the past 52 years, world had already witnessed about 20 oil disruptions of various sizes and durations affecting the oil importing economy. Supply disruptions adversely affect the developing economies like India with oil import dependence of around 69%. This rate of 8.1% is maintained.

5.23 ENERGY SCENARIO

Primary commercial energy demand grew almost three-fold at an annual rate of 65% between 1981 and 2001 to reach 314.7Mt of oil equivalent (MTOE). India's commercial energy demand for the next decade is projected to be among the highest in the world, spurred by sustained economic growth, rise in income levels and increased availability of goods and services.

India's commercial energy demand is expected to grow even more rapidly than in the past as it goes down the reform path in order to raise standards of living. A large part of India's population does not have access to commercial energy. The 479KGOE per capita total energy consumption is only about 20% of the global average in 1997 and compared poorly with the per capita consumption of Thailand (319 KGOE), Brazil (1051 KGOE) and China (907 KGOE).

5.24 TRENDS OF ECONOMIC GROWTH AND ENERGY DEMAND

The average annual world economic growth in the period 1997–2020 is projected at 3.2% while the energy growth rate is estimated at 2.1% per annum. This yields an elasticity of energy consumption at about 0.68%. In the case of India, the elasticity was more than unity for the period 1953–2002. However, the elasticity for primary commercial energy consumption for the period of 1991–2000 is less than unity. This could be attributed to several factors, such as, the improvement in efficiency of energy use and the consequent lowering of the overall energy intensity of the economy and the higher share of hydrocarbons in the overall energy mix.

The projected requirement of commercial energy is estimated at about 412 MTOE (in 2007 AD) and 554 MTOE (in 2012 AD) in the terminal years of the Tenth and Eleventh Five- Year plans, respectively. Based on the inputs of various working groups, the commercial energy demand during the Tenth Five-Year Plane and the Eleventh Five – Year Plan is estimated to grow at an average rate of 6.6% and 6.1%, respectively. **TABLE -5.1** indicates the estimated energy demand in the terminal years of the Tenth and the Eleventh Five-Year Plans. However, the demand may be less by 5% and 10% during 2006–2007 and 2011–2012, respectively due to efficient demand side management, increasing use of information technology (IT) and prevalence of e-commerce, which will affect mainly the demand of energy in transport sector.

5.25 AVAILABILITY OF COMMERCIAL PRIMARY ENERGY RESOURCES

India's energy use is mostly based on fossil fuels. Although the country has significant coal and hydro resource potential, it is relatively poor in oil and gas resources. As a result, it has to depend on imports to meet its energy supplies. The geographical distribution of available primary commercial energy sources in the country is quite skewed with 77% of the hydro potential located in the northern and north-eastern region of the country. Similarly, about 70% of the total coal reserves are located in the eastern region while most of the hydrocarbon reserves lie in the west.

The geographic spread of various energy resources available in the country is given in **TABLE 5.2.** The highlights of individual resources are also given in **TABLE 5.2.**

5.25.1 Coal

The geological coal reserves of the country are estimated at 220.98 billion tones (Bt) as on January 2001. Out of this, proven reserves are 84.41 Bt. While 98.55 Bt are indicated reserves and 38.02 Bt are inferred reserves, coal continues to remain the principle source of commercial energy accounting for nearly 50% of the total supplies. About 70% of the powers generated are coal and lignite based and this trend is likely to continue in the foreseeable future.

India has an estimated 1000 billion cubic meters (BCM) of Coal Bed Methane (CBM), which is likely to emerge as a new source of commercial energy in the country. A demonstration project is under implementation with financial support from the Global Environment Facility (GEF) and the United Nations Development Programme (UNDP). In April 2001, the Government of India announced a programme for exploration and production of CBM. Under the first round of bidding, five CBM blocks have been awarded to private companies.

5.25.2 Lignite

The current estimates of geological lignite reserves in the country are 34.76 Bt spread over Tamilnadu and Pondicherry (87.5%), Rajasthan (6.9%), Gujarat (4.9%), Kerala (0.31%) and Jammu and Kashmir (0.37%). The lignite deposits in the southern and western regions have emerged as an important source of fuel supply for states like Tamilnadu, Rajasthan and Gujarat. Over the years, considerable emphasis has been placed on the development of lignite for power from 24.3 Mt in 2001-2002 to 44.96 Mt in 2006–2007.

5.25.3 Oil and Natural Gas

The latest estimates indicate that India has around 0.4% of the world's proven reserves of crude oil. AS against this, the domestic crude consumption is estimated at 2.8% of the world's consumption. The balance of recoverable reserves as estimated in the beginning of 2001 is place at 733.70 Mt of crude and 749.95 BCM of natural gas. The share of hydrocarbons in the primary commercial energy consumption of the country estimated at 44.9% (36.0% for oil and 8.9% of natural gas). The demand for oil is likely to increase further during the next two decades. The transportation sector will be the main driver for the projected increase in oil demand. Consequently import dependence for oil, which is currently about 70%, is likely to increase further during the Tenth and Eleventh Five-Year Plans.

India has about 0.4% of world's natural gas reserves. Initially, the gas reserves had been developed largely for use as petrochemical feedstock and in the production of fertilizers, but gas is increasingly being used for power generation, industrial applications and more recently in the transport sector. At present, the share of power generation capacity, based on gas, is about 10% of the total installed capacity.

5.25.4 Hydroelectric Potential

The key advantage of hydroelectric power is the ability to store energy and the flexibility of its use during peak load periods. India is endowed with economically viable hydro potential. The Central Electricity Authority (CEA) has assessed India's hydro potential to be about 148 700 megawatt (MW) of installed capacity. The hydroelectric capacity currently under operation is about 26 000 MW and 16 083 MW is under various stages of development. The CEA has also identified 56 sites for pumped storage schemes with an estimated aggregate installed capacity of 94 000 MW. In addition, a potential of 15 000 MW in terms of installed capacity is estimated from small, mini and micro hydel schemes.

5.25.5 Nuclear Resources

Nuclear energy has the potential to meet the future needs of electricity demand in the country. The country has developed the capacity to build and operate nuclear power plants observing international standards of safety. The current installed capacity of nuclear power is 1860 MW accounting for 2.8% of the total installed capacity of the country. The Nuclear Power Corporation of India Ltd (NPCIL) proposes to increase the installed capacity to 9935 MW by the period 2011–2012.

5.25.6 Renewable Sources of Energy

India is endowed with abundant natural and renewable resources of energy, namely, solar, wind and bio-mass. The country has been able to achieve significant capacity addition of 1367 MW through wind farms and ranks fifth in the world after Germany, United States, Spain and Denmark so far the generation of wind energy is concerned. The available renewable resources need to be exploited by giving a commercial orientation, wherever possible. It may be necessary to continue with subsidies in the case of socially-oriented programmes to meet the energy requirements of rural areas, particularly, remote villages, which may be difficult to service through the conventional power grid in the near future. **TABLE 5.3** gives the available potential and the actual potential exploited till August 2001 for various renewable sources of energy.

TABLE 5.1. Estimated Energy Demand

Primary Fuel	*Unit*	*Demand (in Original Units)*		*Demand (MTOE)*	
		2006–07	*2011–12*	*2006–07*	*2011–12*
Coal	Mt	460.50	620.00	190.00	254.93
Lignite	Mt	57.79	81.54	15.51	22.05
Oil	Mt	134.50	172.47	144.58	185.40
Natural Gas	BCM	47.45	64.00	42.70	57.60
Hydro Power	BkWh	148.08	215.66	12.73	18.54
Nuclear Power	BkWh	23.15	54.74	6.04	14.16
Wind Power	BkWh	4.00	11.62	0.35	1.00
Total Commercial Energy				**411.91**	**553.68**
Total Non- Commercial Energy				**151.30**	**170.25**
Total Energy Demand				**563.21**	**723.93**

Mt : Million BCM : Billion Cubic Metres; BkWh : Billion Kilo Watt Hours
Source : Tenth Plan Document published by Planning Commission

TABLE 5.2. Geographical Spread of Primary Commercial Energy Resources in India

Region	*Coal (Bt)*	*Lignite (Bt)*	*Crude Oil (Mt)*	*Natural Gas (BCM)*	*Hydro Power (TWH)*
Northern	1.06	2.51	0.03	0.00	225.00
Western	56.90	1.87	519.47	516.42	31.40
Southern	15.46	30.38	45.84	80.94	61.80
Eastern	146.67	0.00	2.19	0.29	42.50
North-Eastern	0.89	0.00	166.17	152.00	239.30
Total	**220.98**	34.76	**733.70**	**749.65**	**600.00**

Bt :Billion Tonnes; BCM : Billion Cubic Metres; TWH : Trillion Watt Hours; Mt : Million Tonnes
Source : Tenth Plan Document published by Planning Commission.

TABLE 5.3. Renewable Energy Sources Potential

Source/Technology Potential	*Units*	*Available Potential*	*Actual*
Bio-gas	Million	12	~32T~
Bio-mass-based Power	MW	19500	384
Efficient Wood Stoves	Million	120	33.86
Social Energy	MW/km^2	20	1.74
Small Hydro	MW	15000	1398
Wind Energy	MW	45 000	1367
Energy Recovery from	MW	17000	16.2

5.26 TRENDS IN COMMERCIAL ENERGY PRODUCTION

Coal production is likely to grow at an annual rate of 4.46% in the Tenth Five-Year Plan period (as compared to 2.4% annual growth rate during the Ninth Five-Year Plan period) to touch 405 Mt in the terminal year 2007. As against this, the coal demand in that years is estimated at 460.50 Mt. Part of this gap is proposed to be met through import of both coking and non-coking coal. About 70% of the projected demand is for public sector utilities. A substantial expansion in the domestic coal production is, therefore, needed to meet the requirements of the targeted generating capacity additions envisaged during the Tenth and Eleventh Five-Year Plans. The trends in commercial energy production during the last 40 years are given in **TABLE 5.4** which depicts the requirement of various sectors of energy during the period 2006–2007.

Coal production from public sector mines has stagnated and not many private sector mines are in operation now. Some joint sector interventions have added only marginal capacity. The immediate growth pattern is quite dismal. Country is likely to go in for large imports shortly.

The current domestic production of crude oil caters to nearly 30% of the demand and is likely to marginally increase from 32.03 Mt in 2001–2002 to 33.97 Mt in 2006–2007. As against this, the demand for petroleum products, projected as 99.13Mt in 2001–2002, is estimated to grow at the rate of 5.7% a year to touch 134.6 Mt in the terminal year of the Tenth Five-Year Plan and 172.5 Mt in the terminal year of Eleventh Five-Year Plan. The New initiatives are likely to yield larger crude oil share for new finds overseas.

India's natural gas production reached a level of 29.69 BCM in 2001-2002. The projected domestic production of natural gas in the year 2007 is 37.62 BCM. The country has been able to meet the demand with the available domestic production till recently. However, the demand is likely to grow rapidly in the near future. A number of projects for setting up of LNG terminals have been approved by the Government of India to bridge the demand-supply gap. Four LNG terminals at Dabhol, Dahej, Hazira and Cochin are in advanced stages of development and are likely to be completed by the end of the Tenth Five-Year Plan period.

Significant hydro and nuclear generation capacity is likely to be added during the Tenth Five-Year Plan period. The capacity addition programme includes 16083 MW from hydel power plants and 1300 MW from nuclear power plants. In addition, 2000 MW of energy is planned to be harnessed from wind farms.

TABLE 5.5 indicates the trends in primary commercial energy supply from various sources between 1953–1981 to 325 Mt in 2001–2002, the share of coal in total energy supplies has declined from a level of 58.9% to 51.1%. This could be partly due to the increase in the share of inferior grade coal in over-all production. The primary reason, however, is that the share of hydrocarbons in the total energy consumption of the country has been increasing over the years and is currently estimated at 44.9% as compared to 37.2% in 1980-1981. Net energy related imports of 87.85 MTOE in 2001-2002 include the import of 75.43 Mt of crude and petroleum products, 19.60 Mt of coal and 1.4 BkWh of electricity form Bhutan. The share of non-commercial sources in the total primary energy supply is 31.8% in 2001–2002, down from 53.1% in 1980–1981.

TABLE 5.4. Trends in Commercial Energy Production During the Last Four Decades

	Unit	*Production in Various Periods*					
		1960–61	*1970–71*	*1980–81*	*1990–91*	*2001–02**	*2006–07***
Coal	Mt	55.67	72.95	114.01	211.73	325.65	405.00
Lignite	Mt	0.05	3.39	4.80	14.07	24.30	55.96
Oil	Mt	0.45	6.82	10.51	33.02	32.03	33.97
Natural Gas	BCM	-	1.44	2.35	1.79	29.69	37.62
Hydro Power	BkWh	7.84	25.25	46.54	71.66	82.80	103.49
Nuclear Power	BkWh	-	2.42	3.00	6.14	16.92	19.30
Wind Power	BkWh	-	-	-	0.03	1.70	4.00

*: Anticipated, **: Projections for the terminal year of the Tenth Five-Year Plan

TABLE 5.5. Trends in Supply of Primary Commercial Energy During the Last Four Decades

Source of Energy	*Production in Various Period*					
	1953–54	**1960–61**	**1970–71**	**1980–81**	**1990–91**	**2001–02***
Coal	23.62	35.64	36.48	56.96	94.68	133.89
Lignite	-	0.01	0.81	1.23	3.34	6.52
Crude Oil	0.19	0.46	7.01	10.79	33.92	32.03
Natural Gas	-	-	0.60	1.41	11.73	26.72
Hydro Power	0.24	0.67	2.17	4.00	6.16	6.37
Nuclear Power	-	-	0.63	0.78	1.60	5.15
Wind Power	-	-	-	-	-	0.15
Total	**24.05**	**36.78**	**47.67**	**75.19**	**151.43**	**210.82**
Net Imports	2.20	6.04	12.66	24.63	31.69	87.85
Commercial Energy Supply	26.25	42.82	60.33	99.82	183.12	298.67
Total Primary Energy Supply	**64.13**	**74.38**	**86.72**	**108.48**	**122.07**	**139.02**
Total Primary Energy Supply	**90.38**	**117.20**	**147.05**	**208.30**	**305.19**	**437.69**
* : Provisional						

5.27 ENERGY IMPORTS

India is emerging as a large importer of crude and is planning to import LNG during the Tenth Five-Year Plan period. If the present trend continues, India's oil import dependency is likely to grow beyond the current level of 70%. Future strategies should focus on increasing exploration activities to enhance the level of recoverable reserves of the country.

Coal imports account for only about 5.6% of the total domestic consumption in the current year. The steel sector has been importing coking coal mainly for blending with domestic coal to obtain the desired quality of steel production. The cement industry and coastal power stations are importing non-coking coal. The share of net energy imports in primary commercial energy sector during the last two decades is given in **TABLE 5.6.**

TABLE 5.6. Share of Net Energy Imports in Primary Commercial Energy Sector

Year	*Coal*	*POL**	*Electricity*	*Total*
1980–81	0.25	25.45	-	25.70
1990–91	2.22	15.56	0.07	17.85
2001–02	4.12	26.25	0.04	30.41
* : Petroleum oil and lubricants				

The share of primary energy imports in the total commercial energy supply is currently estimated at 29.41% and the same is likely to increase by the end of the Tenth Five-Year Plan. This is a matter of concern from the point of view of energy security.

5.28 ENERGY EFFICIENCY/CONSERVATION

Energy efficiency and energy conservation involve all the sectors of the economy. Although energy efficiency and energy conservation have been discussed for nearly ten years now, the efforts on the ground have been fragmented and half-hearted. There has also been a lack of adequate focus on the institutional arrangements which will devise suitable incentives and disincentives appropriate to each sector, backed by adequate statutory power enforcement.

During the Ninth Five-Year Plan, it was realized that an Energy Conservation Act to be enforced. Accordingly, the Government of India has enacted the Energy Conservation Act, 2001 to meet the legal requirement needed to enforce energy efficiency and conservation measures. The Act provides a long range consequence which is appended below.

- The establishment of Bureau of Energy Efficiency (BEE) in place of the existing Energy Management Centre (EMC).
- Declaration of a user or class of users of energy as a designated consumer.
- To lay down minimum energy consumption standards and labeling for identified appliances/equipment and norms for industrial processes for energy-intensive industries.
- Formation of energy consumption code.
- Establishment of Energy Conservation Fund both at the central and state levels.
- Provision of Penalties and adjudication. No penalties would be effective during the first five years as the focus during this period would be on promotional activities and creating the infrastructure for implementing the Act.
- The BEE would act as a facilitator for the evolution of a self-regulatory system and organizations would regulate on their own with a view to save energy and thereby bring the commercial concept in the organization.

The central government has established the BEE with effect from March 1, 2002. Further, the provisions of Sections 1–29 and Sections 46–62 of the Energy Conservation Act relating to this have come into force from the same date.

The target for energy savings potential during the Tenth Five-Year Plan is given in **TABLE 5.7.** The Tenth Five-Year Plan programme on energy conservation would address issues like pricing of electricity for different categories of consumers, generating awareness on the necessity of energy conservation and the need for an institutional arrangement to co-ordinate the different programmes of energy conservation. It will also deal with mobilizing resources for funding the energy conservation programmes.

TABLE 5.7. Energy Savings Potential During The Tenth Five – Year Plan

End-use Type	*Potential Energy Saving, MkWh*
Motors and drive systems (Industry and agriculture sector)	80 000
Lighting (domestic, commercial and industrial sector)	10 000
Energy intensive industries	5 000
Total	**95 000**

5.29 IMPACT OF POWER SECTOR REFORMS

Reforms were introduced and pursued vigorously in the power sector. The impact was slow but gradually the country's own 'Reforms Model' has taken over and is working. Like the mixed economic model that we pursue, even the successful power sector reforms model is a hybrid between the US model and Indian practice. Success of privatization of distribution, currently in vogue, is a major achievement.

A comparative look of statistics for the years 1998 and 2003 provides the true picture of growth in this sector. **TABLE 5.8** gives the details of the present energy availability scenario of the country which underscores a lot of promise to develop the power scenario of India in future.

The growth pattern reflected in the power sector recently indicates that the system has got into the groove and there is an opportunity to grow as desired.

TABLE 5.8. Present Energy Availability Scenario As On January 2004

Sector	*1998*	*2003*	*Increase, %*
Generation, kWh	422	531	25.85
Capacity, MW	89 167	1 079 73	21.0
Power Load Factor, %	64.7	72	7.4
Transmission Line, circuit km	1 23 267	1 58 484	28.5
Trade, MW	2 600	8 000	207.0
Outlay, Rs. (in crores)	8 180	14 667	79.0
Consumers, Nos. (in crores)	9.3	11.4	22.5
Pump Sets	1 18 49 406	1 37 92 420	16.5
Investment, Rs. (in crores)	1 24 526	2 36 625	90.0
Energy Shortage, %	-	8.0	-
Peak Shortage, %	-	14.0	-
Electrification			
Overall	***Urban***	***Rural***	
42.7%	75.8%	30.4%	
Rural Electrification			
Bihar :		5 % Villages	
Jharkhand :		10% Villages	
Uttar Pradesh :		20% Villages	
Source : Lecture delivered by Minister of State for Power on the National Workshop on 'Rural Electrification Policies' held at New Delhi on January 30, 2004			

5.30 CLUB OF ROME — LIMITS TO GROWTH

The Club of Rome is a global think tank that deals with a variety of international political issues. It was founded in 1968 and raised considerable public attention with its report 'Limits to Growth'.

The Limits to Growth is a 1972 book, modeling the consequences of a rapidly growing world population and finite resource supplies, commissioned by the Club of Rome.

Its authors were Donella H. Meadows, Dennis I. Meadows, Jorgen Randers, William W. Behrens III. The book used world's model to simulate the consequences of interactions between the Earth's human systems. The purpose of the model was not to make specific predictions but to explore how exponential growth interacts with finite resources.

The world model was built specifically to investigate five major trends of global concern;

- Accelerating industrialization
- Rapid population growth
- Widespread malnutrition
- Depletion of nonrenewable resources
- A deteriorating environment.

The model constructed is, like every model, imperfect, oversimplified and unfinished.

The conclusions are:

1. If the present growth trends in world population, industrialization, pollution, food production, and resource depletion continue unchanged, the limits to growth on this planet will be reached sometime within the next one hundred years. The most probable result will be a rather sudden and uncontrollable decline in both population and industrial capacity.
2. It is possible to alter these growth trends and to establish a condition of ecological and economic stability that is sustainable far into the future. The state of global equilibrium could be designed so that the basic material needs of each person on earth are satisfied and each person has an equal opportunity to realize his individual human potential.

If the world's people decide to strive for this second outcome rather than the first, the sooner they begin working to attain it, the greater will be their chances of success.

All five elements basic to the study reported here-population, food production, and consumption of nonrenewable natural resources - are increasing. The amount of their increase each year follows a pattern that mathematicians call exponential growth.

Criticism of Limits to Growth

The Limits to Growth attracted controversy as soon as it was published. The main criticisms are as under;

1. There is insufficient evidence for many of the variables used in the model.
2. The authors have never revealed the equations used in the model.
3. A weak base data has been used in the Limits to Growth predictions.
4. The Limits to Growth predicted Oil running out in 1992 among other natural resources but this date has already expired.

QUESTIONS

1. Explain the energy scenario in India.
2. What are the different sources of energy? Explain in detail.
3. Explain Global Energy scenario.
4. Justify the necessity of developing non-conventional sources of energy.
5. Enumerate the various drawbacks that limit the exploitation of solar energy as an important energy source.
6. Explain the use of solar energy for the purpose of water heating, space heating and drying of agricultural produce.
7. Discuss the use of Bio-energy as a non-conventional source of energy.
8. What is hydropower and tidal energy?
9. Write short notes on;
   (*a*) Hydrogen energy
   (*b*) Wind Energy
   (*c*) Geothermal energy
   (*d*) Nuclear energy
10. What is Club of Rome and explain its report on limits to growth.

6

ENVIRONMENTAL DEGRADATION AND POLLUTION

6.1 ENVIRONMENTAL DEGRADATION

Environmental degradation is an undesirable change in the environment and its surrounding which include air, water and soil and affect the life and create potential health hazards for living organism. Environmental degradation is caused by many man-made activities and pollutant are responsible for environmental degradation; they are listed below:

1. Deposited matter (soot smoke, dust and grit)
2. Acid droplets (Sulphuric and nitric acid)
3. Gases (NO, NO_2, SO_2)
4. Metals (Mercury, lead and iron)
5. Agro-chemicals (biocides, pesticides herbicides, fungicides, nematocides, weedicides etc.)
6. Solid waste
7. Complex organic substances (Benzene, ether, benzopyrenes etc.)
8. Photo chemical oxidants (PAN-Peroxyacetyl nitrate etc.)
9. Fluorides
10. Radioactive waste
11. Global warming
12. Oil spills from tankers
13. Unwanted noise
14. Deforestation and forest fire
15. Disposal of industrial effluents and waste in river and streams.

All these pollutants can be classified as air pollution, water pollution, Soil and Land pollution and Noise pollution.

6.2 BIOSPHERE CYCLE AND CO_2 CYCLE

The biosphere, which is also called eco-system, represents a stable equilibrium of various physical and biological factors, which have been operating in the past. The organic continuity of the system rests on a delicate network of interdependent relationship. The eco-system has a rhythm and movement of its own. All living organisms-microbes, plants, animals, man has survived by adjusting themselves to the environment and attune their lives to its rhythms. It therefore becomes very important that these cycles work. What keeps the biosphere to work is solar energy. This energy comprises 99.98 per cent of the total energy supply of the biosphere. The process by which by which solar energy is transferred to molecular energy is called photochemical process. The most important photochemical activity in the atmosphere is photosynthesis in plants. The light is absorbed by chlorophyll molecules and by other pigments in plants and is transferred to electrons in such a way as to create strong oxidants. These oxidants assist plants in producing carbohydrates and oxygen from molecules of carbon dioxide and water. Plant respire (give out) oxygen, retain 12000 calories and release 7500 calories when broken.

From the preceding line it is clear that the biosphere is the sum of all living matter on earth. All the living matter is highly specialized organism. This organism has adapted (fit in) to the extreme boundaries of the uppermost atmosphere and lowermost ocean depth.

The biosphere is interconnected with there other spheres of physical environment. These three shears are lithospheres, the hydrosphere and atmospheres. A brief description of this sphere is given below one by one.

6.2.1 Lithosphere

The lithosphere is the solid outer layer of the Earth's crust. It includes rock, sand and soil. The lithosphere of the earth is a dynamic area. It goes on processing erosion and earth earthquakes. It also alter plate tectonics (relating to deformation of earth's crust) constantly, altering and forming or destroying the surface. The tectonic cycle describes the formation of new crust in some areas and destruction in others.

Plate tectonics is geological theory that propose the lithosphere is composed of tectonics plates that are in constant motion relative to each other. The plate essentially "floats" on a semi-molten lower crustal layer. Plate motion rate range between 2 and 15 centimeters per years. Tectonic processes have formed new continents, mountain ranges, and moved existing continents across the surface of earth, causing the evolution and extinction of species from resulting climate and environmental changes.

The December 26, 2004 earthquake that killed more than two lakh people in the Southeast Asia (Indonasia, Malasia and A.P.) The latest data available through the Global Positioning System (GPS) has shown the under its impact, that almost all country has moved eastward by few centimeters.

South India moved most on the day of the quake – by more than 15 mm. on the average, Thiruvananthapuran moved by 26 mm, Banglore by 15 mm and Hyderabad by 10 mm. the displacement decreases northwards- Dehradun moves by one about 2 mm.

The data also reveled that the quake caused deforestations on the earth's surface across a radius of 4, 500 km from the epicenter "the kind of impact is astonishing." we also learnt through GPS data that the quake continued for two hours. A recent earthquake in the month of May 2008 rocked the Sichuan province of Chine has left more than 40000 people dead.

6.2.2 Hydrosphere

The hydrosphere includes all water at or near Earth's surface. Water is very important to a number of biological and geological processes. The hydrocarbon cycle is the continuous recycling of water between the oceans and atmosphere. Evaporation is the movement of water from an ocean or a lake to the atmosphere. Transportation by plants contributes to atmosphere water content. Precipitation rainfall or snow is the movement of water from the atmosphere to land or ocean. It changes the environment sometimes quite cold. In the year (2005, Dec.,) BIG CHILL was experienced. Minimum temperature in Churu (Rajsthan). was 1.7^{0}C, Bhunter, (H.P.) 0.8 and Amritsar 2^{0}C. Water on the land can be either surface runoff or filter through soil to groundwater.

Atmosphere: The atmosphere is the envelope of the gases that surrounds the earth. Nitrogen and oxygen compose 99% of the modern atmosphere. The atmosphere becomes progressively thinners with the increasing attitude.

6.2.3 Connectivity of the Sphere

These spheres are connected and relate to each other at a variety of interface. The geological process of volcanism transfer water from the lithosphere to atmosphere, where it precipitates in to the hydrosphere. The interface between water and air likewise allows transfer of material from one sphere to another.

6.3 MONITORING THE ENVIRONMENT

It is periodic checking or testing of environment for knowing levels of different pollutants, identifying the sources and devising strategy for their control. National Environment Engineering and Research Organization (NEERI) is situated at Nagpur.

Birds play an important role in monitoring (informing) the environment. Every spices (living creature) has its own significance and its presence or absence in a particular habitat (local environment of a group of organism) has a specific importance. The birds have freedom to move wherever they like and political boundaries have no meaning for them. The only thing that matter is the suitability of the environment. The birds of the polar region are different from those found in tropics or found in temperate regions. Similarly, the hill species are different from those found in the plains. Cyclone Baaz predict environmental change and wreck havoc.

It is noticed that the metabolic (pertaining to change of food into living tissues) rate of birds is much higher than that of humans, and as a result they suffer from pollution much before it effects humanity. Therefore, the birds effectively forecast an environmental setup of the region. Some bids are intentionally kept in the coal mines and in the case of release carbon dioxide exceed the permissible limits; the birds die and forecast the danger of coal mines explosion. The appearance of vultures, owls, bats *etc.*, are symbolic of bad weather or unfavorable environment, while on the other side, the appearance of a peacock is symbolic of rain, good weather and good environment. Such examination can be multiplied and reflect how the birds monitor the environment. Birds have a close connection with the environment. The sudden fall in the number of aquatic (water foul) organism due to spraying of poisonous pesticides and other chemical also shows their connection with environment. The spraying of DDT destroys the aquatic fauna (animals) on which the marine organism depends. The birds eating fish die on account of consuming fish that were affected by DDT and/or other poisons gases or chemicals.

When the birds living on the banks of lake , ponds or streams start dieing on consuming the polluted water, it clearly indicate that there are industries adjoining these water bodies and there waste are polluting water. The appearance of smog (mixture of fog and smog) during 1956 in London caused a great loss of birds. Acid rains, oil slicks (film off crude oil on sea surface) and atmosphere pollution are responsible for the destruction of birds. A close relationship between environment and the birds had also been established by their presence in specific regions. The presence of parrots will indicate the existence.

6.4 CAUSES OF ENVIRONMENTAL DEGRADATION

1. Environmental degradation is results of the dynamic inter play of socio-economic, institutional and technological activities. Environmental change may be driven by many factors including economic growth, population growth, urbanization, intensification of agriculture, rising energy use and transportation. Poverty still remains a problem at the root of several environmental problems.
2. Population is an important source of development yet it is a major source of environmental degradation when it exceeds the threshold limits of the support systems. Unless the relationship between the multiplying population and the life support system can be stabilized, development programmers, however innovative are not likely to yield desired results. Population impacts on the environment primarily through the use of natural resources and production of wastes and is associated with environmental stresses like loss of biodiversity, air and water pollution and increased pressure on arable land.
3. India supports 17 % of the world population on just 2.4% per cent of world land area. It's current rate of population growth at 1.85 per cent continues to pose a persistent population challenge. In view of the linkage between population and environment, a vigorous drive for population control need hardly be over emphasized.
4. Poverty is said to be both cause and effect of environmental degradation. The circular link between poverty and environment is an extremely complex phenomenon. Inequality may foster un-sustainability because the poor who rely on natural resources more than the rich, deplete natural resources faster as they have no real prospects of gaining access to other types of resources. Moreover degraded environment can accelerate the process of impoverishment again because the poor depend directly on natural assets. Although there has been a significant drop in the poverty ratio in the country from 55 percent in 1973 to 36 percent in 1993–94 the absolute number of poor have however remained constant at around 320 million over the years. An acceleration in poverty alleviation is imperative to break this link between poverty and the environment urbanization poverty and the environment.
5. Lack of opportunities for gainful employment in villages and the ecological stresses is leading to an ever increasing movement of poor families to towns. Mega cities are emerging and urban slums are expanding. There has been an eightfold increase in urban population over 1901–91, India's urban population has doubled from 109 million to 218 million.
6. Such rapid and unplanned expansion of cities has resulted in degradation of urban environment. It has widened the gap between demand and supply of infrastructural service such as energy, housing, transport, communication, education, water supply and sewerage and recreational amenities thus depleting the precious resource base of the cities. The result is the growing trend in deterioration of air and water quality, generation of wastes, the proliferation of slums and undesirable land use change, all of which contribute to urban poverty.
7. To a large extent environmental degradation is the result of market failure that is the non existent or poorly functioning market for environmental goods and services. In this context environmental degradation is a particular case of consumption or production externalities

reflected by divergence between private and social costs or benefits. Lack of well defined property rights may be one of the reasons for such market distortions created by price controls and subsidies may aggravate the achievement of environmental objectives.

8. The level and pattern of economic development also affect the nature of environmental problems. India's development objectives have consistently emphasized the promotion of polices and programmers for economic growth and social welfare. Between 1994–95 and 1997–98, the Indian economy has grown a little over industrial production and manufacturing averaging higher at 8.4 per cent and 8.9 percent respectively during these years. The manufacturing technology adopted by most of the industries has placed a heavy load on environment especially through intensive resource depletion (Fossil fuel, minerals, timber), water, air and land contamination, health hazards and degradation of natural eco-systems. With high proportion fossil fuel as the main source of industries such as iron, steel, fertilizers and cement growing, industrial source have contributed to a relatively high share in air pollution. Large quantities of industrial and hazardous wastes brought about by expansion of chemical based industry has compounded the wastes management problem with serious environmental health implications.
9. Transport activities have a wide variety of effects on the environment such as air pollution, noise from road traffic and oil spills from marine shipping. Transport infrastructure in India has expanded considerable in terms of network and services. Thus road transport account for a major share of air pollution load in cities such as Delhi. Port and harbor projects mainly impact on sensitive coastal eco systems. Their construction affect hydrology, surface water quality fisheries coral reefs and mangroves to varying degrees.
10. Direct impacts of agricultural development on the environment arise form farming activities which contribute to soil erosion, land salination and loss of nutrient. The spread of green revolution has been accompanied by over enforcement capabilities of environmental institutions both at the center and the state. There is no effective coordination amongst various Ministries/institutions regarding integration of environmental concerns at the inception/ Planning stage of the projects. Current polices are also fragmented across several Government agencies with differing policy mandates. Lack of trained personal end comprehensive database delay many projects. Most of the state Government institutions are relatively small, suffering from inadequacy of technical staff and resources. Although overall quality of Environmental impact assessment (EIA) studies and the effective implementation of the EIA process have improved over the year institution strengthening measures such as training of key professionals and staffing with proper technical persons are needed to make the EIA procedure a more effective instrument for environment protection and sustainable development. Shifting cultivation has also been an important source of contamination of water bodies. Intensive agriculture and irrigation contribute to land degradation particularly salination, alkalization and water logging.

6.5 POLLUTION

Pollution is any change in physical, chemical or biological characteristics of the environment that has the potentiality to harm human life, life of other desirable species and natural resources, cultural assets and industries.

Contamination is presence of harmful organisms or their toxins that cause discomfort or disease.

Natural pollution is pollution caused by natural sources, *e.g.*, volcanic eruptions, release of methane by paddy fields and cattle, release of carbon monoxide by plants and animals, emission of natural gas, ozone, nitrogen oxides, soil erosion, dust storms, cosmic rays, ultra-violet rays, etc.

Man-made pollution is pollution resulting from human activities like burning of fossil fuels, deforestation, mining, sewage, industrial effluents, pesticides, fertilizers, etc. Amount of man made atmospheric pollution is hardly 0.05% of total but is more dangerous because of its concentration in certain localities. Pollution can be atmospheric, soil and water. On the basis of emission of pollution, it can be

(*i*) **Point Source Pollution.** From a single point, *e.g.* chimney, municipal sewer.

(*ii*) **Line Source Pollution.** Pollution is passed along a narrow belt, *e.g.*, roads due to automobile exhausts.

(*iii*) **Area Source Pollution,** *e.g.*, mining area, industrial estate.

(*iv*) **Diffuse Source Pollution.** It is over a large area, *e.g.*, sprayed pesticides or fertilizers through run-off.

6.5.1 Pollutants

It is a substance, chemical or factor which has the potentiality to adversely affect natural characteristics of environment, natural resources and assets. Pollution may be a substance in wrong place, *e.g.*, fertilizer causes water pollution.

Primary Pollutant. Pollutant persisting in the environment in the form it is produced e.g., carbon monoxide.

Secondary Pollutant. Pollutant formed from a primary one through change or reaction. Nitrogen oxides and hydrocarbons react photochemically to produce peroxyacyl nitrates and ozone. The secondary pollutants may be more toxic than the primary ones. The phenome[illegible] is called **synergism.**

Qualitative Pollutant. It causes harm due to its nature. It is often produced due to human activity, *e.g.*, insecticides.

Quantitative Pollutant. It is component which becomes pollutant o[illegible] when its concentration reaches beyond a threshold value, *e.g.*, CO_2, CO.

Degradable Pollutant. The pollutant degrades after some time either automatically (*e.g.*, heat) or through the agency of microorganisms (=biodegradable, *e.g.*, sewage).

Nondegradable Pollutant. The pollutant does not get degraded or broken down into harmless material, *e.g.*, DDT, plastics.

6.5.2 Air Pollution (Atmospheric Pollution)

It is the addition of particulate matter, gases and other ingredients into air which can produce an adverse effect on human beings, animals, vegetation, human assets and resources. Major sources of atmospheric pollution are (*i*) Combustion of fossil fuels in homes, factories, thermal plants, automobile, aircrafts, railways, etc. (*ii*) Mining and procession (*iii*) Chemical industries (*iv*) Procession industries etc.

1. **Particulate Matter.** It consist of soot, fly ash, dust of various types, fur, hair, spores, pollen grains, etc. Particulate matter is differentiated into **settleable** (large than 10 μm) and **sus-**

pended (less than 10 µm). **SPM** (suspended particulate matter) is maximum in Calcutta (Tokyo is the most polluted city of the world). It is differentiated into aerosol (less then 1 µm), **dust** (more than 1 ìm) and **mist** (liquid, more than 1 µm). Particles of 2–4 µm size are mostly deposited in the respiratory tract.

(*i*) **Soot** (incomplete burning of carbohydrates), **smoke, flyash** (fine particulate matter passed out along with gases during burning of coal) and dust deteriorate the quality of articles, clog stomata, cover leaf surface, produce allergic reactions, especially bronchial asthma.

(*ii*) Particulate matter from procession industries (*e.g.*, cotton dust, iron mill dust, flour mill dust, gem grinding) causes pneumoconiosis, **byssinosis, emphysema, siderosis** and other pulinonary problems.

(*iii*) Dust and smoke produce **smog.**

2. **Carbon Monoxide.** It is produced due to incomplete combustion, and naturally by plants as well as animals. 70% emissions are from automobiles. Carbon monoxide combines with haemoglobin, produces **carboxyhaemoglobin** or COHb. At 50 ppm, CO converts 7.5% of haemoglobin into carboxyhaemoglobin within 8 hours. It impairs oxygen transport resulting in giddiness, headache, decreased vision, cardiovascular malfunction and asphyxias.
3. **Carbon Dioxide.** It is a green house gas, the concentration of which is constantly rising (0.033 from 0.029 and expected to be doubled in 2020). In excess it causes headache and nausea.
4. **H_2S.** It is a product of putrefaction, treatment of sulphur containing ores, refineries, chemical plants and bituminous fuels. It causes mottled chlorosis and defoliation in plants, decolouries paints, produces eye irritation, throat irritation and nausea.
5. **Hydrocarbons.** They are produced naturally (*e.g.*, marsh gas) as well due to incomplete combustion. Hydrocarbons are carcinogenic, cause irritation of eyes and mucous membrane and bronchial constriction. There is increased mucus secretion and tearing of alveoli. They give rise to secondary pollutants with nitrogen oxides. Methane (marsh gas) has the potential to destroy ozone.
6. **Sulphur Dioxide.** It is produced during combustion of fossil fuels and smelting of sulphur containing ores. Sulphur dioxide produces smog. Maximum SO_2 pollution is found in Calcutta. It causes membrane damage, destruction of chlorophyll (changed to phaeophytin), necrosis and water soaked appearance. Lichens are the most sensitive to SO_2 pollution. Mossess and Garden Pea are also destroyed. SO_2 corrodes metals, equipment, damages buildings, marble, paper and textiles. It produces eye irritation, damages respiratory tract, produces asthma bronchitis and emphysaema. Sulphur dioxide (alongwith nitrogen oxides) produces acid rain that destroys vegetation and degrades various articles. Threat to Taj Mahal from Mathura refinery is due to pollutant gases comprising SO_2, H_2O and nitrogen oxides. They would convert $CaCO_3$ (marble) into calcium sulphate and calcium nitrate.
7. **Nitrogen Oxides.** They are formed electro-photo-chemically in air, burning of fossil fuels and denitrifying bacteria. Maximum NO_2 pollution is recorded from Baroda. Nitrogen oxides causes necrosis, defoliation, dieback and death of plants. Like SO_2, they corrode metals and deteriorate paints as well as various articles. At less than 0.1 ppm nitrogen oxides impair sensation of smell. The oxides produce eye irritation, lung oedema, dilation of arteries, injury to liver and kidneys. It is Carcinogenic and form secondary pollutants.
8. **Fluorides.** They are given out during refining of minerals (*e.g.*, aluminium, also from ground water). Fluorides cause flourosis. In plants there is chlorosis and necrosis of leaf tips and leaf

margins, followed by abscission. In animals, fluorides bring about abnormal calcification of bones and teeth (making them weak), frequent diarrhea and swelling of knee bones. In humans there is mottling of teeth, weak bones, boat-shaped posture, knocking knees, gastrointestinal and neuromuscular disorders.

9. **Chlorofluorocarbons/Chlorofluoromethane/Freon/Aerosols.** There are chemicals used as refrigerants, propellants and solid plastic foams. The chemicals are released as aerosol by jets flying at high altitudes. Alongwith nitrogen oxides, chlorofluorocarbons react with ozone of ozonosphere and depletes the same. Hole in the ozone shield over Antarctic region has widened from 129 to 133 Dobson units in 1994 alone. This can increase the amount of ultraviolet radiations reaching the earth.
10. **Photochemical Oxidants.** They are secondary pollutants (ozone, peroxy-acyl nitrates, aldehydes and phenols) produced due to photochemical reactions between nitrogen oxides and unsaturated hydrocarbons. (i) **Ozone.** It destroys chlorenchyma, produces necrosis, hardens rubber, damages textiles injures mucous membranes, dry throat, haecomorrhages and eye irritation (ii) **Peroxy-Acyl Nitrates (PAN,** also PBN, PPN). Young and spongy parenchyma cells are destroyed resulting in silvering glazing, bronzing and necrosis in leaves. In human beings they cause respiratory distress and eye irritation. (iii) **Aldehydes.** Irritation in gastro-intestinal and respiratory tracts. (iv) **Phenols.** Damage to kidneys, live, spleen and lungs.
11. **Other Atmospheric Pollutants.** Mercury burning of coal, smelting), methyl isocyanate (pesticide manufacture), phosgene (pesticide manufacture, dye industry), ammonia (fertilizer, dye and lacquer industries) and lead (automobile exhausts) are pollutants added to atmosphere though the may not reside in them (*e.g.*, lead in soil). **Bhopal gas tragedy** (Dec, 1984) was due to release of phosgene and methyl isocyanate.
12. **Automobile Exhausts.** Automobiles burn petroleum inefficiently causing 80% of air pollution and 75% of noise pollution in urban areas. They release hydrocarbons (13.7%), carbon monoxide (77.2%), nitrogen oxides (7.7%), sulphur oxides, ammonia, aldehydes and lead (90% of total lead poisoning). Lead is present in petroleum in the form of Pb $(CH_3)_4$ and $Pb(C_2H_5)_4$ as antiknock agent.
13. **Smog** (Des Vocux, 1905). Smog is opaque or dark fog having condensed water vapours, dust, smoke and gases (SO_2, H_2S, NO_2, etc) It causes silvering/glazing and necrosis in plants, allergies and asthma/bronchitis in humans. Some famous smog of the world are Los Angeles smog (1946, London smog (1952), Tokyo (Rome, New York, Sydney 1970).
    Smog is of two types:
    (*i*) **Classical (London) Smog.** It occurs at low temperature, contains sulphur gases (hydrogen sulphide, sulphur dioxide), smoke and dust particles.
    (*ii*) **Photochemical (Los Angeles) Smog.** It occurs at high temperature over cities and towns due to still air, emission of nitrogen oxides and carbohydrates from automobile exhausts and solar energy. Nitrogen dioxide splits into nitric oxide and nascent oxygen. Nascent oxygen combines with molecular oxygen to form ozone. Ozone reacts with carbohydrates to form aldehydes and ketones. Nitrogen oxides, oxygen and ketones combine to form peroxy-acyl-nitrates (PAN).
14. **Pollen/Spores.** They cause allergic reactions or hay fever in certain persons.

6.5.3 Control of Air Pollution

1. **Automobiles.** (*i*) Two-stroke engines fitted in two wheelers (waste fuel 20–30%) be changed to either four-stroke engines or fitted with catalytic converters specially designed for them. A low

cost catalytic converter has been developed at BHU (*ii*) Leaded petrol be replaced with unleaded one and diesel with low sulphur diesel. (*iii*) Tune-ups (for high air-fuel ratio) and catalytic converters (for oxidising $CO \rightarrow CO_2$ and reducing $NO \rightarrow N_2$) be fitted in automobiles. (*iv*) Periodic check up of pollution control for all vehicles.

2. **Begasse and Rice Husk.** Begasse should not be used as fuel. Rice husk should be first converted into briquetts.
3. **Fly Ash.** About 38% fly ash is produced by coal based thermal plants. It should be removed through wet method and used in building material.
4. **Industrial Pollution:**
   (*i*) ***Tall Chimneys.*** They disperse smoke more thoroughly.
   (*ii*) ***Gravity Settling chambers.*** Particles larger than 50 μm settle down
   (*iii*) ***Wet Scrubbers.*** A fine spray of water or alkaline fluid is used to remove soluble gases and particles.
   (*iv*) ***Bag Filters.*** Porous bags of teflon or polyester filter out particulate matter.
   (*v*) ***Electrostatic Precipitators.*** (ESPs). They are electrically charge plates or electrodes which remove most of the particles present in chimney exhausts.
5. **Vegetation.** A broad strip of vegetation along road and around industrial area reduces particulate pollution. Vegetation can also metabolise toxic gases like CO (*e.g.*, Ficus variegate, Coleus, Daucos, Phaseolus) and nitrogen oxides (*e.g.*, Vitis, Pyrus, Robinia, Rhamnus).

6.6 ACID RAIN

The term was coined by Robert Augus (1872). Acid rain is rainfall and other forms of precipitation with a *p*H of less than 4.5. It is caused by large scale emission of acidic gases into the atmosphere from thermal power plants, industries and automobiles. The common ones are sulphur dioxide, nitrogen oxides and hydrogen chloride. Sulphur dioxide and nitrogen oxides are changed in the atmosphere into sulphuric acid and nitric acid by combining with oxygen and water.

Acid rain damages plants by direct effect on foliage and growing points-chlorosis, necrosis, defoliation, dieback. It causes leaching of essential minerals of soil. Toxic minerals left in the soil further kill the plants. 50% of natural forests have been destroyed by acid rain in Germany. Sweden, north east U.S.A., Romania, Poland, etc. Acid rain has also ruined fresh water reservoirs of most industrialized countries, *e.g.*, 80% in Norway, 25% in Sweden, 20% in U.S.A. Acidity dissolves toxic metals Hg, Pb, Zn, Al. Both acidity and toxic metals kill all types of aquatic life except some algae and fungi. Acid rain corrodes metals, marble, painted surfaces, slate, stone, etc. The phenomenon is called stone leprosy.

6.7 OZONE DEPLETION

Ozone layer or shield is present in the stratosphere. It protects the earth from short-wave ultraviolet rays (below 300nm) by changing the same into infra-red rays. A large hole has appeared in ozone shield over Antarctica (first detected by Farmer, 1982) and a smaller one over North Pole. Size of the holes varies with the seasons. Thinning of ozone shield has also been reported elsewhere (*e.g.*, 8% between 30^0–50^0N). Depletion of ozone layer allows harmful ultra-violet radiations to reach earth. It is the major cause of skin cancer, cataract, dimming of eye-sight, decrease in immune system and increased susceptibility to herpes.

Thinning of ozone shield is being caused by a number of pollutants like chlorofluorocarbons (14% of total depletion), nitrogen oxides (3.5% depletion), sulphur dioxide, halon, carbon tetrachloride, methyl chloroform, chorine, etc. Many of these are being released by jets flying in the stratosphere and rockets being fired into space. Others are persistent in the troposphere and gradually pass into stratosphere. Maximum ozone depleting potential or ODP is of chlorofluorocarbon due to release of chlorine by it. A single chlorine atom coverts 1 lakh molecules of ozone into oxygen. Consequently, chlorofluorocarbons (CFCs) are being replaced by hydro fluorocarbons (HFCs) and hydrochlorofluorocarbons (HCIFCs). Carbon tetrachloride, halon and methyl chloroform also deplete ozone by a similar method. Nitric oxide (NO) and other gases released by jets directly react with ozone to form oxygen.

$$NO + O_3 \rightarrow NO_2 + O_2$$

6.8 WATER POLLUTION

It is the degradation of quality of water due to addition of substances (*inorganic, organic, biological, radiological*) or factors (*e.g.*, heat)

1. **Organic Wastes/Sewage.** It is waste water having food residue, animal and human excreta, detergents, discharges from commercial and industrial establishments.
   (*i*) Raw sewage contains a number of pathogens typhoid, dysentery, diarrhea, cholera, giardiasis, amoebiasis, helminth eggs, coliforms and enterococci.
   (*ii*) It stimulates the activity of several decomposer organisms collectively called ***sewage fungus.*** The property of becoming decomposed through microbial activity is known as ***putrescibility.*** Degree of impurity of water due to organic matter is measured in terms of B.O.D. ***Biochemical*** oxygen demand or B.O.D. is the oxygen in milligrams required for five days in one liter of water at 20^0C for the microorganisms to metabolise organic waste. Low pollution-below 1500mg/l, medium pollution-1500–4000mg/l and high organic pollution-above 4000mg/l.
   (*iii*) Water has a brown colouration and an unpleasant odour due to formation secondary pollutants like CH_4, NH_3, H_2S.
   (*iv*) Scum and sludge (H_2S + metallic ions).
   (*v*) there may be algal blooms and eutrophication. The latter is also due to phosphates present in detergents.
   (*vi*) Animals may die due to depletion of oxygen by sewage fungus.

   Sewage pollution can be prevented by treating sewage before passing into water course:
   (*a*) **Primary Treatment.** This consists of shredding, churring, floatation, screening and sedimentation.
   (*b*) **Secondary Treatment.** The organic matter is decomposed through sewage fungus by either ***trickling filter method*** (passing through thick bed of gravel) or ***Activated sludge method*** (aeration/oxidation tanks). The cleared water is now **chlorinated** to kill pathogens. The water can be passed into fields as manured water or into water courses where it may cause eutrophication.
   (*c*) **Tertiary Treatment.** Where water is to be recycled, the chlorinated waste water is mixed with alum, ferric chloride and lime for precipitation of salts. Zirconium is also a good precipitant. Another technique is reverse osmosis.

2. **Fertilizer.** Part of fertilizers added to crop fields are passed down to water bodies during rain through surface run-off. Presence of extra nutrients brings about dense growth of plant and animal life. The phenomenon is called ***eutrophication.*** Eutrophication leads to organic loading, depletion oxygen, death of animals and fouling of water.
3. **Pesticides.** Pesticides sprayed over crops also pass into water bodies due to surface run-off. In excess they cause immediate and mass scale deaths of aquatic animals. **Presistent pesticides** (*e.g.*, organochlorine or chlorinated hydrocarbons like DDT) pass into food chain and increase in amount per unit weight of organisms with the rise in trophic level due to their accumulation in fat. The phenomenon is called ***biomagnification***/bioconcentration/biological amplification, e.g., 0.01–0.05 parts per billion in water, 10 parts per billion or 0.001 ppm in phytoplankton, 0.04 ppm in zooplankton, 0.4 ppm in clan haemorrhage, softening of brain, liver cirrhosis, hypertension, thinning of egg shells, malfunctioning of sex hormones. Population of Bald Eagle has declined due to it.
4. **Silt.** Silt from soil erosion makes water muddy and unfit for plant growth. Silt may clog gills of fishes.
5. **Thermal Pollution.** Hot effluents and hot water (*e.g.*, thermal plants/atomic reactors) bring about rise in water temperature. Warm water contains less oxygen, has lower rate of putrescibility resulting in increased organic loading, replacement of green algae by blue-green ones. Many animal fail to reproduce *e.g.*, Trout, Salmon.

   El Nino is warm ocean or **Peru current** that recurs after 5–8 years. It kills marine animals and plankton over large tracts along Peru and Ecudor coasts.
6. **Oil Pollution.** Oil spills and refinery discharges cause oil pollution. It reduces oxygenatic, inhibits plankton growth and kills animals.
7. **Industrial Effuents.**
   (*a*) **Mercury.** It is changed to water soluble dimethyl mercury which undergoes biomagnification. Eating poisoned animals causes deformity known as ***minamata disease*** which is characterized by diarrhea, hemolysis, impairment of various senses, meningitis and death.
   (*b*) **Copper Hypertension,** uremia, occasional fever and coma.
   (*c*) **Lead** (also common from automobile exhausts). Anaemia, vomiting, convulsions, loss of appetite, damage to liver, kidneys and brain.
   (*d*) **Zinc.** Vomiting cramps, renal damage.
   (*e*) **Cobalt.** Diarrhoea, hypotension, bone defects and paralysis.
   (*f*) **Chromimum** Gastro-intestinal ulcers, nephritis and nervous system disorders.
   (*g*) **Cadmium.** Anaemia, hypertension testicular atrophy, damage to liver and kidneys, diarrhea and skeletal deformities called ***itai-itai*** (ouch-ouch, first reported in 1947 in Toyoma city of Japan)

6.9 SOIL POLLUTION

It is alteration in soil caused by removal or addition of substances and factors which decrease its productivity, quality of plants and ground water. **Negative soil pollution** is reduction in soil productivity due to erosion and over-use. **Positive soil pollution** is reduction in soil productivity due to addition of undersirable substances (*e.g.*, pesticides, fertilizers, industrial wastes, air pollutants wasted down by rain, faulty sanition) **Landscape/third pollution** is converting fertile land into barren one dumping wastes (*e.g.*, ash, sludge, garbage, rubbish, industrial wasters, broken cans, bottles, etc.) over it.

1. **Pesticides.** They include insecticides, fungicides, algaecides, rodenticides and weedicides. Pesticides are generally broad spectrum and function as **biocides.**
   (*i*) **Organochlorines/Chlorinated Hydrocarbon.** The important ones are DDT (dichloride diphenyl trichloroethane), BHC (benzence hexachloride), aldrise dieldrin and endrin. They are **persistent,** fat soluble and show **biomagnification** whence they prove harmful to higher trophic level organisms. Hence, their use is being restricted

   (*ii*) **Organo-Pesticides.** Degradable but toxic to workers, *e.g.*, malathion, parathion, carbamates.

   (*iii*) **Inorganic Pesticides.** They contain arsenic and sulphur. They are persistent hence, their use is highly restricted.

   (*iv*) **Weedicides.** Often persistent and harmful.
2. **Fertilizers.** Excessive use causes soil deterioration through decrease in natural microorganism leading down causes pollution of underground water (third poison). Salts entering crop plants in excess may prove harmful. For example, nitrate rich leaves, fruits and water produce nitrite in eliminatory canal that enters blood, combines with haemoglobin forming **met-haemogiobin** and reduces oxygen transport. It may prove fatal in infants. **Organic farming** involves use of biofertilisers, manures, pesticides of organic origin, biological control and resistant varieties.
3. **Industrial Effluents/Wastes.** They contain a number of toxic substances including cyanides, chromates, acids, alkalies and metals like mercury, copper, zinc, lead, cadmium etc.
4. **Mine Dust.** It destroys vegetation and produces many deformities in animals and human beings.
5. **Soil Salination**

   It is increase in salt concentration of soil making the latter **halomorphic.** Salination is often accompanied by white incrustation over the surface. If makes the soil barren. Only halophytes can grow over it. Soil salination is caused by;

   (*i*) Parent rock.

   (*ii*) Poor drainage and elevated water table.

   (*iii*) Salt rich ground/canal water.

   (*iv*) Excessive fertilizers.

   (*v*) Salts blown from rocks/sea.

   (*vi*) Plants that absorb acidic ions. India has about 6 million hectares of saline soils. They can be reclaimed by addition of gypsum/related chemicals, trenches and proper drainage, growing salt tolerant plants and blue-green algae.
6. **Recycling of Waters**

   Sludge from sewage treatment plants is burnt alongwith coal. Organic wastes be composted or employed in biogas plants, plastics and metals are now recycled. Agricultural wastes are used in manufacture of paper, hard board, animal feed, etc.

6.10 RADIOACTIVE POLLUTION

It is degradation of environment due to release of radioactivity (emission of α-particles, β-particles and gamma rays) by changes in **nuclides** (atomic nucleus) of unstable/radioactive elements causing short-range and long range harmful effects on living beings including humans. Radioactivity is measured in units called roentgens or r.

1. **Background Radiation.** It is radiation level found naturally in biosphere due to cosmic rays reaching earth and radio-nuclides found in earth's crust. The naturally occurring radioactive elements are Uranium-238, Uranium-235, Radium-224, Radon-222, Potassium-40 and Carbon-14. **Half life** is time required for one half of radioactive atoms to disintegrate, *e.g.*, 5000 years in C-14. Maximum background radiation is found in Kerala beach where 75% of thorium deposits of the world are found.
2. **Man-Made Radiations.** They are due to mining and refining of radioactive elements like Plutonium, Uranium and Thorium, nuclear power plants and fuels, preparation of radio-active isotopes, production and explosion of nuclear weapons.

6.11 NUCLEAR WEAPONS

They are Uraniun-239, Plutonium-239 for fission and hydrogen or lithium as fusion material. A nuclear explosion produces;

(*i*) Uncontrolled chain reactions.

(*ii*) Tremendous heat

(*iii*) Neutron flux that changes other elements to radioactive state, *e.g.*, Barium-142, Cesium-137, Iodine-131, Krypton-91, Strontium-90, Cobalt-60, Phosphorus-32

(*iv*) Unused explosive and activation products.

(*v*) A lot of radioactive and other gases forming mushroom type cloud. The radioactive particles are carried to all places polluting air, water, soil and food chains. Nuclear bombing of Nagasaki and Hiroshima (1945) not only destroyed the two cities, killed and maimed thousands of humans and animals but have left a chain of genetic deformities. Iodine-131 forms thyroxin and damages spleen, lymph nodes, leucocytes, bone marrow, produces lung tumor, skin cancer as well as sterility. Strontium-90 causes bone cancer, blood cancer and tissue degeneration. Cesium-137 brings about nervous, muscular and genetic charges. Cattle, dairy milk, drinking water, vegetables, fruits and grains all pass on radioactivity to humans. Certain animals accumulate radioactive elements. Oysters-Zn-65, Fish-Fe-55, other marine animals-Sr-90.

Atomic Reactors: They employ controlled radioactive fission/fussion for liberation of energy;

(*i*) The coolant water causes thermal pollution.

(*ii*) Small amount of radioactivity enters coolant water which undergoes biomagnification to some 75000 times in birds.

(*iii*) They release halogens and inert gases.

(*iv*) A mishap can be dangerous as an atomic explosion, *e.g.*, Chernobyl in Ukraine.

(*v*) Radioactive waste is highly pollutant and its dumping requires several precautions.

Radio-Isotopes. They are used in research (*e.g.*, metabolic pathways), induction of mutations in plants and microorganisms, radiotherapy, *etc. e.g.*, Co-60, P-32, C-14, I-125.

***Harmful Effect*:** They were first recorded in 1909 in uranium miners as skin burns and cancers. Many plants are killed even at low level of radioactivity. Young and recently divided cells are more easily damaged.

Short Range Effects. Loss of nails and hair, subcutaneous bleeding, changed proportion of blood cells, changed metabolism, damage to all organs, death in high dose.

Long Range/Delayed Effects. Tumours, cancers, mutations, genetic deformities, shorter life span.

Non-ionising Radiations

They are ultra-violet rays (100–300 nm) which have low penetration. They damage eye sight due to damage of corneal cells, injury to germinative layer, rupture of subcutaneous capillaries, blisters and reddening of skin (sunburn), mutations, skin and other cancers due to inactivation of organic molecules and formation of pyrimidine dimmers in DNA. Damage to ozone layer will bring all these hazards. Thick cuticle, sporopollenin and carotenoids protect plants from UV radiations, especially at higher altitudes. Human beings can protect themselves by wearing sun glasses. They have an enzyme system for repair or UV damage. Absence of this protective system produces a deficiency called ***xeroderma pigmentosum.***

6.12 NOISE POLLUTION

It is a physical form of pollution that affects the receiver directly. Noise pollution is due to released of unwanted sound generally with a value of 80 dB and above. Frequency of sound is measured in Hz (Hertz) while unit of sound is dB (decibel). Range of human hearing is 20 Hz to 20000 Hz. ***Infrasonic*** vibrations (below 20 Hz) can be felt because some body parts resonate at this frequency. ***Ultrasonic*** vibrations (above 20000 Hz) is used in imaging, cleaning, drilling, cutting, welding and sealing. Moderate conversation produces 60 dB sound, loud conversation 70 dB, plying of truck/bus 90 dB, jet aero plane 150 dB, rocket 180 dB. Noise becomes uncomfortable above 100 dB. Sound is produced by loud speakers/public address system, transport vehicles, defense equipment, industries, domestic gadgets and out door equipment.

Effects. Noise brings about;

(*i*) Damage to ear drum and impairment of hearing (a ten year exposure to 80 dB impairs hearing by 15 dB.

(*ii*) Interference in conversation and hearing.

(*iii*) Emotional disturbance, development of anxiety and stress (first effect).

(*iv*) Damage to eye sight, colour perception, night vision, etc.

(*v*) Hypertension, decreased heart output and gastric problems.

(*vi*) Startle reaction.

Control.

(*i*) Delimitation of acoustic zoning

(*ii*) Use of cotton plugs or ear muffs in occupational exposure.

(*iii*) Development of quieter machines.

(*iv*) Restricted use of loud speakers

(*v*) Acoustic furnishing and low voice radio/TV.

(*vi*) Regulation of noise on road.

(*vii*) Green muffler.

Global Warming

Rise in the temperature of globe (earth) is called global warming. The average global temperature reached 15.39 Celsius in 1995 breaking the previous record. Coming of every year will be hottest. For example, Vishakhapattnam recorded a day temperature of 39.6, the highest in the last 100 years on July, 1977 more or less, same is the case with almost all coastel cities of Andhra Pradesh. Global warming is being felt due to the specific characteristics of the of the greenhouse gases (GHE). Greenhouse gases are three: (*i*) Carbon dioxide CO_2 (*ii*) Water vapour, (H_2O) and (*iii*) Ozone, O_3, Out of theses three gases , only carbon dioxide contributes largely to the greenhouse effect in the earth's atmosphere, because carbon dioxide is much more uniformly distributes in atmosphere. The major suppliers of carbon dioxide gas to the atmosphere are:

(*i*) Eruption of volcanoes

(*ii*) Respiration of living organism

(*iii*) Decay of dead organisms

(*iv*) Burning of fossil fuels and

(*v*) Cultivation of land. Contribution to global warming is 24% by industrial processes; 14% by deforestation; 13 by agriculture and 49% by energy use.

Before explanation of greenhouse effect and its result on earth; we should recall the composition of sun-light. The sunlight consists of three types of radiations: ultra violet radiation, visible light and infra-red radiation. Infra-red radiations are heat rays. They heat the earth and atmosphere. Infra-red rays emitted from hot bodies like sun, these waves are short in wavelength which is not absorbed by either carbon dioxide or even glass sheet. So, the short wave-length infra-red rays are emitted by the sun can pass through a blanket of carbon dioxide gas or even from a glass sheet.

The infra- red rays emitted by less hot objects like the earth's hot surface. They are long wave length which can be absorbed by carbon dioxide molecules. They cannot escape by a blanket of carbon dioxide in the atmosphere. The long wave length infra- red rays cannot go out even through a glass sheet. This blanket of carbon dioxide gas in the atmosphere allows the sunlight to come freely but does not allow the infra-red radiation reflected by the earth's surface to go out. It is just because the sun-light can come in freely but infra red rays cannot go out freely that the temperature of the earth is raised. This results in global warming. It be noted that, there is a direct relationship between carbon-dioxide in the temperature and the temperature of earth's atmosphere. So, if the proportion of the carbon dioxide gas in the atmosphere increases, then the temperature of the earth's atmosphere will also rise further. This is the main cause of global warming. During the 1980, scientist, governments and public became quite concerned about the feeling that the world may beget warmer. Global warming brings major changes in the world climate. Sea level can rise because the warmer water in seas would expand, and water stored in polar glacier would melt.

Effect of global warming:

1. Increase in global warming will lead to dislocation of suitable land for agriculture. This may adverse effect in food production. Wheat growing areas shift from fertile land to poor soils. In case of India wheat production will drop in the fertile northern belt. This is what soil experts predict.

2. Dislocation of some biological species along with degradation and disturbance in eco-system.
3. Global warming result into acid rain which affects eco-system 39.9 C was hotteset temperature of 15th April, 2003 in the last 5 years.
4. Depletion of Ozone layer.
5. Increase in global warming increase average temperature which increase the amount of water vapour in the atmosphere why because saturated vapour pressure increase with temperature thereby increase the long wave optical depth, trapping more long wave radiation and increase the temperature further.
6. More evaporation in tropics and alternation in existing precipitation patterns.
7. Flooding of many costal areas due to rising sea level resulting from thermal expansion of oceans caused by global warming because glaciers and ice-sheets melt.
8. due to global warming caused by (GHE) *Green House Effect* mean temperature will increase by 1.5 to 4.5 on earth after 50 years, and alter the climate, humidity and weather facilitating human diseases such as malaria, influenza, bronchitis, pneumonia and heat strokes.
9. more warming up at higher latitudes during late autumn and winter, than in tropics.
10. increase salinity of many Americans estuaries.

Future impacts of Global Warming:

Global temperatures are like bank rates. A small change can bring big difference. A 1% rise in surface temperature can cause major disruptions in weather patterns that could produce flash floods and unexpected droughts. Impacts of global warming anticipated by scientist and environmentalists are stated below:

Vanishing wild life:

One tenth of the world's known species at high altitudes plants and animal that are found in the Himalaya will be under threat.

Drought attack:

In much of the world, drought prone zones are likely to expand dramatically. In the sub-continent, Pakistan and Indian scientists argue that global warming may result in more rainfall over India.

Sudden floods:

Rapidly melting glaciers will cause floods resulting in wide spread damage to crops and livestock. Bangladesh and Pakistan will be in trouble. University scientist find ocean levels are rising twice as fast to day as they were before the industrial revolution. Studies show 2mm a year rise in sea level.

Health Hassles: Warmer and wetter conditions will increase the level of infectious diseases.

Crop out:

Wheat crops are sensitive to an increase in maximum temperature. Production will hit by 15 percent. Crop like soyabean could fall by 30%.

The saline invasion:

The rising sea is expected to increase salinity all across India's coast line causing a Kutch-like barrenness.

The marauding seas:

38 Indian districts face the danger of being partially sub-merged by rising sea levels, with Mumbai and Goa likely to be worst impacted. Internationally too, cities such Asmaini, New York and Venic face serious sea erosion problems.

Controlling Global Warming:

There are strong evidence linking planet's temperature and co2 changes. So in order to solve global warming problem, attempts must be made to use those sources of energy that do not produce carbon dioxide.

We should take steps right now to avoid underway disaster of global warming then it will be too late to make change. Climatologists suggest that following measures to cut emissions or increase absorption.

(*i*) industrialized countries should phase out Chlorofluro-carbons Because they damage stratospheric ozone as well as absorb heat.

(*ii*) Increase energy efficiency in homes, automobiles, factories and offices and cut CO_2 emissions.

(*iii*) Pass a carbon tax to discourage use of fossil fuels and encourage renewable energy sources.

(*iv*) Reduce speed levels, raise CAFÉ (corporate average fleet efficiency) for new cars and trucks. This is being implemented in USA, Canada and some European countries by committing themselves to a 20–25 percent decrease in CO_2 emission.

(*v*) Plant trees, investment in energy-efficient appliances, encourage mass transit and recovery of methane from landfills as these can reduce greenhouse gases.

(*vi*) Nuclear power industry points out that their reactors are an alternative to fossil fuels and produce no co2.

Two broad reasons have been ascertained that contribute that contribute to global warming. One is human activity and the other anthropogenic (human-caused) releases of gases into atmosphere. Statistics placed clearly show that amongst human activities maximum contribution to global warming is from burning fossil fuels whereas contribution to global warming is from burning fossil fuels whereas contribution of carbon dioxide is released through human caused (anthropogenic) activities.

Human Activities		*Human caused Activities Gases produce*	
Burning fossil fuels	49%	Chlorofluorocarbons	20%
Industrial Processes	24%	Methane	16%
Deforestation	14%	Ozone	8%
Agriculture	135	Nitrogen oxide	6%

(Sources: World Resources Institute and the United Nations Environment Programme)

James Baker, former administrator of the US National Oceanic and Atmospheric Administration while being in India (New Delhi) some time back, commented that the world will see a definite impact of global warming within the next 10 years. Increase in global warming will make society more vulnerable to climate change. "policy-makers all over the world will either have to learn to do with lesser greenhouse gases or live with climate change". According to him global temperature had steadily increased over the past 100 years. He further added "more scientist are convinced that we are seeing human influences on the climate....what we are worried about is the enhanced Greenhouse Effect, with carbon dioxide levels in the atmosphere growing"

Baker admitted that "Developed countries have no right to tell developing countries to slow down their economic growth. "

Developed countries need to find solutions and help developing countries leapfrog over polluting technologies in their quest for raising standards.

Predictions of Global Warming:

(*i*) Global warming will not be distributed evenly around the globe. Someplaces will experiences high temperature whiles a few quite less. Dallas may have 162 days per year that hot then 100 as it does now. Kolkatta where temperature is mostly hot, will hot, will not get much hotter.

(*ii*) Change in precipitation patterns can be one of the most serious consequences of the greenhouse effect.

(*iii*) Ecological disruption could be fatal for many wild plants and animals and they would be forced out of current ranges.

(*iv*) There shall be a rise in sea level due to global warming. A few oceanographers are of the opinion that thermal expansion alone could raise the sea level by one meter or more in the next country.

(*v*) Global warming might cause increase in violent storms such as Hurricane.

(*vi*) Many scientist states that the global change in environment has already started. They point to the fact that four of the five warmest years on record were in the late 1980s and early 1990s.

(*vii*) A few approaches to gauge earth's climate has found that the atmosphere would witness more warming during 21st century. It is said the global warming would be faster than what has been anticipated till now.

The approach, developed at the Handley centre for climate prediction and research in Bracknell in UK and published in nature, said average global temperature could be 5.5 degree Celsius higher by 2100, which is around 1.5 degree Celsius higher then forecast accepted worldwide.

Contrary to earlier estimates which are based on limited factors only, the new approach says that the warming of earth's atmosphere would depend on a host of other factors like volcanoes, fluctuations in son's activity and changing levels of greenhouse gases and ozone. It also allows oceans to affect the land and vice-versa.

Global Catastrophes

Climatologists predict that climate change can destroy earth. According to them climate change over the next 20 years could result in a global catastrophe costing millions of lives in war and natural disasters. Disruption and conflict will be endemic features of life according to pentagon analysis.

According to one report following predictions are revealed:

1. In 2005, widespread flooding by rise in sea levels will create major upheaval for millions.
2. By 2020 "catastrophic shortage" of water and energy supply will become increasingly harder to overcome, plunging the planet into war.

3. by the same year Britain will have a "Siberian" climate . Nuclear conflict, mega droughts, famine and widespread rioting will erupt across the world.

Ecological Effects of Acid Rain

Any atmospheric precipitation (rain snow falling on earth) with pH value less than 5.6 is acid rain. Past observation of environmentalist reveal that acid precipitation is usually varies between 5.6 and 3.5 and even less than in many cases. Acid rain is cause due to deposition of acid in air. Acid deposition in air is either dry or in wet form. Acid rain is due to pollution in air emitted by factories and vehicles. Wet forms of acid deposited in air constitute acidified rain, snow, fog and cloud vapour. Pollutions like NO_2 (Nitric acid) and so2 (Sulphuric acid) emitted in air also assist in the formation of acid rain. Soil is polluted when acid rain fall on earth. pH scale measure strength of acids and alkalis in atmosphere precipitation. Pure water in the form of rain water is neither acidic nor alkaline. It has a ph value of 7 which is neutral. Solutions having a ph value of less then 7 are acidic while those having a pH value more then 7 are alkaline.

The lesser the pH value, the more acidic is the solution and acids rain mainly contains sulphuric acid and nitric acid.

Most people in the united sates became aware of the problems associated with acid precipitation. But credit goes to English Scientist Report Angus Smith for coining the term "ACID RAIN" in 1850s, fear was developed latter that precipitation of the acids in the rain.

Causes of Acid Rain:

(*i*) **H_2SO_4 and HNO_3**: out of many the primary cause of acid rain is sulphuric acid, created by the reaction of sulphur dioxide with oxygen and water in the atmosphere. Once sulphur dioxide, so2, reaches the atmosphere, it reacts with water vapour in the air. First it is oxidised, taking up two atoms of oxygen to become a sulphur ion (SO_4). This then combines with the hydrogen to form sulphuric acid. Most of the atmosphere, sulphure dioxide is produced by burning coal with a high sulphur content.

There is another cause of acid rain – nitrogen dioxide NO_2, and nitric oxide, NO. They are given off by power stations and automobile exhausts. In the atmosphere, nitric oxide, which is itself oxidized and changed into nitric acid, HNO_3.

(*ii*) Burning of fossil fuels in various industries, electric power plants, smelting plants, oil refineries, Electric generation plants, and industrial boilers release oxides of sulphur and nitrogen into atmosphere which produce the acid rain. Coal contains a lot of sulphur and its burning is a major cause of SO_2 production. SO_2 is the major contributor accounting for nearly 50% of its originated sources. War between USA and Iraq (April 2003) caused acid rain because nearly 1000 oil wells were burnt by Iraqi rulers.

Natural Sources of Acid Rain

Volcanic eruptions, forest fires and bacterial decomposition of organic matter produce oxidize of sulphur and nitrogen. They mix in the air and finally in the atmosphere released through precipitation in the form of acid rain.

Impact of Acid Rain:

(*i*) Damage to ecosystems: forests. It is difficult to have a quantitative effect of acid rain on ecosystem. However, intense sulphur dioxide pollution around smelters is known to cause death of many kinds of trees and other vegetation.

The typical effects are;

(*a*) Yellowing and browning of needles and leaves.

(*b*) Decreased annual income

(*c*) Death of herbaceous vegetation

(*d*) Active shedding of needles and leaves while still green

(*e*) Death and diseased trees.

(*ii*) Acidification can be natural as well as anthropogenic. It kills the plankton, fish and adversely affects productivity of aquatic ecosystem. Acidity in soil and water due to acid rain leads to increase in dissolved matters specially aluminum. Aluminum affects the grills of fish. On account of respiratory failure they die due to heavy metals as Al, Zn, Mn, Cd, Pb, Cu added in water increases acidity beyond permissible limits.

(*iii*) Effects on soil: Acid rain leach out soil nutrients, adversely affects the properties of soil, causes damage to soil fertility, damages vegetation including cropland and forests, and makes the leaves of plants yellow and brown. This is how acid rain affects the productivity of forests, grasslands and crops.

(*iv*) Effects on microorganisms and wildlife:

Proliferation of microbial species is governed by Ph value. There are three ph value of water: acidic, neutral, alkaline. Acidic is extremely smog; neutral reflect normal rain and alkaline is beyond 7.0 pH value contain ammonia. Many bacteria and protozoa have pH optima near neutrality; most fungi prefer an acid environment and many blue green bacteria prefer an alkaline environment. So ph determines whether any microbial species can proliferate in a particular environment and the rate at which it can reproduce.

It is believed that acid rain affect wildlife also. Acid rain carry metals bound in soil and sediment released into the aquatic environment. The toxic substances so discharged are ingested by wildlife, like birds that feed in such environment. This is a direct effect of acid rain. The indirect effects of acid rain are loss of food and habit resources. The elimination of fish from lake during the breeding season is one example. Over 10,000 lakes in Sweden have become acidified. Fish population has decreased tremendously, and there are deaths of salmon trout etc. the fish-less area are now fish grave yards. Many bacteria and blue algae are killed due to acidification, thus disrupting ecological balance.

(*v*) Effect on human health:

Human are vulnerable to acid rain. In 1952 in London greatest air pollution took place which shrouds the city for 5 days. Though pH measurement of fog was not recorded yet fog was highly diluted with diluted sulphuric acid mist with a pH of about 1.4 to 1.9. Assessment of disaster is described below:

(*a*) 50,000 premature deaths are caused by sulphate laden air due to acid rain annually.

(*b*) 40 million people in United States or about one in six are drinking water with lead levels above 20 ppb (parts per billion. Lead (Pb) is one of the factors responsible for high blood pressure and heart attacks in adults and brain damage in children.

(*c*) Direct effects of acid rain are: chronic bronchitis, pulmonary emphysema, (related to lungs' cancer.

***Effects of building*:**

A brief review of acid rain effects on buildings and materials are given below:

1. Taj Mahal is affected by "*Stone Cancer*" due to acid rain.

2. The pigments in paint make paint work susceptible (doubtful work) and humidity encourages metal corrosion.
3. Urban building having stone carved into complex shapes tends to erode more quickly than flat. Vertical stone placed on buildings retain pollutants limply. Acid rain brings physical location and shape of the material.
4. Leather and leather shoes weaken due to residual acids introduced in manufacture.
5. Metals are corroded and tarnished due to sulphur oxides and other acidic gases.
6. Paper is discolored due to sulphur oxides, moisture, and acidic material introduced in manufacturer.
7. Rubber cracks due to ozone, sunlight and physical wear.

The green India 2047 project of TERI has estimated that India is already losing 11% to 26% of agricultural output on account of soil degradation. Acid rain would of streams, soil and rest of environment. However, some of the measure described below:

1. Buffering Methods used having deadening impact is the meaning of buffering. This is done by adding neutralizing agent to the acidified water to increase the pH value. Practice show use of lime in the form of calcium oxide and calcium carbonate.
2. Hatchery fish in acidified waters:
   This is done to maintain an ecological balance within acidified lakes. The procedure of breeding hatchery fish in acidified water curb fish population from further declining.
3. Controlling excess emission of Sox and Nox. Strict laws should be designed and effectively implemented by the government to regulate the emission. Some measure are:
   (*i*) Ordering power stations manager to burn less fossil fuels as it will decrease emission of Sox. They should be asked to use alternative sources of energy such as tidal waves, wind and hydro-power.
   (*ii*) Using low sulphur fuel.
   (*iii*) desulphurization
   (*iv*) modification in engines.

6.13 ENVIRONMENTAL ISSUES AND ACTS: THE INDIAN SCENARIO

6.13.1 Meaning of Environmental Issues

Environmental issues have assumed significance in the recent years. Commensurate (in relation) with industrial expansion, a strong regulatory mechanism is needed to ensure minimum environmental standards. There are 17 highly polluting industries which are. Aluminium, Caustic, Cement, Copper, Distillery, Dyes and D.I., Fertilizer, Iron and Steel, Leather, Pesticides. Petrochemical, Phrama, Pulp and Paper, Refinery, Sugar, TPP and Zinc. These industrial units along the major rivers and lakes in the country have been found discharging their untreated or partially treated effluents into the rivers and lakes. Opening new industrial units is developmental activity and nobody can deny this fact too, but such activities should not result in stress on various eco-system and quality of life. This is a hard fact that most of the development activities use natural resources in some or other form and generate waste leading to pollution in various environmental media like air, water, soil etc. ***Realizing the air and waste quality, increasing vehicular emission and higher noise levels, Ministry of Environmental and Forests, Govt. of India have adopted policy for abatement of***

pollution which provides multi-pronged strategies in the form of regulations, legislations, agreements, fiscal incentives and other measures. These rules and regulations; issues and acts are stated below with the sole objective of developing and promoting cleaner technologies and waste-minimization.

6.13.2 Objectives

The broad objectives of Ministry of Environment and Forests are:

(*i*) Conservation and survey of flora, fauna, forests and wildlife.

(*ii*) Prevention and control of pollution.

(*iii*) Afforestation and regeneration of degraded areas.

(*iv*) Protection of the environment.

(*v*) Ensuring the welfare of animals.

The above mentioned objectives are being fulfilled by the ministry primarily through:

(*a*) EIA (Environmental Impacts Assessment)

(*b*) Eco-generation.

(*c*) Assistance to organization engaged in environmental, forestry and animal welfare programmes.

(*d*) Promotion of environmental and forestry research, extension, education and training.

(*e*) Dissemination (spreading) of environmental information.

(*f*) International cooperation.

(*g*) Creation of environmental awareness.

6.13.3 Rules and Legislation

There objectives are well supported by a set of regulatory and legislations as mentioned below:

1. Water Prevention and Control of Pollution Act 1974.
2. The Air Pollution Control Act 1981.
3. The Environment Protection Act 1986.
4. The Indian Forest Act, 1927.
5. The Wildlife Act (Protection) 1972 and 2003.
6. The Public Liability Insurance Act 1991.
7. The Prevention of Cruelty Tribunal Act 1960.
8. The National Environment Appellate Authority Act 1997.
9. The National Environment Tribunal Act 1995.
10. Biological Diversity Act 2002.

6.13.4 Activities of Environmental Ministry

An overview of the activities undertaken by ministry as well as subordinate offices is as follows:

(*i*) Flora and Fauna : Germplasm of 254 species of rare, endangered, wild relatives of ornamentals and medicinally important plants have been conserved. Sixty five extensive faunal surveys were undertaken by Zoological survey of India. In includes important ecosystems and some selected conservation/protected areas.

(*ii*) A programme has been initiated for the regular and systematic monitoring of wild elephants under the **MIKE** (Monitoring of illegal killing of elephants.)

(*iii*) Assistance for construction of shelters, houses, dispensaries etc., are being provided for stay, of infirm and abandoned animals. This job is done by Animal Welfare Division. **NBWL** (National Board for Wildlife) and **AWBI** (Animal Welfare Board of India).

(*iv*) Work and activities are in progress to bring about awareness regarding the harmful effects of Ozone Depleting Substances (ODS). Ozone friendly products are developed. Refrigerators can be operated very efficiently with hydro-carbon as a refrigerant. (**Source:** Annual Report 2003–2004, MOE, GOI) CFC phase out project in foam, commercial refrigeration and aerosol have been approved.

(*v*) Cleaner technologies and cleaner production strategies can ensure resolution of conflicts between growth and environment through economic benefits via increased resource efficiency, innovation and reduced cost for environmental management. Various activities mentioned below have been undertaken to achieve the said objectives :

(*a*) Commercial production of horizontal and vertical laminates have been made using Bambusa bamboo species. Construction of demonstration houses has been taken up in Bamboo growing states.

(*b*) Bricks and tiles from marble slurry are being made in Udaipur (Raj). To utilize wastes arising out of marble cutting and processing for the purpose of improving the local environment. Awareness workshops are organized to make the products made out of marble slurry popular and to promote the usage.

(*c*) In Maharashtra (Thane) bio-remediation technique has been used. The activities include dredging and physical cleaning of lake and application of bio-clean product.

(*d*) Natural dyes from forest wastes have been identified in Uttranchal by FRI (Forest Reasearch Institute), Dehradun.

(*e*) Small utencils manufacturing units have been installed in Jagadhri. These utencils are being manufactured from cut pieces and wastes of steel, brass and aluminium.

(*f*) Industrial Waste Exchange Bank have been established in Karnataka in order to provide necessary links to create a close cycle of material reuse and recycling among business houses.

(*g*) IIT, Delhi has continued project on development, demonstration and dissemination of bio-degradable emulsion technology for increasing the shelf life of Fruits and Vegetables.

(*h*) RRL (Regional Research Laboratory), Bhopal is aimed at developing suitable process for making active chemical manganese dioxide from the wastes generated in the cell house in Zinc plant and to separate lead from these wastes. The manganese dioxide so generated will be used in battery manufacture.

(*i*) Development of a domestic unit employing a complexing resin for removing arsenic and other pollutants from water. It comes under the promotion of clean technologies.

(*j*) Waste-minimisation is one of the important strategies adopted for prevention of industrial pollution. The main objective of the waste minimization is to optimize the consumption of raw material and reduce waste generation by adopting cleaner production techniques in the existing units without necessarily changing the production process or unit operations.

(*k*) Several notifications have been issued to check noise levels. The general standards for industrial, commercial, residential areas and silence zones are notified based on the guidelines issued by WHO. The source of noise has been notified which include standards for firecrackers, vehicles, diesel generator sets. NPL, Delhi has been entrusted with the responsibility to check noise. The standards for generator sets of different sizes have been notified.

(*l*) Vehicular emissions is the major cause for deterioration of urban ambient air quality. It is proposed that PUC check shall be done every six months or any lesser period as may be specified by the state Government form time to time.

In 11 mega cities who have already introduced Bharat Stage-II equivalent to EURO-II norms during April 2000 to 2003 would introduce EURO-III and EURO-IV equivalent norms by April 1, 2005 and April 1, 2010 respectively. EURO-II and EURO-III norms would be introduced in the entire country by April 2005 and April 2010 respectively. (**Source :** Annual Report, Ministry of Environment and Forests GOI).

6.13.5 Ozone Regulation Control and Task Forces

Ozone is an extremely rare component of earth's atmosphere and in every 10 million molecules of air, only three are ozone. Most of the Ozone(90%) is found in the upper atmosphere (the stratosphere) between 10 and 50 kms. above the earth's surface. The ozone layer absorbs all the harmful UV radiation emanating from the sun. Ozone layer protects plant and animal life from UV radiation. UV radiation has the potential (of course in the absence of ozone layer or its becoming thin) to cause skin cancer, eye damage, suppress body's immune system, decrease crop yield, cause damage to forests and affects marine organisms. The Govt. of India has entrusted the work relating to the ozone layer protection to Ministry who has set up an Ozone Cell as a national unit to look after and to render necessary services to implement the Protocol and its Ozone Depleting Substances (ODS) phase out programme in India. To ensure compliance of the provisions of the Montreal Protocol measures have been taken up urgently to discourage use of Ozone Depleting Substances (ODS) and encourage ODS substitutes. Licensing system has been adopted to regulate import and export of ODS. A ban has also been imposed on trade of ODS with non-parties. The notified ozone depleting substances (Regulation and Control) Rules, 2000, provides a legal framework for ensuring compliance of the Montreal Protocol.

6.13.6 Regulation and Control:

(*i*) Deadline is fixed to phase out various ODS.

(*ii*) Prohibiting the use of CFCs in the manufacture of various products beyond 1st January, 2003

(*iii*) Use of halons is prohibited after 1st, Jan, 2001 except for essential uses.

(*iv*) Compulsory registration of ODS based products.

(*v*) Maintenance of periodic reports for monitoring production and use of ODS.

(*vi*) Those producers who have switch overed to non-ODS technology must declare that the equipment used for ODS has been destroyed.

(*vii*) Expansion of capacity of manufacturing facilities of ODS has been prohibited.

(*viii*) Various activities relating to wild life conservation and implementation of the wild life (protection) Act, 1972 were carried out. During the year ban on export of 29 species of plant, plant portions and their derivatives obtained from wild were continued. Export of six species of exotic birds were continued subject to preshipment inspection and CITES (Convention on International Trade in Endangered Species) permit wherever required. Several cases of poaching and illegal trade in wildlife products are detected.

6.13.7 Animal Welfare

Various schemes with regard to animal welfare have been implemented. These are scheme for shelter houses for looking after the animals, scheme for provision of ambulance services to animals in distress, scheme for relief to animals during natural calamities and unforeseen circumstances, control and supervision of experiments on animals, prevention of cruelty to animals (Transport of Animals on Foot) Rules 2001, control and supervision of experiments on animals.

"Project Tiger" has been launched to ensure maintenance of a viable population of tiger in India for scientific, economic, aesthetic, cultural and ecological values. Project Tiger is undisputedly the custodian of major gene pool.

'Project Elephant' was launched in 1992 with main activities such as-protection of wild elephants from poachers, conservation of elephant habitats, eco-development and veterinary care, Eight-elephant reserves have been set.

In India, keeping in view the degradation of coastal ecosystems of the country, work relating to identification and mapping of the coastal systems, for the purpose of sustainable management of these ecosystems has been initiated.

Eco-Task Forces : ETFs is an on-going scheme. Ecological Task Forces of ex-servicemen are employed in remote and difficult areas to undertake restoration, soil conservation and water resources management techniques. Four battalions of such forces are operating in India primarily two in Uttaranchal, one in Rajasthan and one in J & K National action programme to combat desertification has been designed. **June 17, is World Day to combat Desertification.** Desertification means saving the land from getting deserted.

6.14 HAZARDOUS SUBSTANCE MANAGEMENT

Rule 1989 and 1996 pertain to manufacture, storage and import of hazardous chemicals. These rules and their implementation is being constantly pursued and monitored except islands and high hill areas as well as Chandigarh. The Harardous Wastes (Management And Handling) Amendment Rules, 2003 were notified. Development and demonstration of process technology for remediation of polychlorinated biphenyls (PCBs) in oils and paints by Radiolysis" was initiated. The objective of the project is to develop and optimize a radiolytic process for decomposition of PCBs in oils, capacitor oils and paint scrapings.

6.15 THE BIOLOGICAL DIVERSITY ACT, 2002

Whereas India is rich in biological diversity and associated traditional and contemporary knowledge system relating thereto.

And whereas India is a party to the United Nations Convention on Biological Diversity signed at Rio de Janeiro on the 5th day of June, 1992.

And whereas the said Convention came into force on the 29th December, 1993.

And whereas the said Convention reaffirms the sovereign rights of the States over their biological resources.

And whereas the said Convention has the main objective of conservation of biological diversity, sustainable use of its components and fair and equitable sharing of the benefits arising out of utilization of genetic resources.

And whereas it is considered necessary to provide for conservation, sustainable utilization and equitable sharing of the benefits arising out of utilization of genetic resources and also to give effect to the said Convention.

The Act extends to whole of India.

6.15.1 Definitions

In this Act, unless the context otherwise requires :

(*a*) ***Benefit claimers*** means the conservers of biological resources, their byproducts, creators and holders of knowledge and information relating to the use of such biological resources, innovations and practices associated with such use and applications

(*b*) ***Biological diversity*** means the variability among living organisms from all sources and the ecological complexes of which they are part, and includes diversity within species or between species and of eco-systems

(*c*) ***Biological resources*** means plants, animals and micro-organisms or parts thereof, their genetic material and by-products (excluding value added products) with actual or potential use or value, but does not include human genetic material

(*d*) ***Bio-survey and bio-utilization*** means survey or collection of species, subspecies, genes, components and extracts of biological resource for any purpose and includes characterization, inventorisation and biossay

(*e*) ***Commercial utilization*** means end used of biological resources for commercial utilization such as drugs, industrial enzymes, food flavors, fragrance, cosmetics, emulsifiers, oleoresins, colours, extracts and genes used for improving crops and livestock through genetic intervention, but does not include conventional breeding or traditional practices in use in any agriculture, horticulture, poultry, dairy farming, animal husbandry or bee keeping

(*f*) ***Fair and equitable benefit sharing*** means sharing of benefits as determined by the National Biodiversity Authority under section 21

(*g*) ***Local bodies*** means Panchayats and Municipalities, by whatever name called, within the meaning of clause (1) of article 243Q of the Constitution and in the absence of any Panchayats or Municipalities, institutions of self-government constituted under any other provision of the Constitution or any Central Act or State Act

(*h*) ***National Biodiversity Authority*** means the National Biodiversity Authority established under section 8;

(*i*) ***Research*** means study or systematic investigation of any biological resource or technological application, that uses biological systems, living organisms or derivatives thereof to make or modify products or processes for any use;

(*j*) ***State Biodiversity Board means*** the State Biodiversity Board established under section 22;

(*k*) ***Sustainable use*** means the use of components of biological diversity in such manner and at such rate that does not lead to the long-term decline of the biological diversity thereby maintaining its potential to meet the needs and aspirations of present and future generations

(*l*) ***Value added products*** means products which may contain portion or extracts of plants and animals in unrecognizable and physically inseparable form.

6.15.2 National Biodiversity Authority

Seven ex officio members to be appointed by the Central Government to represent respectively the Ministries of the Central Government dealing with

(*i*) Agricultural Research and Education
(*ii*) Biotechnology
(*iii*) Ocean Development
(*iv*) Agriculture and Cooperation
(*v*) Indian Systems of Medicine and Homoeopathy
(*vi*) Science and Technology
(*vii*) Scientific and Industrial Research.

Five non-official members to be appointed from amongst specialists and scientists having special knowledge of, or experience in, matters relating to conservation of biological diversity, sustainable use of biological resources and equitable sharing of benefits arising out of the use of biological resources, representatives of industry, conservers, creators and knowledge-holders of biological resources.

6.15.3 Duties of the Central and the State Governments

Central Government to develop National strategies plans, for conservation, *etc.*, of biological diversity:

1. The Central Government shall develop national strategies, plans, programmes for the conservation and promotion and sustainable use of biological diversity including measures for identification and monitoring of areas rich in biological resources, promotion of in-situ, and ex-situ, conservation of biological resources, incentives for research, training and public education to increase awareness with respect to biodiversity.
2. Where the Central Government has reason to believe that any area rich in biological diversity, biological resources and their habitats is being threatened by overuse, abuse or neglect, it shall issue directives to the concerned State Government to take immediate ameliorative measures, offering such State Government any technical and other assistance that is possible to be provided or needed.
3. The Central Government shall, as far as practicable wherever it deems appropriate, integrate the conservation, promotion and sustainable use of biological diversity into relevant sectoral or cross-sectoral plans, programms and policies.
4. The Central Government shall undertake measures:
   (*i*) wherever necessary, for assessment of environmental impact of that project which is likely to have adverse effect on biological diversity, with a view to avoid or minimize such effects and wherever appropriate provide for public participation in such assessment;
   (*ii*) to regulate, manage or control the risks associated with the use and release of living modified organisms resulting from biotechnology likely to have adverse impact on the conservation and sustainable use of biological diversity and human health.

5. The Central Government shall endeavour to respect and protect the knowledge of local people relating to biological diversity, as recommended by the National Biodiversity Authority through such measures, which may include registration of such knowledge at the local, state or national levels, and other measures for protection, including **sui generic** system.

   Explanation : For the purpose of this section

   (a) ***ex situ conservation*** means the conservation of components of biological diversity outside their natural habitats;

   (b) ***in situ conservation*** means the conservation of ecosystems and natural habitats and the maintenance and recovery of viable populations of species in their natural surroundings and, in the case of domesticated or cultivated species, in the surroundings where they have developed their distinctive properties.

6.15.4 Biodiversity Management Committees

1. Every local body shall constitute a Biodiversity Management Committee within its area for the purpose of promoting conservation, sustainable use and documentation of biological diversity including preservation of habitats, conservation of land races, folk varieties and cultivars, domesticated stocks and breeds of animals and micro organisms and chronicling of knowledge relating to biological diversity.

   Explanation :- For the purposes of this sub-section,

   (a) ***Cultivar*** means a variety of plant that has originated and persisted under cultivation or was specifically bred for the purpose of cultivation;

   (b) ***Folk variety*** means a cultivated variety of plant that was developed, grown and exchanged informally among farmers;

   (c) ***Landrace*** means primitive cultivar that was grown by ancient farmers and their successors.

Legislation And International Treaties To Protect Endangered Species & Biodiversity.

Following steps have been taken in this direction:

(i) Recognition of Ecological conservation

(ii) Wildlife protected in a network comprising 84 national parks and 447 wildlife sanctuaries.

(iii) The Endangered species Act and CITES represent a new attitude towards wildlife in which we protect organisms just because they are rare and endangered.

(iv) U.S. Endangered species Act of 1973 and COSEWIC (Committee on the Status of Endangered Wildlife in Canada) 1976 provide powerful approaches to wildlife protection.

(v) The Biological Diversity Bill 2000, (India) awaiting Parliamentary sanction and presidential approval, is a landmark legislation aimed at protecting India's bio-assets and ensuring equitable benefit-sharing by the users and the owners. The Bill No. 93 for the year 2000 *Provide for conservation of biological diversity, sustainable use of its components and equitable sharing of the benefits arising out of the use of biological resources and for mattes connected there with or incidental thereto.*

Carbon Credits

Carbon credits are a key component of national and international attempts to mitigate the growth in concentrations of greenhouse gases (GHGs). Carbon trading is an application of an emissions trading approach. Greenhouse gas emissions are capped and then markets are used to allocate the emissions among the group of regulated sources. The idea is to allow market mechanisms to drive industrial and commercial processes in the direction of low emissions or less "carbon intensive" approaches than are used when there is no cost to emitting carbon dioxide and other GHGs into the atmosphere. Since GHG mitigation projects generate credits, this approach can be used to finance carbon reduction schemes between trading partners and around the world.

There are also many companies that sell carbon credits to commercial and individual customers who are interested in lowering their carbon footprint on a voluntary basis. These carbon off setters purchase the credits from an investment fund or a carbon development company that has aggregated the credits from individual projects. The quality of the credits is based in part on the validation process and sophistication of the fund or development company that acted as the sponsor to the carbon project. This is reflected in their price; voluntary units typically have less value than the units sold through the rigorously-validated Clean Development Mechanism (CDM).

Burning of fossil fuels is a major source of industrial greenhouse gas emissions, especially for power, cement, steel, textile, fertilizer and many other industries which rely on fossil fuels (coal, electricity derived from coal, natural gas and oil). The major greenhouse gases emitted by these industries are carbon dioxide, methane, nitrous oxide, hydro fluorocarbons (HFCs), etc, all of which increase the atmosphere's ability to trap infrared energy and thus affect the climate.

The concept of carbon credits came into existence as a result of increasing awareness of the need for controlling emissions. The IPCC (Intergovernmental Panel on Climate Change) has observed that:

Policies that provide a real or implicit price of carbon could create incentives for producers and consumers to significantly invest in low-GHG products, technologies and processes. Such policies could include economic instruments, government funding and regulation,

while noting that a tradable permit system is one of the policy instruments that has been shown to be environmentally effective in the industrial sector, as long as there are reasonable levels of predictability over the initial allocation mechanism and long-term price.

The mechanism was formalized in the **Kyoto Protocol**, an international agreement between more than 170 countries, and the market mechanisms were agreed through the subsequent Marrakesh Accords. The mechanism adopted was similar to the successful US Acid Rain Program to reduce some industrial pollutants.

Emission Allowances

The Protocol agreed 'caps' or quotas on the maximum amount of Greenhouse gases for developed and developing countries. In turn these countries set quotas on the emissions of installations run by local business and other organizations, generically termed 'operators'. Countries manage this through their own national 'registries', which are required to be validated and monitored for compliance by the UNFCCC. Each operator has an allowance of credits, where each unit gives the owner the right to emit one metric tonne of carbon dioxide or other equivalent greenhouse gas. Operators that have not used up their quotas can sell their unused allowances as carbon credits, while businesses that are about to exceed their quotas can buy the extra allowances as credits, privately or on the open market. As demand for energy grows over time, the total emissions must still stay within the cap, but it allows industry some flexibility and predictability in its planning to accommodate this.

By permitting allowances to be bought and sold, an operator can seek out the most cost-effective way of reducing its emissions, either by investing in 'cleaner' machinery and practices or by purchasing emissions from another operator who already has excess 'capacity'.

Since 2005, the Kyoto mechanism has been adopted for CO_2 trading by all the countries within the European Union under its European Trading Scheme (EU ETS) with the European Commission as its validating authority. From 2008, EU participants must link with the other developed countries who ratified Annex I of the protocol, and trade the six most significant anthropogenic greenhouse gases. In the United States, which has not ratified Kyoto, and Australia, whose ratification came into force in March 2008, similar schemes are being considered.

Kyoto's 'Flexible Mechanisms'

A credit can be an emissions allowance which was originally allocated or auctioned by the national administrators of a cap-and-trade program, or it can be an offset of emissions. Such offsetting and mitigating activities can occur in any developing country which has ratified the Kyoto Protocol, and has a national agreement in place to validate its carbon project through one of the UNFCCC's approved mechanisms. Once approved, these units are termed Certified Emission Reductions, or CERs. The Protocol allows these projects to be constructed and credited in advance of the Kyoto trading period.

The Kyoto Protocol provides for three mechanisms that enable countries or operators in developed countries to acquire greenhouse gas reduction credits

Under Joint Implementation (JI) a developed country with relatively high costs of domestic greenhouse reduction would set up a project in another developed country.

- Under the Clean Development Mechanism (CDM) a developed country can 'sponsor' a greenhouse gas reduction project in a developing country where the cost of greenhouse gas reduction project activities is usually much lower, but the atmospheric effect is globally equivalent. The developed country would be given credits for meeting its emission reduction targets, while the developing country would receive the capital investment and clean technology or beneficial change in land use.
- Under International Emissions Trading (IET) countries can trade in the international carbon credit market to cover their shortfall in allowances. Countries with surplus credits can sell them to countries with capped emission commitments under the Kyoto Protocol.

These carbon projects can be created by a national government or by an operator within the country. In reality, most of the transactions are not performed by national governments directly, but by operators who have been set quotas by their country.

Emission Markets

For trading purposes, one allowance or CER is considered equivalent to one metric tonne of CO_2 emissions. These allowances can be sold privately or in the international market at the prevailing market price. These trade and settle internationally and hence allow allowances to be transferred between countries. Each international transfer is validated by the UNFCCC. Each transfer of ownership within the European Union is additionally validated by the European Commission.

Climate exchanges have been established to provide a spot market in allowances, as well as futures and options market to help discover a market price and maintain liquidity. Carbon prices are normally quoted in Euros per tonne of carbon dioxide or its equivalent (CO_2e). Other greenhouse gasses can also be traded, but are quoted as standard multiples of carbon dioxide with respect to their global warming potential. These features reduce the quota's financial impact on business, while ensuring that the quotas are met at a national and international level.

Currently there are five exchanges trading in carbon allowances:

- The Chicago Climate Exchange
- European Climate Exchange
- Nord Pool
- PowerNext
- The European Energy Exchange.

Recently, NordPool listed a contract to trade offsets generated by a CDM carbon project called Certified Emission Reductions (CERs). Many companies now engage in emissions abatement, offsetting, and sequestration programs to generate credits that can be sold on one of the exchanges. At least two private electronic markets have been established in 2008: CantorCO2e and Preserval Marketplace.

Managing emissions is one of the fastest-growing segments in financial services in the City of London with a market now worth about 30 billion, but which could grow to 1 trillion within a decade. Louis Redshaw, head of environmental markets at Barclays Capital predicts that *Carbon will be the world's biggest commodity market, and it could become the world's biggest market overall.*

Carbon Credits Versus Taxes

Carbon credits and carbon taxes each have their advantages. Credits were chosen by the signatories to the Kyoto Protocol as an alternative to Carbon taxes. A criticism of tax-raising schemes is that they are frequently not hypothecated, and so some or all of the taxation raised by a government may be applied inefficiently or not used to benefit the environment.

By treating emissions as a market commodity it becomes easier for business to understand and manage their activities, while economists and traders can attempt to predict future pricing using well understood market theories. Thus the main advantages of a tradable carbon credit over a carbon tax are:

- the price is more likely to be perceived as fair by those paying it. Investors in credits have more control over their own costs.
- the flexible mechanisms of the Kyoto Protocol ensure that all investment goes into genuine sustainable carbon reduction schemes, through its internationally-agreed validation process.
- if correctly implemented a target level of emission reductions is achieved with certainty, while under a tax the actual emissions would vary over time.
- it provides a framework for rewarding people or companies who plant trees or otherwise sequester carbon.

The advantages of a carbon tax are:

- possibly less complex, expensive, and time-consuming to implementat. This advantage is especially great when applied to markets like gasoline or home heating oil.
- perhaps some reduced risk of certain types of cheating, though under both credits and taxes, emissions must be verified.

- reduced incentives for companies to delay efficiency improvements prior to the establisment of the baseline if credits are distributed in proportion to past emissions.
- when credits are grandfatered, this puts new or growing companies at a disadvantage relative to more established companies.
- It is clear what effect the policy has on the price of energy.

Criticisms:

1. Environmental restrictions and activities have been imposed on businesses through regulation which are uneasy at managing emissions.
2. U. S. A., Austrialia and China have avoided mandatory caps.
3. A question has been raised over the grand fathering of allowances.
4. There have been concerns raised over the validation of CDM credits.

6.16 ROLE OF ENGINEERS IN ENVIRONMENTAL PROTECTION

People are demanding as a fundamental right, not only clean air and pure water and food, but also decent housing, an unpolluted land, freedom from excessive noise, adequate recreation facilities, and housing in communities that provide comfort, privacy, and essential services. The engineers and planners must ensure at the planning stage itself that the projects are so designed as to be pleasing and enhancing (increasing) the well being of the community. Before any project is undertaken, a site is selected wherein following considerations are important:

1. The site selected should be good from the geological point of view. It should have suitable foundation soils, and be safe from earthquakes. The use of limestone, dolomite or gypsum should be avoided.
2. Special care should be taken in matters of drainage and flooding. The site selected should be above the 100-year flood level before erection of the structure is started. Access to roads, well drained and above flood level, is equally important.
3. Access roads, proper designs, safe entrance and exit lanes, noise, dust and other air pollution on transportation routes should be assessed.
4. It should be seen that the site is screened and landscaped.

6.17 ECO-FRIENDLY TECHNOLOGY

The phrase environmentally friendly is used to refer to goods, services and/or practices considered to inflict little harm on the environment. The phrase has been in common usage at least 20 years and is often added to product advertising or packaging to promote a sale.

It also means being friendly to the environment. For example you can be environmentally by recycling or being 'green' as people call it. There are also many organizations that associate themselves with the term, such as Greepeace.

In pest control, integrated pest management is regarded as more environment friendly than traditional pesticide use, as its goal is to reduce the use of pesticides to a minimum by using a variety of less impactive means, with pesticides only as a last resort. More recently, the development of biological pest control methods are an effort to go a further step to being environment friendly.

In waste management, recycling and composting are viewed as more environmentally friendly than bury and burn practices. For example Edmonton, Alberta and Canada has adopted large scale composting to deal with its urban waste.

Some Eco-friendly Technologies

There are various eco-friendly technologies existing today. Some of them are as under:

- Fluorescent light bulbs
- Fuel efficient cars
- LCD Monitor
- Energy efficiency building materials
- Smart traffic lights
- Turbo charged engines
- Solar lights.

Eco-friendly Products:

1. ***Hand made paper***
   It reduces pollution by 70%. It also helps in saving trees as it used non wood raw material.
2. ***Hand made paper products***
   More the selling of hand made paper products, more the saving of trees. It also provides solution to unemployment.
3. Eco Wheels such as electric Vehicles can help in decreasing the air pollution.
4. ***Eco Furniture***
   It good alternative to wooden furniture. It uses lass wood and more of waste products.
5. ***Eco Fans***
   It does not need electricity or any external batteries to work. Its thermoelectric module works on a small generator to power the fan. It is capable of distributing the warm air in whole of your house.
6. ***Eco Shoes***
   These are made of contents such as recycled rubber recycled plastic, textiles and synthetic waste products which are recyclable
7. ***Eco Friendly Paints***
   They carry non toxic elements, which have lesser chemical element. They are made of naturally derived raw materials. They are in great need for healthier environment.
8. ***Solar Products***
   They are without any air or noise pollution. It gives a positive boost to save our environment. The most lucrative feature of solar product is its energy conservation capability.
9. ***Hand Craft Products***
   They are mostly handmade and are environment friendly, which use less energy. These items help in eradicating unemployment.
10. ***Paper Bags and Note Books***
    The eco-friendly alternative to plastic bags are paper bags. There are eerily biodegradable. The notebooks made of hand made paper are quite popular these days.

11. ***Jute Bags***

    Jute is 100% biogradable and thus environment friendly. Known for its course character, jute has come to acquire the centre stage as an eco-friendly alternative.

12. ***Organic Food***

    These are the foods made without the use of non-organic pesticides, insecticides, herbicides and fertilizers.

13. ***Natural Beauty Products***

    Natural beauty products are ; natural soap, natural hair care, organic skin care.

Eco-friendly technology looks to improve in two major areas ;

- Energy Efficiency
- Reduction of harmful waste

QUESTIONS

1. What is pollution? Explain the types of pollution.
2. What is Environmental degradation? What are the main causes of Environmental degradation?
3. What are the main causes of air of pollution?
4. Explain the various types air pollutants and their effects.
5. Discuss the methods of controlling Ozone depletion.
6. State the effects of Global Warming and predictions about it.
7. What are the major water pollutants? Explain briefly.
8. Explain various sources of water pollution.
9. Explain Soil Pollution and various causes of soil pollution.
10. What is eco-friendly technology? List some eco-friendly technologies.
11. What is role of engineers in environmental protection?
12. What are effects of noise pollution? Explain the methods of control of noise.

7

NUCLEAR THREAT AND ARMS RACE

7.1 THE NUCLEAR THREAT

Humanity at the moment may be compared to a colony of bacteria in a Petri dish with a constant food supply, with no way to remove the wastes of their metabolism, the bacteria eventually poison themselves.

The atomic bombings of Hiroshima and Nagasaki are a tragic chapter in the history of the human race. These bombings not only demonstrate the cruelty that humanity can inflict upon itself, but they also foreshadow a terrifying future if we do not halt nuclear proliferation and embrace nuclear disarmament. It has been 61 years since nuclear weapons were first used in war. We wish we could say that the world has learned the lesson that the survivors, the Hibakusha, have been trying to teach us since then: The lesson that humans and nuclear weapons cannot coexist.

Unfortunately, all evidence points to the contrary. The world is a far more dangerous place than it was 63 years ago. The world is a far more dangerous place than it was 10 or even 5 years ago. The likelihood that countries will seek nuclear weapons and the likelihood that countries will use nuclear weapons has increased.

To begin, nuclear weapons have become far more lethal since 1945. The 21 kiloton bomb used at Nagasaki is considered miniscule by modern standards. Today, there are thousands of missiles tipped with nuclear warheads hundreds of times more powerful. A full nuclear war would likely bring about the end of the human race. But, even the amount of suffering and destruction that would result from the detonation of just one of these nuclear weapons over a populated area is unprecedented. A book published in 2004, titled **'Whole World on Fire'** by **Lynn Eden**, details the heat and blast effects of a moderate-sized 300 kiloton weapon detonated over the Pentagon.

It would create a fireball more than a mile in diameter producing temperatures of more than 200 million degrees Fahrenhiet-about four to five times the temperature at the centre of the sun.

In Pentagon City, asphalt and metal would melt, paint would burn. Offices and cars would explode into flames. The blast wave would create 750 mile per hour winds tossing burning cars into the air.

On the edge of the Potomac the fireball would be 5,000 times brighter than a desert sun at noon. It would melt the marble at the Lincoln and Jefferson memorials. Four seconds after detonation, these structures would collapse from the blast wave that followed.

On Capitol Hill, the House and Senate office buildings would burn. The blast would shatter exterior windows and level surrounding buildings.

Within tens of minutes, everything within approximately three-and-a-half to four-and-a-half miles of the Pentagon would be engulfed in a massive fire. The fire would extinguish all life and destroy almost everything else.

For decades, the international community has tried to prevent countries from causing this level of destruction. The cornerstone of these efforts has always been international agreements that encourage arms control. The most important of these is the nuclear Non-Proliferation Treaty (NPT).

The NPT has three key provisions. It guarantees countries the right to use nuclear energy for peaceful means. It prohibits the passing of nuclear weapons technology to, and the development of nuclear weapons by, non-nuclear weapons states. Most important, it requires countries with nuclear weapons to negotiate nuclear disarmament. The treaty establishes an effective framework discouraging more countries from developing nuclear weapons. When the treaty was signed in 1968, there were five nuclear weapons states. The NPT has not been perfect and currently there are nine, but without it there would likely be more.

Today, the international anti-nuclear framework set forth in the NPT is unraveling. There are countries such as Israel, India and Pakistan that have never signed the treaty, and developed large-scale programs. North Korea has broken away from the NPT in order to develop nuclear weapons. Iran may be in the early stages of a weapons program. These countries are endangering themselves and their neighbors, as are the original five nuclear weapons states.

One of the most dangerous recent developments is in Russia. Russia is currently building up its own nuclear arsenal, in significant part, in response to a US missile shield. A recent article in the '*Nation and Foreign Policy*' magazines argues that, given the state of Russia's infrastructure, such a build-up is extremely dangerous. The Russian government is not investing in proper safety mechanisms to prevent catastrophes such as accidental launches. There has already been a near miss. In 1995, the world came within minutes of nuclear Armageddon when the Russian early warning systems confused the launch of a Norwegian weather rocket with a preemptive nuclear attack by the United States. Boris Yeltsin had nuclear launch codes in front of him and would have retaliated had the mistake not been caught at the last minute. Russia's early warning system has only further deteriorated since then. There are massive holes in its detection capabilities. Russian commanders rely on antiquated radar rather than satellite technology to detect possible launches.

Together, the United States and Russia have 26, 300 nuclear weapons. They possess the ability to carry out precision nuclear strikes anywhere in the world. They have hundreds of nuclear missiles on hair-trigger alert, pointed at each other, that could be fired in a matter of minutes. An accidental nuclear launch by Russia and the retaliatory response by the United States would result in the deaths of millions of people.

But let's not forget the biggest nuclear player and the destabilizing effect it has on non-proliferation regimes. In 2002, the United States placed increased emphasis on the role that nuclear weapons play in its foreign policy. The Bush administration's Nuclear Posture Review states:

1. Nuclear weapons play a critical role in the defense capabilities of the United States, its allies, and friends.
2. Nuclear weapons can be used to achieve political or strategic goals.
3. US policy now supports preemptive attacks, possibly nuclear, on countries with Weapons of Mass Destruction or hardened targets.

The United States is relying now, more than ever before, on nuclear weapons. It also has lowered the threshold for the use of nuclear weapons.

In the past five years, the Bush administration has ignored many important international arms control treaties. It has failed to ratify the Comprehensive Test Ban Treaty, which prohibits all forms of nuclear testing. It has withdrawn from the Anti-Ballistic Missile Treaty. It has invested billions of dollars into a missile shield program; an action seen by many other countries as an aggressive gesture.

In violation of the Non-Proliferation Treaty, it has attempted to develop nuclear weapons that can be used more readily in combat, such as the *bunker buster*. It has also attempted to upgrade the US nuclear arsenal with the implementation of the Reliable Replacement Warhead Program. Most recently, it has negotiated a *deal* with India that allows the exchange of nuclear technology.

By steering around international treaties that encourage arms control, attempting to build new weapons and then seeking to use them for political or strategic goals, the United States is encouraging other countries to do the same. As more countries go down this road, the likelihood of nuclear weapons use will only increase.

Although the only way to ensure that nuclear weapons are never used again is through total disarmament, there are ways to stop proliferation and minimize the risk of nuclear weapons use. In order to be effective, these efforts must have the support of the United States. As the world's most powerful nation in possession of thousands of nuclear weapons, and as the only country that has used nuclear weapons as an instrument of war, the United States is ethically obligated to pursue non-proliferation, arms control and disarmament. This effort must begin with the following three steps:

1. Altering US current nuclear policy. The US must de-legitimize the idea that nuclear weapons are an effective way to achieve political or strategic goals by declaring that it will never be the first to use nuclear weapons in war.
2. Ratifying and complying with the provisions set forth in international treaties such as the Non-Proliferation Treaty, the Comprehensive Test Ban Treaty, the Anti-Ballistic Missile Treaty, the Strategic Arms Reduction Treaties, and the Strategic Arms Limitations Treaties that promote non-proliferation and disarmament.
3. Abandoning our policy of preemptive attacks, which further emboldens countries like Iran to pursue nuclear weapons.

Despite the dangers that we now face and despite all that needs to be done to make the world a safer place, historically, anti-nuclear activism in the United States has been incredibly effective. Anti-nuclear activism was a significant factor in bringing an end to nuclear testing in the US and around the world. Activism was influential in slowing the nuclear arms build up in the 1980s. History has shown that our government listens to the public about nuclear weapons. If the people of the United States work together to tell our government that the creation and use of unclear weapons is not acceptable, we can actually change nuclear policy to make the world safe. We strongly encourage you to find a way to get involved in anti-nuclear work.

Nuclear weapons are the most significant threat to the future of the human race. As long as they exist, no human being is safe. Today, more and more countries are adopting dangerous nuclear policies. It is imperative that we pressure our government to bring the world back from potential nuclear anarchy. Only then, can we prevent proliferation and prevent future Hiroshimas and Nagasakis.

7.2 NUCLEAR WARFARE AND NUCLEAR PROLIFERATION

Nuclear Warfare and Nuclear Proliferation, dissemination and use of nuclear weapons and military nuclear technology. The first use of nuclear weapons in war took place in August 1945, when the United States dropped atomic bombs on Hiroshima and Nagasaki in Japan. Atomic or fission bombs involve a self-sustaining atom-splitting chain reaction in a mass of uranium or plutonium, causing the release of a huge amount of energy in a very short time. Atom bomb design and construction is very precise, but it is now widely accepted that a competent nuclear physicist could glean all the information needed from readily available scientific literature. Most modern nuclear weapons are scientifically more advanced and belong in the second category of hydrogen or thermonuclear weapons. These weapons employ a fission explosion to create sufficient energy for the hydrogen fusion process to take place. There is then a massive release of energy, much larger than that from an atomic bomb. There is theoretically no limit to the size of a thermonuclear explosion.

From the moment nuclear weapon technology was revealed in 1945, it has been widely, though not universally, understood that these are not "weapon" in the traditional sense of conferring a viable military advantage on the possessor. When the use of nuclear weapons threatened to annihilate not just the belligerents in any conflict, but also to destroy and contaminate much of the Earth's surface, it seemed that weapons development had finally gone beyond the bounds of proportionality and utility. The British strategist Basil Henry Liddell Hart commented that *with the advent of atomic weapons, we have come either to the last page of war or the last page of history*.

A nuclear weapon involves a delivery system as well as the bomb or warhead. The first delivery system was the long-range bomber. As the Cold War progressed, so a wide variety of delivery systems were developed: missiles of various ranges launched from the air, ground, and sea, battlefield artillery, shipborne depth charges and torpedoes, and landmines. Nuclear weapon technology is now so diffuse that an effective means of delivery could be as simple as the boot of a car or the proverbial "suitcase bomb". The yield or explosive force of nuclear weapons has also increased dramatically. The Hiroshima and Nagasaki bombs had a yield of about 13 kilotons-equivalent to 13,000 tonnes of TNT explosive. The yield of some US and Soviet intercontinental ballistic missiles (ICBMS) was in the region of 1.5 megatons, or 1,500,000 tonnes of TNT. In 1962 the Soviet Union exploded a thermonuclear bomb with a staggering yield of 58 megatons. Nuclear weapons have also become extremely accurate: the "circular error probable" of a US MX missile is just 100 metres, meaning that there is a 50 per cent change of the 11,000-kilometre-range missile delivering a warhead within 100m of the target.

The first effects of a nuclear explosion are an extremely bright and hot wave of thermal radiation, and a massive blast, resulting in widespread destruction and fires. A series of Electro Magnetic Pulses (EMP) are also released which, while not damaging to humans or buildings, could destroy communications systems. Nuclear radiation, extremely harmful to all forms of life, comes in the form of direct radiation at the time of the explosion and radiation from radioactive fallout, the dirt and debris sucked up and irradiated during the explosion and falling back to Earth. It is possible to design weapons which augment one or other effect. Weapons designed to be used against military units might produce a high EMP in order to break down the military communications network. The best known special design was the so-called *neutron bomb*, or enhanced radiation/reduced blast bomb, which maximized lethal direct radiation to kill tank crews but which kept the blast effect to a minimum.

The United States had the monopoly in atomic bombs from 1945 until the first Soviet test in 1949. The US tested a thermonuclear bomb in 1952, and the Soviet Union followed suit one year later. In 1957 the Soviets launched the Sputnik satellite into the first global orbit using an intercontinental ballistic missile, prompting US fears of a "missile gap". Both sides then raced to produce bombers, missiles, and other means of delivery. By the end of the Cold War, each had well over 10,000 strategic (*i.e.*, intercontinental) warheads, with many more in the sub-strategic range. Other acknowledged possessors of nuclear weapons are Britain, China, and France. In 1998 both India and Pakistan conducted nuclear tests, confirming long-held suspicions that they too possessed a nuclear capability. Several states, such as Israel, are widely suspected of possessing nuclear weapons. Other suspected "threshold" states, those thought to be capable of developing nuclear weapons, include Iran and North Korea. The dismasting of Iraq's nuclear weapons programme after the Gulf War revealed it to be well on the way to developing a nuclear device.

The "proliferation", or spread of nuclear weapons has taken place in two ways, "vertical" and "horizontal". Vertical proliferation is the expansion and development of existing arsenals and is controlled through arms control agreements between the possessors. The prevention of horizontal proliferation is the object of the nuclear non-multilateral export controls, inspection and verification agencies, bans on weapon testing and most recently, an attempt to prohibit further production of fissile material.

The key to the regime is the Nuclear Non-Proliferation Treaty (NPT) signed in 1968. In essence, the NPT is a bargain in which non-nuclear states agree to renounce weapons-related research, development, and acquisition in exchange for access to civil nuclear technology. Non-signatories, however, include Israel, India, and Pakistan, three states that that fought several wars in recent history. North Korea's threat to withdraw from the NPT in 1993, to avoid opening its nuclear facilities to inspection by the International Atomic Energy Agency (IAEA), brought the counter-threat of a pre-emptive strike on the facilities by US bombers. The NPT, with 187 signatories as of December 2002, was the subject of a major international review conference in 1995 to meet its official expiry in May. The major powers possessing nuclear weapons, with permanent seats on the United Nations Security Council, secured indefinite continuation of the NPT regime by, among other proposed policies, promising concerted retaliation against any state guilty of nuclear attack or threats against an NPT signatory.

Many states criticize the NPT as discriminatory, and demand that the nuclear "haves" do more to meet their part of the bargain by working more conscientiously towards nuclear disarmament and complete test bans. The non-proliferation regime faces several other challenges. There is a fundamental difficulty in distinguishing between civil and military use of nuclear technology, and it seems increasingly difficult to control the traffic in key components and materials. It may also be that the taboo against the viewing of nuclear weapons as acceptable weapons of war has been eroded in various parts of the world. And it is not beyond the realms of possibility that a terrorist organization might eventually acquire and use its own bomb. In January 2003 North Korea announced its intention to become the first nation to withdraw from the NPT, having admitted to the pursuit of a secret nuclear programme and expelling United Nations nuclear inspectors the previous month.

In an important benchmark decision, 'the International Court of Justice of the United Nations ruled in July 1996 that the use of nuclear weapons in warfare was contrary to the rules of war, except in "extreme circumstances". A ruling was initially requested in December 1994 by a large majority vote of the Un General Assembly, reflecting the concern of the non-nuclear majority at the possession of nuclear weapons by the "nuclear club" of permanent UN Security Council members. The decision rendered the use of tactical or the nuclear weapons illegal in any circumstances. Without an effective mechanism for enforcement, the decision is significant chiefly as a moral precedent.

7.3 ARMS RACE

An arms race is a competition between two or more countries for military supremacy. Each party competes to produce superior numbers of weapons, larger armies, or superior military technology in a technological escalation.

Most arms races have occurred in the modern era. One of the first arms races occurred in the pre-World War I era from the 1890s to 1914 where the five great powers of Europe (Germany, Australia, Hungry, the Russia Empire, France, and the United Kingdom) were looked in an all-out military buildup, ranging from land armies, conscription, and artillery to battleships and competition between each country's mobilization speed. France reached a mobilization speed of just three days. At one point, it was estimated Germany could become fully war ready in only two days. Perhaps the best known arms race of this era was the naval one between the United Kingdom and the German Empire, during which the Royal Navy and the High Seas Fleet spent millions on the development and construction of Dreadnoughts and other types of warships.

The term, "arms race" is used generically to describe any competition where there is no absolute goal, only the relative goal of staying ahead of the other competitors. Evolutionary arms races are common occurrences, *e.g.*, predators evolving more effective means to catch prey while their prey evolves more effective means of evasion. This is sometimes called the **Red Queen effect**. In addition to predators, parasites can force their hosts into an arms race.

In technology, there are close analogues of the arms races between parasites and hosts, such as the arms race between computer virus writers and anti-virus software writers or spammers against Internet Service Providers and E-mail software writers.

Lewis Fry Richardson made an arms race model, trying to retroduct World War I, where he showed how two countries would go to war if more money was spent in the arms race than in trade. There is an interesting economic idea here, but since the model was made after the war, it did not have the prophetic power it might have had.

7.4 NUCLEAR WEAPONS IN INDIA

Background

India's nuclear weapons program was started at the Bhabha Atomic Research Center in Trombay. In the mid-1950s India acquired dual-use technologies under the *Atoms for Peace* non-proliferation program, which aimed to encourage the civil use of nuclear technologies in exchange for assurances that they would not be used for military purposes. There was little evidence in the 1950s that India had any interest in a nuclear weapons program, according to Joseph Cirincione of the Carnegie Endowment for International Peace. Under the "Atoms for Peace" program, India acquired a Cirus 40 MWt heavy-water-moderated research reactor from Canada and purchased from the U.S. the heavy water required for its operation. In 1964, India commissioned a reprocessing facility at Trombay, which was used to separate out the plutonium produced by the Cirus research reactor. This plutonium was used in India's first nuclear test on May 18, 1974, described by the Indian government as a ***peaceful nuclear explosion***.

According to the Bulletin of Atomic Scientists, India began work on a thermonuclear weapon in the 1980s. In 1989, William H. Webster, director of the CIA, testified before the Senate Governmental Affairs Committee that *indicators that tell us India is interested in thermonuclear weapons capability*. India was purifying lithium, producing tritium and separating lithium isotopes. India had also obtained pure beryllium metal from West Germany.

Testing

After 24 years without testing India resumed nuclear testing with a series of nuclear explosions known as "Operation Shakti." Prime Minister Vajpayee authorized the tests on April 8, 1998, two days after the Ghauri missile test-firing in Pakistan.

On May 11, 1998, India tested three devices at the Pokhran underground testing site, followed by two more tests on May 13, 1998. The nuclear tests carried out at 3:45 pm on May 11th were claimed by the Indian government to be a simultaneous detonation of three different devices - a fission device with a yield of about 12 kilotons (KT), a thermonuclear device with a yield of about 43 KT, and a sub-kiloton device. The two tests carried out at 12:21 pm on May 13th were also detonated simultaneously with reported yields in the range of 0.2 to 0.6 KT.

However, there is some controversy about these claims. Based on seismic data, U.S. government sources and independent experts estimated the yield of the so-called thermonuclear test in the range of 12-25 kilotons, as opposed to the 43-60 kiloton yield claimed by India. This lower yield raised skepticism about India's claims to have detonated a thermonuclear device.

Observers initially suggested that the test could have been a boosted fission device, rather than a true multi-stage thermonuclear device. By late 1998 analysts at Lawrence Livermore National Laboratory had concluded that the India had attempted to detonate a thermonuclear device, but that the second stage of the two-stage bomb failed to ignite as planned.

Test	*Device*	*Date*	*Yield claimed*	*Yield reported*
	Fission device	18 May 1974	12-15 kiloton	4-6 kiloton
Shakti 1	Thermonuclear device	11 May 1998	43-60 kiloton	12-25 kiloton
Shakti 2	Fission device	11 May 1998	12 kiloton	??
Shakti 3	Low-yield device	11 May 1998	0.2 kiloton	low
Shakti 4	Low-yield device	13 May 1998	0.5 kiloton	low
Shakti 5	Low-yield device	13 May 1998	0.3 kiloton	low

India's Nuclear Arsenal

Though India has not made any official statements about the size of it nuclear arsenal, the NRDC estimates that India has a stockpile of approximately 30-35 nuclear warheads and claims that India is producing additional nuclear materials. Joseph Cirincione at the Carnegie Endowment for International Peace estimates that India has produced enough weapons-grade plutonium for 50-90 nuclear weapons and a smaller but unknown quantity of weapons-grade uranium. Weapons-grade plutonium production takes place at the Bhabha Atomic Research Center, which is home to the Cirus reactor acquired from Canada, to the indigenous Dhruva reactor, and to a plutonium separation facility.

According to a Jan. 2001 Department of Defense report, *India probably has a small stockpile of nuclear weapon components and could assemble and deploy a few nuclear weapons within a few days to a week.* A 2001 RAND study by Ashley Tellis asserts that India does not have or seek to deploy a ready nuclear arsenal.

According to a report in Jane's Intelligence Review, India's objective is to have a nuclear arsenal that is "strategically active but operationally dormant", which would allow India to maintain its retaliatory capability "within a matter of hours to weeks, while simultaneously exhibiting restraint." However, the report also maintains that, in the future, India may face increasing institutional pressure to shift its nuclear arsenal to a fully deployed status.

Doctrine

India has a declared nuclear no-first-use policy and is in the process of developing a nuclear doctrine based on "credible minimum deterrence." In August 1999, the Indian government released a draft of the doctrine which asserts that nuclear weapons are solely for deterrence and that India will pursue a policy of "retaliation only." The document also maintains that India "will not be the first to initiate a nuclear first strike, but will respond with punitive retaliation should deterrence fail" and that decisions to authorize the use of nuclear weapons would be made by the Prime Minister or his 'designated successor(s).'

According to the NRDC, despite the escalation of tensions between India and Pakistan in 2001-2002, India remains committed to its nuclear no-first-use policy. But an Indian foreign ministry official told Defense News in 2000 that a 'no-first-strike' policy does not mean India will not have a first-strike capability.

India has not signed the Comprehensive Test Ban Treaty (CTBT) or the Non-Proliferation Treaty (NPT). India is a member of the International Atomic Energy Agency (IAEA), and four of its 13 nuclear reactors are subject to IAEA safeguards.

Despite promoting a test ban treaty for decades, India voted against the UN General Assembly resolution endorsing the CTBT, which was adopted on September 10, 1996. India objected to the lack of provision for universal nuclear disarmament "within a time-bound framework." India also demanded that the treaty ban laboratory simulations. In addition, India opposed the provision in Article XIV of the CTBT that requires India's ratification for the treaty to enter into force, which India argued was a violation of its sovereign right to choose whether it would sign the treaty. In early February 1997, Foreign Minister Gujral reiterated India's opposition to the treaty, saying that "India favors any step aimed at destroying nuclear weapons, but considers that the treaty in its current form is not comprehensive and bans only certain types of tests."

QUESTIONS

1. Explain the nuclear threat to humanity.
2. Give heat and blast effects of a modern nuclear weapons.
3. What are the first effects of nuclear explosion?
4. What is nuclear proliferation? What is NPT?
5. Explain the term Arm's race.
6. Explain the Nuclear program of India.

8

APPROPRIATE TECHNOLOGY

8.1 INTRODUCTION

We want appropriate technology to feed the rapidly growing population. The application of technology shall ensure optimum use of natural resources which may remain unutilized or underutilized in the absence of the same. No doubts in the past quarter of a century the developing countries like, India have made progress in terms of increasing per capita incomes of their rapidly growing populations. Of course progress has also been made in the fields of health, education and infrastructure (electricity, water, transportation and communication) but despite this growth record, poverty has not been lessened appreciably. In India the levels of unemployment and income inequality are increasing.

According to the World Bank almost half of the total population of the developing world lives in absolute poverty. Such people are denied basic components of average living, adequate clothing and housing and access to basic services such as clean water, energy, schools, and health facilities which are taken for granted by most people in developed countries.

An Appropriate technology means method of doing a given job or works in a befitting manner and style as per the liking of users.

'An appropriate technology is one which is best suited to the environment in which it is most suitable'. The idea of appropriate technology was first welcomed and received by third world countries. The third world countries are underdeveloped developing countries.

The idea of appropriate technology appeared first in the mind of great thinker **Schumacher**. It is he who popularized the term and we can find its reference is his famous work ***Small is Beautiful.***

8.2 IMPORTANCE OF APPROPRIATE TECHNOLOGIES

The appropriate technology has the following advantages:

(*i*) Appropriate technologies involve low capital cost and are quite useful for capital poor countries including India.

(*ii*) Use of local material is made available where ever possible.

(*iii*) Create job opportunities and employ local skill and labour.

(iv) The scale of production is so small that even marginal and small group of farmers are able to use without much hesitation and hindrance.

(v) It does not need any high level western style education and is quite easy for villagers to understand, control and maintain.

(vi) This technology can be produced out of a small metal working shop if not in a village itself.

(vii) Suppose that people have desire to work together collectively with a view to bring improvements and the objectivity as well as recognization, many important decisions are made by groups rather than individuals alone, then this technology is found useful appropriate and suitable.

(viii) Appropriate technology involve renewable energy sources, such as wind, solar energy water power, methane gas, animal power and pedal power such as bicycle etc.

(ix) It makes technology understandable to the people who are using it and thus suggest ideas that could be used in further innovations.

(x) Appropriate technologies are highly flexible (subject to change) so that they can continue to be used and fitted as per the changed circumstances.

(xi) It do not demand patents, royalties and consultant gees etc.

(xii) No import duty, shopping charges are needed.

(xiii) Practical plans can be obtained free or at a low cost while making use of appropriate technology.

8.3 EXAMPLES OF APPROPRIATE 'TECHNOLOGY' ARCHITECTURE

A Delhi-based architect working with appropriate technology for the last 15 year meets the demand of low-energy or eco-friendly architecture. According to DLF (Delhi) building a house is an expression of thought. By using appropriate technology a low energy and high comfort homes can be produced. The designs should be as follow:

(i) A south facing entrance and thick thermal mass walls ensure sun-soaked bedrooms in chilly winter month; and a central cooling system, sunshades, green patches and as insulated while glazed tile roof keep the house cool during hot summers.

(ii) 'Clerestorey'-a skylight like feature in the dining room to minimize artificial light. This dispense with air-conditioners, room-heaters and electric lights during the day. Apart from cutting costs it also minimize initial construction expenses by limiting the use of steel to mere 9 kg whereas in a conventional home normal steel used is 30 kg/Sq.m.

(iii) In Bangalore, an architect has built a colony of 35 wallet friendly stone houses, procuring the raw materials from local quarries. In this system, stone slabs are used in an interlocking system where they are places horizontally, locked in a grove and supported by a pillar. With this technology, the saving on foundation material is upto 70 per cent and the total savings in cost is upto 20 per cent. Not only does stone work well in Bangalore's climatic conditions, but such buildings also consume less cement, steel and water during construction.

(iv) In hot and humid Bombay, building energy conscious architecture is a tricky exercise. But Space Design Associates, Delhi turned the situation around by designing the country's first energy-saving sky-scraper-CMC building. Since the city requires central air-conditioning around the year, a computerized automatic system was installed to save energy.

Lighting efficiency is maintained technically by a double-glazed continuous window, with reflective Venetian blinds operated by computer-controlled motors, which tilt the slats at appropriate angles depending on the brightness, and a micro-processor, which controls the light switches, according to the time and daylight available. 'Though the owners have not maintained any records of energy consumption, CMC should be saving an estimated 25 per cent on energy.

In a country of homeless millions, the cost efficiency of appropriate architecture is a driving force, while the actual cost-saving cannot be quantified in rupee terms, the Tata Energy Research Institute (TERI) is presently engaged in evaluating the performance level of such architecture. "The cost of constructing a 1,500 sq. ft. higher-income home in a conventional manner could come to Rs.400 to Rs.50 per sq. ft. While a solar passive home of the same size would not only cost the same, but its recurrent expenditure on electricity will be reduced, and the comfort levels will be much higher.

8.4 APPROPRIATE TECHNOLOGY AND CAPITAL INTENSIVE TECHNOLOGY: CAN BOTH EXIST TOGETHER

Capital Intensive Technology is the modern concept developed by industrialized and computerized countries like U.S.A., U.K and Germany etc.

Both the Appropriate technology & Capital Intensive technologies cannot go together simultaneously due to the following differences:

(*i*) Appropriate technology involve low capital cost whereas capital Intensive technology, involve high capital cost.

(*ii*) Appropriate technology as used in underdeveloped economies and is employment oriented, providing more and more jobs. But Capital Intensive Technology when used in Under Developed Countries create unemployment because ratio of man to capital CM: K is reduced.

(*iii*) Appropriate technology can be understood and applied by all those skilled and unskilled workers engaged in a particular working environment. But the application of CIT need higher skill and technical knowledge which is highly cost effective. In other words much more training is needed in case of the application of CIT. This may not be very suitable to underdeveloped countries where infrastructural facilities are meager.

8.5 INDIGENEOUS TECHNOLOGY

Indigenous technology means, all those techniques which are developed and created domestically without any foreign assistance and help.

Technology is developed for increasing the power of human beings over their material environment. Indigenous technology is called home technology a technology which is developed by the nation without any foreign assistance. It is low cost and socially appropriate technology. This technology is well adapted to the local cultural needs. It is the most appropriate technology for under-developed and developing countries of South-Asia., Use of this technology save precious foreign exchange. Other uses of this home made technology are given below:

(*i*) It is need based.

(*ii*) Its skill and capital requirement are low.

(*iii*) It can be managed with local materials.

(*iv*) It is very popular in third world countries.

Advantages of indigenous technology are given below:

1. Indigenous technology besides saving foreign exchange which is precious to India guarantees economic, religious and political independence too. Whenever an effort is made to modify the world through the application of scientific knowledge, the domain of technology starts. This technology is to first economise human labour. Science too makes its contribution to technology by supplying the basic results. Evolution of science and technology is perhaps devoted to the issue when man tried to use it in day to day life. This knowledge first of all, perhaps began in the Nile Valley, when the people began to built cities with the use of metals for building temples and tombs. It is historically recollected that science and technology developed in India when the people of West were still living in the forests. In 1958, Govt, of India announced its science and technology policy. The aims of this policy are to foster, promote and sustain cultivation of science and scientific knowledge.

8.6 TECHNOLOGY IMPORT POLICY OF INDIA

Technology import policy of India is encouraging Multinational Corporation to enter in the field. As the financial and administrative structure of Multinational Corporation is much better than Swadeshi industries, it will harm adversely the Swadeshi movement and our traditionally rural based industries. They will face stiff competition and even may be wiped out.

Recently Govt. of India get cleared from parliament Insurance regulatory bill which permit foreign companies to invest capital in Indian Insurance Sector. This step will open gates for foreign insurance companies.

But however, it is suggested that if Indian cottage, small or big industries keep their quality at the standards and prices of their products much cheaper than offered by MNC's or foreign companies, swadeshi movement will not be harmed much and Indian Industry can sustain.

8.7 INTERMEDIATE TECHNOLOGY AND THIRD WORLD COUNTRIES

What technology should be used by developing countries including India is another point needing attention of millions. E.F. Schumacher in 1973 first brought the concept of intermediate technology to the attention of the public in his best selling book ***'Small is Beautiful'***. He was not only a person to identify the need for intermediate technologies in the third world, Eminent figures like Gandhi and Julius Nyerere have also talked about and propagated such technologies earlier. Schumacher suggested intermediate technology when the problems of the developing countries themselves were facing increasingly severe economic and environmental problems.

It is a difficult task and debatable issue also to define intermediate technology. However attempts have been made to define intermediate technology in terms of equipment cost per workplace. In this respect intermediate technology has been assessed in terms of equipment and finance involved. If we say Pound one technology of underdeveloped developing countries and Pound 100-technology of the developed countries, than the gap between traditional (village type) and modern technology is quite wide from productivity point of view. Intermediate technology is preferred to indigenous (demotic or within the nation) technology as it is quite cheaper than the sophisticated, highly capital intensive technology of modern industry. With Pound 100 technology reasonably good work places could be created within a fairly short time; and the creation of such work places would be within reach for the more enterprising minority within the district not only in financial terms but also in terms of their education, aptitude, organizing skill, and so forth.

8.7.1 Advantages of Intermediate Technology

India and many other under-developed countries have adopted and may be on their way to adopt this technology in view of the following advantages:

(*a*) Intermediate technology fit much more smoothly into the relatively unsophisticated environment in which is to be utilized.

(*b*) The equipment used are fairly simple and are easily adaptable by the people on the job.

(*c*) On the spot maintenance and repair of tools and machinery is possible.

(*d*) Equipment used is normally far less dependent on raw materials of great purity.

(*e*) Men are more easily trained; supervision, control and organization are simpler.

(*f*) Unforeseen difficulties can be easily overpowered.

(*g*) Capitalization is less and very large work places can be created within a fairly short time.

8.7.2 Types of Technologies

The main types of technologies in vogue are discussed below:

1. ***Alternative technology:*** This technology describes new types of equipment or new organization forms which show a possible native to the existing 'main stream' technologies today. ***Example:*** Self housing schemes instead of conventional urban development programmes, or small scale organic farming instead of large scale energy intensive cultivation techniques.
2. ***Capital saving technology:*** It is also called light capital technology. Capital saving technology is a technology characterized primarily by its low cost in capital and the small size of the investment needed to create jobs. It is widely used by the US Agency for International Development for building roads with efficient labour intensive methods and not bulldozers embodying light capital technologies.
3. ***Community technology:*** It is a small scale technology which does not require a complex infrastructure (electricity, transport and heavy equipment) and is specifically developed to the needs and capabilities of small urban or rural communities. The term community technology is widely used in American culture. ***Example:*** Small scale co-operative industrial activities or decentralized water supply and waste disposal system.
4. ***Environment sound and appropriate technology:*** This is a concept developed by the United Nations Environment Programme and Amulya K. Reddy of India. It is an appropriate technology which is particularly well adopted to the local social and economic environment, and which uses renewable resources other than non renewable resources. Examples in the energy field are biogas plants or biomass conversion systems. Hardware, a term borrowed from the computer industry and now used by the Appropriate technology community is the physical embodiment of technology that is tools, implements, machines, devices and equipments.
5. ***Low-cost technology:*** It is a technology whose main feature is the low cost of the final product or service, or the low cost on the investment required to provide this product or service. ***Example:*** Stabilization ponds for sewage treatment.
6. ***Socially appropriate technology:*** It is a technology which is likely to have beneficial effects on income distribution, employment, work satisfaction, health and social relations. ***Example:*** A vaccine against Malaria or Schistosomiasis ***(A type of disease).***

7. ***Soft technology:*** It is a technology which is well adapted to the local cultural and social environment, which uses renewable rather than non-renewable resources, and which does as little damage as is possible to the surrounding eco-system. ***Example:*** Windmills, small hydro power plants. Software is the non-material dimensions of technology for example knowledge experience, organizational forms, managerial tools, institutional structures, legal provisions and financial incentives.
8. ***Village technology:*** It is small scale technology aimed primarily at meeting the basic needs of rural population in the developing countries. This concept was developed by UNIFEF (United Nations Children's Fund) ***Examples:*** Small scale on farm storage systems for food, low cost dryers.
9. ***One world technology:*** This technology is associated with the equilibrium economy. The idea that capital intensive technology is appropriate for the advanced countries and intermediate technology for the developing countries will not apply in an equilibrium world economy. The idea of this technology is one which is good to work with, sparing its use of resources, produce a good end product and kind to the environment.

8.8 ADVANTAGES OF TRANSFER OF TECHNOLOGY

Technology transfer is unavoidable so far as development of nation is concerned. Some of the advantages of technology transfer are discussed below:

1. It contributes significantly in widening and strengthening the industrial base of recipient countries. Once the industrial base is developed then one will go on a long way in furthering the developing and enhancing the GNP (Gross National Product), decreasing poverty, increasing standard of living and reducing unemployment.
2. It increases employment rate due to expansion in industries; reduction in poverty and enhanced purchasing power in turn results in more industrial production and brisk industrial activities.
3. When a developing country procures a technology from another developing country, it gets chance of setting buyers market which is not only advantageous but a step forward in the development of economy.
4. The developed countries develop and design a technology in such a manner and style that in many cases it becomes quite useful and suitable to developing countries. It appears less capital and more labour intensive suitable to the climatic condition of developing countries. This reduces the chances of closing down or shutting down the plants.
5. Technology transfer taking place through consultancy has much more advantages because the transferees (who transfers) have vast experience of running a number of industries and commissioning of plants. Under those condition which are suitable to recipient countries, such technology when used show concrete results in a short period of time.
6. Technology transfer when properly adopted check brain drain problem.
7. To poor countries technology transfer is a great source of saving.
8. The main objective of technology transfer is to speed up the process of development.

8.9 APPROPRIATE TECHNOLOGY SYNDROME (CHARACTERISTICS)

Availability of proper and appropriate technology is a major problem in the field of production. The use of right technology can always reduce cost of production besides improving the quality of produce and increasing the level of productivity. Whatever technology is developed, the same is used by big farmers and industrialists. Adequate attention is not paid to develop technologies suitable for resource poor farmers in dry land areas. The marginal farmers and small factory-owners are not able to use technology due to of number of reasons such as;

(*i*) High costs of improved technology

(*ii*) Lack of extension efforts and methodology directed toward small farmer

(*iii*) Problems regarding availability of physical inputs

(*iv*) Problems of institutional credit and problems of institutional credit and

(*v*) Problems of risks and uncertainties.

A number of studies have been conducted which reveal that technologies developed have been found to be unsuitable to the conditions of many small farmers.

8.10 CRITERIA FOR SUCCESS OF TECHNOLOGY TRANSFER

Transfer of technology is more important than other functional areas, may it be research or education. There are a large number of farm technologies which have not reached the farmers, particularly small and marginal farmers who constitute the bulk of our farming population. The lab-to-lab projects were launched with a view to cater to the needs of such resource poor farmers. Low-cost technologies suiting to the pockets of marginal farmers diversified agricultural practices necessary for economic viability and appropriate technology can bring great success in the field of transfer of technology. Various programmes, measures and criteria as developed from time to time for making the transfer of technology as a great success are discussed below:

1. **Development of Low Cost Farm Technologies:** Considerable attention should be paid by agricultural scientists to develop low cost farm technologies since high cost of technologies hinder most of the small farmers from adopting them. Low cost technologies are given as follows:

   (*a*) The use of neem as an ***insect repellent, antifeedant*** and insecticide has proved experimentally correct. The locally available neem can be used in plant protection by small farmers. Mixing neem leaves with grains for storage purposes is common in rural India.

   (*b*) Use of ***rhizobium inoculants*** in place of high cost fertilizers, especially nitrogenous, can be made use of. The inoculants are available in the form of charcoal based culture. About 10 to 70 percent increase in yield over control due to Rhizobium treatment has been reported. The use of ***Azoprillium*** as seed inoculants is recommended for crops like Jowar, Bajra, Ragi, Barley and Oats.

2. **Low Cost Farm Storage Technology:** The loss of large quantities of grain during storage in rural households has attracted farm scientists to develop farm level low cost storage structures especially to suit the conditions of small farmers, who do not have access to the costly storage structures. Scientists have developed a low-cost improved storage structure called the ***Pusa Bin*** which involves simple changes in the existing structures or constructing new structures to achieve moisture content of the store grains is an important factor which determines

the development of insects or other pathogens and biochemical. They determine the development of insects or other pathogens and biochemical deterioration. By ensuring airtihghtness the level of oxygen content is reduced which is harmful to the life of insects. The storage structure ***Pusa Bin*** is a modification of the ordinary mud storages structure used in villages. Commonly available polythene sheeting of 700 gauge thickness is used at the top as well as on all sides of the mud structure to achieve air tightness and water proof conditions. The problems due to rats are avoided by constructing the storage structure with unburnt bricks on a floor made of burnt bricks or concrete. The bin can be made with a storage capacity ranging from 0.5 ton to several tones. The grains or seeds can be stored in this way for more than a year. He cost of construction of a Pusa Bin of two tones capacity will be around Rs.200. The benefits of this storage structure can be seen from infestation after the storage for a period of 6 to 7 months. The Pusa bin will have negligible effect compared to 7 to 35 percent level of grains stored in gunny bags.

3. **Lab to Land Programme:** The aim of the programme was selecting and adopting about 50,000 small farmers and landless agricultural labourers by the scientists spread all over the country. The farmers were selected with a view to educate them and introduce new technologies to them so that their income and production could be increased. An evaluation of the Lab to Land programme has shown that the programme has succeeded in transferring the modern technologies to small farmers.

8.11 SOCIO-ECONOMIC SYSTEM AND TECHNOLOGY ADOPTION

Modern society is passing through the age of technological advancement. Many technological changes have taken places and many have yet to take place in the near future. The Industrial Revolution, equipped with a number of new technologies, ushered in new dimensions in the economic and socio-political outlook of people. Social institutions connected with a number of new technologies, ushered in new dimensions in the economic and socio-political outlook, largely influenced by the Industrial Revolution. Now, underdeveloped and developing economies should try to develop and adopt indigenous technology since procuring foreign technology often poses many hurdles.

Adoption of technology, which means making use of technology, is good for socio-economic prosperity. But in underdeveloped countries, amongst which India is one, adoption and absorption of the 20th century's technological advancement and its transfer is constrained due to a number of socio-economic conditions. Adoption of modern technology and agricultural package programmes has not been sufficient to date to ensure the success of agricultural development plants. This needs appropriate technology suitable to farms and existing industrial conditions. Cultural factors such as caste wise distribution of operational land holdings, farm family characteristics particularly literacy rate, sub-division and fragmentation of holdings, and class-wise inequalities in the distribution of operational holdings are considered important variable (factors that change) in production efficiency in agriculture as they motivate the production efforts of farmers with regard to the choice of production techniques and agricultural as well as industrial enterprises. Critical evaluation of cultural constraints, therefore, becomes quite important in this respect.

It is true that land productivity and total agricultural production can be raised substantially through better inputs and the use of modern technology. But these inputs become more effective and their potential is better utilized if appropriate energy and power sources are made available to the farmer. 60 percent of the total power available to Indian farmers is derived from human and ***bovine*** (related to oxen) sources. This dependence on human and bovine sources does not go in harmony with new technology. Besides this, there are a number of other reasons also which come in the way of technology use and these are stated below:

1. It is agreed that most farmers having small landholdings cannot afford to spare capital out of their own saving to buy machines. But we cannot accept this argument. If we accept it, then it would mean that we are not ready to avail the advantages of new technology. The success of technology strictly depends upon the availability of improved inputs including capital and technology.
2. The small size of farms created due to social and institutional factors and a number of socio-economic changes make the operation of the transfer of technology developed in the laboratory impossible. For full mechanization of the type prevalent in most advanced countries, large farms are necessary. In the USA, for example, the average size of the farm is about 123 hectares and in Canada it is 188 hectares. As against this, in India, the average size the farm is 2.3 hectares. Most of the farms in India are, however, even below this average size. There are a few large farms in the country, but their number is quite small.
3. The tiny size and irregular shape of fields make it impossible to carry out field operations with any degree of efficiency and with the use of mechanization and technology. So the engineers, scientists and technologists should be engaged in evolving those machinery and technologies useful to our small farms. Japan has already shown a way by introducing wheeled tractors, known as ***garden tractors,*** that can be applied with equal efficiency on small farms. An NCAER study reveals that 31 percent of the total tractor stock is owned by the small farmers with les than 4 hectares of land.
4. The new method of cultivation as extended in the ***Green Revolution*** is both expensive and risky. The modern technology can no doubt give wonderful results and benefits, only if inputs are applied in optimal doses and on time. Only big farmers with large resources and vast credit at their command can secure the necessary inputs and the knowledge to use them. The growing of new ***hybrid crops*** requires careful handling and supervision from the first to the last operation. All this often necessitates of money, time and other non-monetary resources is possible for big farmers. But this is certainly not he case with small farmers, who have very limited resources of their own and have still less command over institutional finance.
5. ***Cultural thinking*** and the physical conditions also influence the type of implements and technology to be used on farms. Where a major pat of the land is leveled it is possible to use farm machinery and other equipments, including iron made implements, provided other conditions such as size of holding, irrigation facilities etc. are favorable. In addition, the degree of intensiveness in farming, the commercial character of agriculture and the farmer's way of life further determine the kind of farm implements to be used. The farmer operating sub marginal and small-sized holding (piece of land) are conservative. It is often the small size of holdings, ***subsistence farming*** and the limited economic means of the farmers that stop them from the use of modern implements and technology as it inhibit the transfer of technology from the laboratory to the field. There is little possibility of changing their attitudes unless the whole fabric of farm economy is changed.
6. The Green Revolution, which is merely a ***technical revolution,*** could not spread to our rural areas in the way it was expected on account of a number of reasons, the first being technology itself. As has happened elsewhere, in other countries, the new technology has been mostly of such a character that it requires sufficient and stable sources of water to absorb large doses of

fertilizers. In India, where the bulk of the land is subject to insufficient and unstable water supply from rains, the new strains could not be used everywhere. The second factor is concerned with the inadequate government policy in respect of irrigation and chemical fertilizers.

7. The inducement to invest in modern technology is limited by the size of the market. It is not difficult to find examples and illustrations of the way in which the small size of a country's market can discourage, or even prohibit, the profitable application of modern capital equipment and technology by any individual entrepreneur in any particular industry. In a country, for instance, where a great majority of the people are too poor to wear leather shoes, setting up a modern shoe factory may be a doubtful as a business proposition because the market may be small. Many articles that are in common use in the United States can be sold in a low-income country in quantities is limited because a machine working only a few days or weeks can produce enough for a few year's consumption and would have to stand idle for the rest of the time. In Chile, for example, it has been found that a modern ***rolling mill,*** which is standard equipment in any industrial country, can produce in three hours sufficient supply of certain type of iron shapes to last the country for a year. In these circumstances, the inducement to install such equipment is lacking.

Social and attitudinal constraints in the transfer of technology from laboratory to field.

The constraints in the transfer of technology are:

1. High cost of translating a lab or drawing board level innovation to a pilot plant or prototype level.
2. Most of the R&D ideas being followed are already being practiced.
3. The entrepreneur does not trust this technology and hence wants to know about the ready availability of the inventor in the initial phases of the trial and erection to the technology.
4. Transfer of technology very often gets mixed up with the ideologies such as self-sufficiency, socialistic pattern and freedom of competition.

8.12 TECHNOLOGY TRANSFER—NEED OF TRAINING

Developing countries are very keen to transfer technology (advanced) from developed countries to improve the standards of living and know how. The developing countries are backward due to serious lack of technical education, their institution lack latest equipment on the one hand and qualified and technically trained person on the other hand. Due to this reason they find very difficult to transfer the technology. So training is utmost important for them to understand and use the technology properly and develop their country. For training, qualified person are sent to developed countries to get trained and educated for use of advanced technology. Such training programmes are funded by UNDP or developing countries itself. Sometime developed country's people also come to under-developed countries to provide training. So without proper training, transfer of technology has no meaning.

8.13 HUMAN CAPACITY TO STORE AND TRANSMIT TECHNOLOGY

Transfer of technology means communication of knowledge about techniques of doing things. Understanding the process of transfer of technology involves knowing the manner in which technology develop from the initial idea to the final product. Transfer of technology is an exchange of expertise when the holder of technology passes over to recipient. Transfer of technology to producing unit is not so simple and automatic. A few points need attention. The technology selected for transfer should be appropriate; this technology should not be so much expensive and finally the technology transferred should be provide with the back-up service.

8.14 RECOMMENDATIONS AND CRITERIA FOR TECHNOLOGY TRANSFER

1. The adoption of new technology is hindered by a number of constraints. Some of these are technological, socio-cultural, economic, physical, administrative and the like.
2. There is need for close interaction between scientists and small farmers to reduce technology application gap and develop appropriate technologies. This includes learning about small farmers and integration of small farm family's circumstances into the project process.
3. Adaptive trials must be conducted on the fields of small farms before the experimental recommendations are released for adoption to these farmers.
4. The development of technologies for the small farmers should take into consideration the whole farming system of the small farms.
5. Small farmers have a favourable attitude towards new technology and they have selectively adopted the same with suitable changes or modifications.
6. Small farmers should be associated with the State or National Screening Committee for releasing recommendations.
7. The differences between small and big farmers are significant and their implication in terms of their attitude, value and belief towards the new technologies as well as farming have to be taken into account in the overall developmental programmes of small farms.
8. Formation of small farmers' organizations like Farmers' Groups or Farmers' Service Societies or Farmers' Co-operatives will facilitate adoption of new technologies as well as help in the efficient distribution of essential inputs.
9. All the inputs and credit needs of small farmers should be routed through a single agency and a 'single window approach' is suggested for this purpose.
10. The supply of credit must be accompanied by improved technology, technical guidance and proper supervision.
11. The modern communication technologies such as videos, televisions, cassette recorders, cinema slides etc. have also to be used in proper combination to popularize the new technologies among the small farmers.
12. The small farmers' problems with regard to transportation, grading and marketing of farm produce have to be solved. The farmers have to be educated as well as infrastructure should be developed to help them.
13. Greater attention has to be paid to develop and popularize low-cost farm technologies. Technologies which will be suitable to dryland areas shall also be developed especially to overcome the problems of frequent monsoon failures and lack of adequate irrigation water. Water conservation (*i.e.*, watershed management and water harvesting) technologies should be adopted.
14. Projects to extend irrigation facilities to small farms should be given priorities along with performing soil and moisture conservation activities.

15. A proper rural communication policy should be evolved with emphasis on reaching small farmers with modern technologies.
16. The small farmers should be ensured a profitable market price for their produce.
17. Mixed farming must be given to small farmers on subsidiary enterprises like dairying, goat rearing, sheep rearing, poultry farming and fish culture etc.
18. The training of small farmers in the modern agricultural technologies should receive greater attention, as at present training of small farmers has received only less attention.

8.15 LIST OF TECHNOLOGIES

1. Mobile phones
2. High Efficiency Inverters
3. SPV Modules & System
4. Digital Video Recorder
5. Network Camera
6. Feeder Energy Monitoring system load profiler
7. R&D Innovations from CSIR.
8. Biotechnology based products
9. Herbal based products
10. Oscilloscope and Educational Training Equipment.
11. Metallic Ceiling System
12. Solar Water Heater-Reverse Tube Type
13. Textile and Jute Sizing Processes.
14. Galvanizing Zinc kettle
15. Polypropylene and P-diethyl benzene catalyst.
16. Speciality Polymers-Thermoformable polypropylene, Biodegradable Polylactic Acid, Acrylic super Abserbent polymers.
17. Specially chemicals-Linear Plphan Olefins, Alkoxyl-Alky Benzoates Stereo Regulators.
18. Cement Testing Techniques.
19. Anti corrosive packing
20. Bent Air Purification System for Edible Oil Processing plant.
21. Gas operated Burners for corrugation machines
22. Carbon dioxide Gas Plants
23. Water Treatment Technologies
24. Cotton Seed De-lining plant
25. Dry Lubricants for steel plants
26. Column pipes for submersible pumps.

8.16 HUMAN CENTERED TECHNOLOGY

Human Centered Technology can be understood as being technology that supports human activities in various usage contexts. Human centered technology has applications everywhere and in involves strongly a cross-disciplinary research on several research areas. The technology should functional in different cultures, in different age groups and overall in different contexts. Technology should be considered as tool that helps and simplifies normal human processes.

Main Points

1. The basic objective of technology used in a field is to increase and improve the productivity, qualitatively and quantitatively by involving agricultural technology, mechanical technology, biological technology and chemical technology.
2. History reveals that the development of human societies has always been associated with the technological break-through. Technology which means application of mind and hands associated with techniques for completing a given work efficiently is always recognized.
3. Estimation and assessment of technology means examination of technology form suitably point of view. It is governed from socio-economic environment as well as from societal point of view.
4. Prediction about the future use of technology is called technological forecasting. Observable difference between technological forecasting and upgradation exist.
5. Modes and methods involved in transferring technology are foreign investment agreement, import of technical know-how, government to government transfer of technology and technology transfer through training and education. A number of problems are faced with regard to transfer of technology.
6. An appropriate technology is one which is the best suited to the environment in which it is most suitable. The idea of appropriate technology was first welcomed by third world countries. Shumacher mentions this technology in his work 'Small is Beautiful'.
7. Success of technology transfer depends upon development of low cost and effective programmes from laboratory to land programmes.
8. There are many pitfalls too when technology is transferred from laboratory to field.
9. The adoption of new technology is hindered by a number of constraints. These are technological, socio-cultural, economic, physical and administrative.

QUESTIONS

1. What do you mean by appropriate technology?
2. Discuss the criteria for the assessment of appropriate technology.
3. State the effect of Socio-economic system on technology adoption.
4. State the criteria for success of technology transfer.
5. What is difference between appropriate technology and capital intensive technology?
6. What is intermediate technology and what are its advantages?
7. What are the advantages of transfer of technology?

9

TECHNOLOGY ASSESSMENT AND TRANSFER

9.1 INTRODUCTION

The history of mankind reveals that the development of human societies has always been associated with the technological breakthrough. The development of countries especially in Asia and Pacific regions has been very closely associated with their technological capabilities in agriculture. Countries like India, South Korea have solved their food problems through the application of modern farm technology in their production processes. Man creates technology. Transfer of technology is done for the sake of man to make him healthy and happy in the society.

9.2 DEFINITION OF TECHNOLOGY

Technology is the application of mind for learning methods and techniques of completing a given work or job efficiently. Technology involves collaboration (association) of human hand with human brain to fashion and design tools and devices as well as technical know-how and applying the same for which they have been created. Use of technology examines the ability of man beyond natural capacity of his body. Understanding connection between elements of nature on the one hand and the capacity of manipulating it on the other hand to achieve human ends is necessary for the emergence of technology. So we can say that existence of manipulative capacity of human mind reflecting creativity underlying the emergence of tools, techniques and processes together refer to technology system which involve intellectual comprehension (understanding). There can be both ***hardware*** and ***software technology.***

Hardware: Hardware refers to physical computer equipment and associated devices. Hardware must provide five major functions:

(*i*) Input of entry
(*ii*) Output
(*iii*) Secondary Storage for data and programmes.

(*iv*) Central Processor (Computation, Control and primary storage).

(*v*) Communications.

Software: Software is a broad term given to the instruction that direct the operation of the hardware. The software can be classified into two major sections: (1) **System Software and** (2) **Application Software.**

Database: The database contains all data utilized by the application of software.

Procedure: Formal operating procedures are physical components because they exist in physical form such as manual or instruction booklet.

Three types of procedures are required as mentioned below:

(*i*) User instruction

(*ii*) Instructions for preparation of input

(*iii*) Operating instructions for computer operations personnel.

Operation Personnel: They are computer operators, system analysis, programmers, data preparation personnel, information systems managers and data administrators etc.

Engineering and Technology

The systematic application of knowledge to meet the needs and desires of man, including its embodiment in machines, tools, and product processes and software, is called technology. Software means organization and management of ways of doing things. There can be both capital intensive as well as labour intensive technology.

One who designs and repair machine and tools is called, an engineer. Engineering is a process to perform designing and preparing as well as utilities. Engineering subjects are taught in engineering colleges. Superior technocrats are also called **engineers.**

9.3 TECHNOLOGY TRANSFER

Technology transfer is unavoidable so far as development of a nation is concerned. The various reasons for technology transfer are given below:

1. It contributes significantly in widening and strengthening the industrial base.
2. It increases employment rate both at rural and urban sector because expansion takes place.
3. It establishes buyer's market.
4. It checks brain drain problem.
5. It speeds up the process of development and increase in GNP (Gross National Product).
6. It enhances scientific and technical knowledge.
7. It encourages Research and Development (R & D).
8. It provides market testing and market development.

9.4 NEED FOR TECHNOLOGY TRANSFER

What type of technology is being transferred is an important question. Before importing any technology following points are required to be studied carefully to make the technology transfer a success:

1. Understanding of the user's actual needs.

2. Taking into account country's conditions.
3. Technology must be appropriate and best suited.
4. It must be provided with back up services.
5. It should be cheap and preferably adopted on use of renewable energy resources.
6. The local industries should not suffer from the transfer.
7. Its adoption should not have problems of risk and uncertainties.
8. The technology transferred should be eco-friendly so that its use do not come in the way of ecological development.

9.5 ROLES OF TECHNOLOGIES

The basic objective of technology used in any field is to increase and improve the productivity qualitatively and quantitatively. Various fields have been discussed below where the roles of technologies are recognized in abundance:

1. ***Role of Agricultural Technology:*** The introduction of agricultural technology is directed towards achieving of crops and livestock (animal wealth). and increase in the employment and income of farmers. The introduction of modern farm technology will not only result in increasing the farm produce but also an improving the quality of life and standard of living of the rural population. As reported by ***Mclnerney*** (1978), ideally technology embodied in a rural development project might be expected to bring the following changes:
   (*a*) Raise the aggregate output and production capacity of the agriculture
   (*b*) Provide increased opportunities for the employment of labour
   (*c*) Create resource demands to encourage the development of rural non-farm productive capacity
   (*d*) Require minimal amounts of scarce resources, such as capital or entrepreneurial skills, which may be in short supply or only available to a number of large farmers,
   (*e*) Generate benefits that accrue to small farmers in a measure at least equal to those for the larger farmer
   (*f*) Offer stability and improved nutritional quality in the rural family's food supply
   (*g*) Enchance utensils, experience, level of awareness, and enterprise of the rural producer, and
   (*h*) Engender change conducive to continued innovative activity in social and institutional structures and wealth distribution.

9.6 TYPES OF TECHNOLOGY

1. The recent advances in **molecular biology** and **genetic engineering** have opened up unusual possibilities of developing a number of technologies which will have a direct bearing on production and productivity. Further, efforts are also being made to introduce technology to combat (to control) new problems arising related to agricultural production. Some of the emerging technologies are listed below:
   (*i*) Integrated pest management
   (*ii*) Integrated nutrient supply

(*iii*) Scientific water supply
(*iv*) Soil health care on system basis
(*v*) Farming systems based on integrated principles of ecology and economics, and
(*vi*) Energy efficient farm machinery and equipments.

2. ***Role of Mechanical Technology***: Mechanical Technology refers to the development of different farm machineries and equipment such as tractors, threshers and tillage implements, which help in farm production. Progress has been made in the application of improved implements at the production and post harvest levels.

   Some of the most common mechanical innovations are as follows:
   (*a*) Tractors and tractor drawn implement
   (*b*) Improved plough
   (*c*) Improved seeds and fertilizer drills
   (*d*) Power sprayers, dusters, etc.
   (*e*) Electric or diesel pump-sets
   (*f*) Puddlers and transplanting machines
   (*g*) Harvesting and *winnowing machines*, and
   (*h*) Threshers.

   Apart from the above innovations, improvements pertaining to water management, water harvesting, soil and water conservation, post-harvest technology and bio-gas technology are gaining increased popularity.

3. ***Role of Biological Technology***: One of the important milestones in improving the productivity of crops, during this century is through repatriating of *genetic architecture* of plants. This has resulted in the development of improved varieties and hybrids of various crops. Hybrid maize was commercially released in 1933, followed by the release of dwarf rice (1955), *hybrid sorghum* (1957), hybrid pearls millet (1959), hybrid wheat (1961), hybrid cotton (1973) and hybrid rice (1976). Most of the developing countries especially in Asia and the Pacific region have developed high yielding crop varieties to suit their local environments. Apart from the high yield, some of the characteristics sought after in the new varieties are high productivity per unit of water, high photosynthetic ability, high response to nutrients and multiple resistance to insects and diseases.

   The agronomic and farming systems researches have resulted in improved cropping systems and cultural practices. Some of the innovations in these areas are related to the following:
   (*a*) **Multiple cropping** system in irrigated areas,
   (*b*) **Rainfed farming**
   (*c*) Inter-cropping practices
   (*d*) Mixed farming and
   (*e*) Composts and their benefits.

4. ***Role of Chemical Technology***: The discovery of chemical fertilizers has made it possible to supply crops with needed nutrients in the midst of scarcity of organic manures. Following the chemical fertilizers, a number of related discoveries such as pesticides, fungicides, hormones, herbicides and other growth regulations have been made. These technologies have a remarkable impact in the production and productivity of crops. Researches pertaining to the use of above technology are well established in most of the developing countries and countries like India.

9.7 VARIOUS SOURCES OF TECHNOLOGY

In developed countries, the main agencies of technology are universities, institutions, industries and government. The NRDC in U.K. gets 50% process from the universities, in Japan from Japan research Development Corporation and U.S.A. receive practically all of its new ideas from Universities. The main sources of technology in India are:

1. N.R.D.C. and I.A.R.L.
2. R&D cells within the Corporation and their production departments.
3. NCL, NPL, National Productivity Council and National Physical Laboratory.
4. Laboratories established by public and private sectors.
5. CSIR (Council of Scientific and Industrial Research).
6. Defence Research & Development Organization (DRDO).
7. Department of Atomic Energy.
8. ICAR (Indian Council of Agricultural Research).
9. ICMR (Indian Council of Medical Research).
10. Departments of the Govt. of India like Railways, P&T, AIR etc.

9.8 TECHNOLOGY ASSESSMENT

Assessment of technology means estimation and examination of technology. The one who examines, judge and advice for using a given technology is called assessor or even valuer. Critical evaluation of technology is very essential before it is deemed fit for a transfer. A particular technology may be useful in one type of environment whereas the same may not be in other environment.

So assessment of technology from the suitability point of view is very essential. The recipient of technology should know whether a particular type of technology suits to the prevailing environment of that country (the country who wants to receive) or not.

In view of this, following issues need attention before technology is adopted:

1. **Technology and Employment**: Broadly there are two types of technology; Labour intensive and capital intensive. So before adopting any technology we must exercise difference between these two because our country (India) is a country where rate of unemployment is quite high due to higher population growth rate. If we use capital intensive technology then the problem of unemployment shall become more serious.
2. **Socio-Economic Environment:** Technology should be assessed on the basis of socio-economic environment that prevails in the society. Only that technology should be adopted which is most suitable to the environment and found economically feasible.
3. **Technology and Budget:** Assessment of technology for the purpose of use strictly depends on the budget and cost aspect. Costly technology does not suit to under developed countries amongst which India is one.
4. **Societal Process:** We know that tools and techniques when combined together produce systems of technology. The system of technology is of varying degrees of complexity. So while assessing the technology deemed fit for use a number of social processes are involved which must be taken into account before adopting a technology. A particular technology may be suitable for one country and the same may not be suitable to other country.

9.9 LABOUR INTENSIVE AND CAPITAL INTENSIVE TECHNOLOGIES

The difference between both the technologies is described below:

(*i*) Labour intensive technology is one in which more labour and less capita is used per unit of output.

(*ii*) In labour intensive technology Ratio of Capital (machinery) to labour is less. In other words K: L is low (K<L), or (L>K). K=capital, L= Labour.

(*iii*) In capital intensive technology ratio of capital to labour is more. In other words K:L is high (K>L) or (L<K).

(*iv*) In developed countries like U.S.A, U.K. and Germany capital intensive technology is used because their Science & Technology is advanced as well as financial budget is on the higher side and population growth is low.

(*v*) In underdeveloped countries, the system is otherwise. Here availability of labour both skilled and unskilled is abundant due to high population growth. So labour intensive technology is proactive here. Moreover our budget is also tight.

9.10 PROBLEMS FACED BY TECHNOCRATS AND MANAGEMENT

While selecting proper technology for transfer, technocrats face the following problems:

1. Interferences by non technical people increases beyond limits.
2. Shortage of proper staff under them.
3. Improper training of technical staff.
4. Lack of funds.
5. Bureaucrats disturb technocrats by unwanted interferences.
6. Bureaucracy delays the work schedule.
7. Non availability of modern equipments.
8. Heavy burden of work due to population explosion.
9. Improper Govt. or organizational policies.

9.10.1 Management of Technocrats' Problems

While managing and implementing a particular technology, the Technocrats face the following problems:

(*i*) Cultural thinking and physical conditions are adverse. So a technocrat has to understand the cultural level of people encountered before implementing a designed technology.

(*ii*) Non-financial and non-monetary institutions coming in the way of techniques used by technocrats are so strong and stiff that surmounting these hurdles is not so easy. Technocrats encounter pitfalls before managing the use of technology.

(*iii*) Technocracy and Bureaucracy are two different streams. A technocrat has to face bureaucratic policies which are highly politically coloured and oriented.

9.11 IMPORTANCE OF TECHNOLOGY

Technology may be regarded as operations or procedures to achieve desired outputs from given science, that is, *the systematic application of knowledge to meet the needs and desires of man, including its embodiment in machines, tools, products processor, and software. Software means organization and management of ways of doing things. It also includes the possession of such knowledge by skilled labour, professionals, managers and librarians.* The economists are increasingly accepting the role of technology to achieve economic growth. According to them economic growth does not arise by simply changing the technology but by commercializing know-how through innovation.

Five factors have been suggested for innovation and these are:

1. Understanding user's need.
2. Marketing as well as sales performance.
3. Management structure-well managed and not mismanaged.
4. Communication system.
5. Research and development.

***Importance of Technology*:**

1. Importance of technology lies in its diffusion (spreading).
2. Technology increases GDP (Gross Domestic Product).
3. It improves the product and service qualitatively and quantitatively.
4. It improves standard of living.
5. It makes country self sufficient.
6. It is designed for increasing the power of human beings over their meterial management.

9.12 IMPACT OF HIGHLY CAPITAL INTENSIVE TECHNOLOGY ON EMPLOYMENT SITUATION OF COUNTRY

Capital intensive technology is highly useful and do not cause unemployment in so called developed European countries but in Asian countries amongst which Indian economy is at the top suffer from unemployment of diversified capital intensive technology. Reason being that ours is the second largest populous country in the world having more than 100 crores of people. We are already having disguised, white colour, blue-colour as well as seasonal unemployment. Capital intensive technology can further aggravate it. Any type of mechanization can further cause unemployment. The unemployment shall be caused in all the sectors of the economy come what nature it has. It can and is causing unemployment in primary, secondary and tertiary sector.

9.13 TECHNOLOGICAL FORECASTING AND UPGRADATION

Prediction about the future use of technology is called *technological forecasting*. This is important for maintaining sustainable growth rate. Even substituting one technology for another technology serve this purpose. According to Fisher and Ray substitution proceeds at a rate given by the relation, between forecasting and up gradation. Equationaly it is started below:

$$F/(1-F) = \exp\{a(T-T_0)\}$$

where F is fraction of the new product or service that has substituted at a time T; T_0 is the time when the substitution is half complete; a is a rate constant that can be defined in terms of the time, T, required for the substitution fraction, F = Substituted total to 80 from 0.1 to 0.9, *i.e*, $a = 2$. In (9/DT).

A number of diverse technical advances follow this simple law. Substitution curves (F versus T) may be plotted and used for technical forecasting.

Advancement to higher level of technology is called ***technological up gradation.*** Technological up gradation increases productivity to meet the enhanced needs of the society. But up gradation of technology must be suitable to the society. If technology is upgraded as it will increase productivity then two variables are essential for its implementation. Firstly we require highly skilled manpower and secondly backup service. In the absence of both up-gradation of technology should be avoided as it will create more troubles then results for which we aspire. Accept/reject should be followed while accepting technological forecasting and its upgradation. Technological planners having knowledge of societal position of the country should be consulted.

As already started that technological involves collaboration of the human hand with the human brain to design those tools and techniques the use of which enables us to extend (increase) beyond the natural capacity body. No technology can come up without understanding the connections between the elements of nature on the one hand, and the capacity of manipulating them to achieve human ends on the other.

Transfer of technology means communication of knowledge about techniques of doing things. This has been in existence since the birth of the tools making and domestication of animal. Man being a social animal, can watch, communicate and intimate others. This even other animals do, but the difference between man and animal is that man possesses symbolic systems of communication which animals do not. Technology is prior to the use of both. The techniques are put together to produce systems of technology of varying degrees. Though all technological system involve mindful understanding, yet modern technological system have ended to become increasingly dependent on scientific modalities (methods/procedures) of understanding. Three kinds of things are involved in case of scientific modalities and they are:

1. **Firstly,** logic of organization of technological systems which today depends more on the basic methodology of science than it did in the past. In olden times, technological system could be developed around the impression of thumb, prediction and feeling of master craftsmen and sacred conventions. Modern science and technology do not fully depend upon these, but reed experimentation and testing of the system.
2. **Secondly,** the level of knowledge of the technological system must be known to the engineers so that they are able to design the high level performance and fine tuning of the systems.
3. **Thirdly,** the engineers who design the system have to work not only in association with fellow engineers but also with scientists whose help is essential in understanding the problems found in high-technology system.

Now the question arises what exactly transfer is. In other words what is covered under the concept of technology transfer? To be very brief there are ***three elements of modern technology such as;***

(*i*) Product design

(*ii*) Production techniques

(*iii*) Management of systems.

Product design range from smallest thing like in to ship whereas production techniques include operational methods. Management system constitutes plans, layout, technical equipment maintenance and market research. In a good technical system three things are incorporated.

Firstly, general knowledge which cover information common to an industry pertaining to reading of blue print, tools and fixtures.

Secondly, under system specific information about industrial capability available in a firm. It also includes capability of finding out solutions to problems on the basis of experience obtained in previous manufacturing system related to production.

Thirdly, we have firm specific knowledge which differs from system specific as it cannot be attributed to a particular production item but results form firm's overall activities in some areas. This type of technical knowledge goes beyond the general level of knowledge that an industry possess.

Understanding the process of transfer of technology involves knowing the manner in which technology develops from the initial idea to the final product. It is not necessary that the idea comes only to the mind of an engineer or a scientist. A salesman, a consumer or, in fact, any concerned individual could give birth to an idea, but its materialization and management involves the skills of a host of individuals.

Transfer of technology implies the transfer of knowledge or knowhow. Transfer of technology is an exchange of expertise when the holder of technology deliberately changes the level of exchange with the recipient of technology. Levels of exchange can be of the following types:

(*a*) Where a technologically advanced, supplier sells consumer goods produced in exchange for money.

(*b*) Where an advanced supplier sells machines to produce industrial consumer goods in exchange for primary products or consumer goods.

(*c*) Where an advanced supplier sells machines to produce other machines alongwith the technology and the system of maintenance. The deal then is a turn-key plant for money.

(*d*) Where an advanced supplier sells the know-how of producing a particular industrial product the purchaser is capable of mobilizing the industrial plant level technology to become self-sufficient in the manufacture of the end product.

The 'status' of the recipient of the technology at every successive 'level of exchange' shows improvement vis-à-vis the supplier. His assumption here is that, the more 'science –based' a technology is, the higher is its market value compared to the labour-intensive technologies.

It is obvious that increasing the 'level of exchange' would be possible only if the supplier feels confident that the purchase of the technology will not be a threat to the supplier's market. On the other hand, he may feel that by up scaling the technological demand level of the market for which the technology is being supplied, the supplier of technology may feel that the technology level of the recipient himself is already high enough so that in case he does not sell the technology, it would be developed by the recipient itself in a relatively short period of time. By with holding the sale of the technology, the supplier would only have lost the chance of entering the market of the recipient with his technology even though indirectly and in collaboration with the recipient.

'**Technology**' is designed and developed for increasing the power of human beings over their material environment. Consequently, control over technology gives one advantage in economic exchanges. Marx had understood the central significance of such a control in the process of the establishment of the structure of relations between groups and classes. Control over the hardware aspect of technology is easy to establish and institutionalise in terms of private property. Control over the 'knowledge' aspect as private property was expressed, in pre-modern times, by the master craftsmen either by being secretive about the special craft technologies or through guild (group) and case monopolies. Modern societies have, however, had to evolve a more complex way to deal with the problem of the social need to spread technology, and the private interest of the possessor of technology to maintain personal monopoly over it. The institution of patenting is one of such device. Besides patenting of innovations and inventions, the owners of technologies have evolved other modes of restricting the rapid spreading of their inventions such as:

(*a*) Making the hardware of production technology more complex than it need to be, in order that it may be more difficult to copy it.

(*b*) Keeping the overall lead in the production, distribution and maintenance systems of the products such that their competitors are always at a disadvantage if they merely copy their technologies.

(*c*) When the technology is sold to a potential competition, an attempt is made to arrive at such agreements which tend to perpetuate future dependency in the related development in technology.

(*d*) Cartels (group OPEC is international carte) are created, so that a small number of firms control the distribution of technology.

(*e*) Secretiveness is maintained as regards crucial invention and innovations until developments have taken place all along the innovation chain.

This is not concerned only with the technological development a nation has achieved but with a number of other issues also, such as internationalization of the fundamental values of rationality, punctuality, honesty and commitment to work. Thus, even when the productive capacities of Europe and Japan were shattered (destroved) at the end of World War 11, their levels of techno-culture were by and large was unaffected (had no bad effect). So due to the existence of techno-culture their material aspects were recreated with greater vigour and strength.

The fundamental values underlying he techno-culture levels of different societies are:

(*a*) Care for detail

(*b*) Quality of product

(*c*) Commitment to finish the work on time

(*d*) Risk taking and innovativeness (discovery/invention), and

(*e*) Honesty in dealings while doing exchange or transfer of technology.

Quality control is an important aspect of techno-culture. Quality control depends upon the morals of those who monitor (supervise) it. Societies which do not provide autonomy (independence) to their professionals (technicians, scientists and engineers) engaged in the monitoring of standards may never reach high techno-economic-cultural levels. A constant check is needed to avoid the monopoly of technocrats. Transfer of technology at this simplest level, namely in the transfer of tools and turn-key plants is possible even when two societies are at different techno-cultural levels.

The major transfers of technology occur across the following institutions:

1. Between Research and Development (R&D) institutions and the users, including private and public sector industrial and trading corporations.
2. Between the R&D Cells within a corporation and their production departments.
3. Between different corporations within a country.
4. Between one country and another country.

5. Between the present multinational companies and their subsidiaries in other countries.

These transfers of technology occur because there is a feeling of gain from he deals between the donors or sellers and the recipients or buyers of technologies. When the innovation is not encashable in terms of the sale of the final product, as in the case of an R&D institute or a multinational corporation which may be restrained by the government of a third-world country to sell its hardware or industrial unit on a turn-key basis or finished products, the only alternative may be to sell the technology itself. Some of the possible modalities (methods) of technology transfer are:

(*a*) Lump sum payment,
(*b*) Licensing,
(*c*) Equity participation in the recipient company,
(*d*) Periodic payments for specific consultancy services to the transfer or as and when required.

In return the purchaser can expect the following kinds of facilities:

(*i*) One-shot transfer of design and process specifications of the technologies under the guidance of technical experts or in-house training of the experts of the purchase

(*ii*) Periodic supply of expertise at different stages of the establishment of technology.

In the case of transfer of technology between the buyers and purchasers who are at very different levels of technological development, the transfer of hardware technology or even turn-key plants may occur along with the stage-by-stage transfer of software expertise in the following stage:

(*a*) Know- how to run and maintain the hardware is transferred along with the hardware.

(*b*) Know- how to assemble the hardware is transferred along with the hardware in Semi-Knocked Down (SKD) packages.

(*c*) Machine and process specifications are sent along with expertise to manufacture these and test them for their quality. Instrument for testing are identified and expertise transfers.

(*d*) Quality control and R&D cells are established in the purchasers' factory under the guidance of the experts from the buyers.

(*e*) Training of experts of the buyer is arranged or facilitated by sellers.

9.14 METHODS AND MODES OF TECHNOLOGY TRANSFER

Developing countries of the world are keen to have right type of technology from developed countries because its use helps these countries in removing poverty. The underdeveloped developing countries also want to have that technology which is most suited to their socio-economic need. A number of modes and methods involved in transferring technology are stated below:

1. ***Foreign Investment Agreement:*** Foreign investment is an important mode of technology transfer. Under this an agreement is made and signed by both the countries as suitable to both after discussions and negotiations. This agreement can be of many types. If foreign investor agrees to invest in terms of money then quite possible that the recipient country provides the balance capital in the form of infrastructural facilities and both skilled, trained and untrained manpower. The proportion of equity is and in nature of control between the two countries which is very well mentioned in the agreement backed by law. In such agreements the foreign investor has usually upper hand because it is he who provides technical know-how necessary for

exploiting natural resources and other resources, which local partner cannot provide. A number of such agreements have been made between India and USA and other European countries.

2. ***Import of Technical Know How:*** Under this methods technologically backward countries purchase technical know-how from advanced countries to be used by them in their country. The technology can be straight way bought by one country from another or it can be on the basis of percentage also called royalty depending upon the turn over. This is necessarily done because technology developed by an exporting country is a patented property.
3. ***Turn key Project:*** This is quite popular method of technology transfer being used. Under turn key jobs the foreign firms install plants with local firm. The foreign firm is given the job to plan and engineer the whole plant. Besides installation and erection work, it provides initial training to the local personnel. This method of technology transfer ends responsibility with the ultimately commissioning of the plant. It is different from technical collaboration.
4. ***Franchise:*** This is a form of licensing; however, the source usually provides some type of continual support to the receiver, for example, by supplying materials, marketing support, or training. This channel is commonly used in food chains.
5. ***Joint Venture:*** Two or more entities combine their interests in a business enterprise in which they can share knowledge and resources to develop a technology, produce a product, or use their respective know-how to complement one another.
6. ***Foreign Direct Investment (FDI):*** A corporation, usually a multinational, decides to produce its products or invest some of its resources overseas. This permits the transfer of technology to another country, but the technology remains within the boundaries of the firm (*i.e.*, is still controlled by the firm). This type of investment has advantages for both the investor and the host country.
7. ***Government to Government Transfer of Technology:*** Technology can be transferred from one government to the government of other country. This mode of transferring technology is much more easier as compared to private sector of one country to another. Under this system of technology governments sign agreements with each other and agree to transfer technology on certain terms and conditions already decided and clearly defined. The lending governments try to dictate terms and take political advantages also. Such agreements involve stage financed projects. For example setting up of steel mill, heavy engineering projects, river valley projects, public health services and power stations etc.
8. ***Centralised Purchase of Technology:*** It is another method of technology transfer being used. A state controlled organization in the host country purchases technology rights from other country. It is modified suitably and given to firms in the country. By doing so more than one firm in the host country is benefited. Such technology is usually bought at nominal cost. The biggest advantage of this is that central organization in consultation with local firms can develop research institutes in the country to achieve degree of self-sufficiency over a period of time.
9. ***Technology Transfer Through Training and Education:*** Developing countries are very keen to have advanced technology from technologically rich countries, because the use of advanced technology can end poverty and provide better and higher standard of living. Such countries are technologically quite backward but also lack technical education seriously. To over-power this problem technologically qualified persons can be sent from developed nations to under-developed nations to impart necessary education and training so that after being trained they are in a position to give training to other persons in their country too. Such programmes of

technology transfer are permissible and international agencies like UNDP (United Nations Development Programme) finance these.

10. ***International Conferences, Seminars and Symposia:*** Exchange of useful information and dissemination (spread) of latest technology in itself is no less important in the matters of technology transfer. Such information promote local research and development activities. A number of conferences are being held from time to time in different countries of the world and personnel are invited from a particular region. The expenses are met either by host country or participating countries or shared between both. So seminars and international conferences in this way keep abreast (to keep aware of) of the latest technological developments. A large number of autonomous organizations are engaged in this work. Minutes and proceedings on latest technology developed are published for users on payment basis.
11. ***Technical Consortium and Joint R&D project:*** Here, two or more countries collaborate in a large venture because the resources of one are inadequate to effect the direction of technological change. Typically this type of venture takes place between two countries or two large conglomerates (groups). For example, a consortium (partnership) was firmed between France and England to develop a supersonic plant (the Concorde). Both nations needed to combine their technical and financial resources to develop expensive technology and, in the meantime, to compete with their rivals in the United States.

9.15 PROBLEMS

The supply of R&D innovations from government labs to the private industry is based on the assumption that the transfer of technology from the labs to the industry is a national benefit, and hence it need not be priced in terms of the "unit cost plus" principle which would be the case if the R&D happens to be commercial venture.

When R&D is located within a research laboratory, the transfer of technology takes on varied problems. These problems arise from the following types of situations:

(*a*) The technology developed by the R&D units has to be communicated to the user departments. What is the most appropriate form of linkage between the two to take care of the organizational disturbances which may arise with the introduction of the new technologies?

(*b*) What is the most appropriate system for the communication of user feed-back and forward market demand assessments, so that R&D gears itself to provide for current and future user needs? This involves relations between R&D, marketing and quality control departments.

(*c*) How is one to assess the technological capability of a production set-up to absorb technology brought in from outside the country? This kind of a question arises often in situations where a foreign technology is purchased and has to be adapted or indigenised.

(*d*) Transfer of technology very often gets mixed up with the ideology, self-sufficiency, socialistic pattern and freedom of competition. Why does this happen? To understand this we must have a clear idea of the actors involved in the process of technology transfer along with their interests and ideologies. For example transfer of agricultural technology involves the interests not only of the agricultural scientists and farmers but also of the producers of chemical fertilizers and pesticides, farm machinery, the whole group of ***bureaucracies*** (IAS officers etc.), politicians, the traders and the humble agricultural labourers.

(*e*) Similarly, in the case of transfer of imported technology the interested actors are not only the capitalists who purchase technologies across the international borders but also the policy makers, as they have to take the view points of scientists and engineers as well as the industrialists and capitalists because they feel threatened by the imported technologies. If imported technology is used then the indigenously developed technologies will be put out of use.

(*f*) In matters of transfer of technology, labour unions are also involved because advanced technology and its use threatens employment opportunities and prospects.

(*g*) New and imported transferred technology causes new types of social and ecological disturbances.

9.16 TRANSFER OF TECHNOLOGY FROM LABORATORY TO FIELD

Transfer of technology from laboratory to industry producing unit is not so simple and automatic. A lot of efforts are required in this direction. Following are the factors which are normally considered for the transfer of technology from laboratory to field.

Choice of Technology: Choice has to be made with regard to technology available before it is transferred to industry. The choice has to be based upon the following factors:

(*a*) The technology selected for transfer should be appropriate and the purpose for which it is intended to be transferred.

(*b*) Status of development is another important issue with regard to transfer of technology.

(*c*) The technology selected for use should not be so much expensive. It should be competitive otherwise it will not be commercially accepted.

(*d*) The technology to be transferred should be provided with the back-up services. The user of technology should be provided with the services like planning, selection of equipment, layout of the factory, erection, commissioning and trouble shooting during operation.

(*e*) Evaluation of technology is important before it is transferred from lab to field or industry. This evaluation has to be done to find out whether the claims made are justified by actual facts. If they are not then that very technology should be dropped.

(*f*) Transferring agency may be like NRDC is essential because technology transfer is not only a matter of technology but also a matter of commerce and law.

(*g*) Provision of guarantees with regard to performance of the technology on commercial-scale is equally important.

9.17 PITFALLS OF TECHNOLOGY TRANSFER—FROM LABORATORY TO FIELD

The government of India established the CSIR (Council of Scientific and Industrial Research) in the late forties (1942) and its real functioning started in the sixties. There are today about 40 R&D (Research and Development) labs under the CSIR. Similarly we have the CMERI (Central Machanical Engineering Research Institute) and the NRDC (National Research Development Corporation). These labs are engaged in the transfer of technology form research units to production unit.

During themed-seventies, the NRDC was given enhanced funds to participate financially in joint private and public sector projects based on indigenously developed technologies. It could also underwrite investments related to a certain portion of the pilot plant and prototype level investigations.

The research and development system, thus, had an institutional mechanism for the transfer of technology from the lab level to the level of its final fruition as industrial production set-up. We must, however, keep in mind the fact that only a small proportion of innovations emerging from industrial labs, get ultimately productionised. There are a number of reasons for this phenomenon.

Firstly, the high cost of translating a lab or drawing-board level innovation to a pilot plant or prototype level. This stage of development entails serious financial risks. When the technologies are not proven, the preliminary investigations may turn out to be more expensive and or less efficient than the existing technologies in the market.

Secondly, most of the indigenous R&D depends upon and follow the ideas which are already being practiced abroad at the industrial level. The Indian entrepreneur, therefore, feels unhappy if he is asked to purchase an unproven modification of a technology readily available on payment from aboard.

Thirdly, the entrepreneur would like to the assured of the ready availability of the inventor in the initial phases of the trial and erection of the technology. This cannot be assured by the NRDC or the labs since very often the scientists may no longer be even on the rolls of the labs.

Fourthly, Direct and Indirect cost involved along with restrictive conditions possess serious implications for the developing countries. The direct cost of imported technology is of very high order in the form of royalty and technical fees etc. the high foreign exchange cost of technology transfer is likely to impose heavy burden on the balance of payments unless the transfer make favorable effect on the foreign trade sector.

The heart of the R&D efforts are the research institutes. It is in the institutes that the generation of ideas first take place, leading to their preliminary translation into a series of experiments which again need to be integrated into broader interdisciplinary strategies for the designing of the pre-commercial ***upscaling*** (pilot plant or prototypes). Developmental research involves many disciplines and support facilities and their effective coordination before a project qualifies to advance into the successively more complex stage of the innovation chain.

Care should be exercised and an assessment be made before a development work is undertaken.

Kind of Organization Most Conducive to Innovate Fruitful Research

From the very nature of developmental research, it is obvious that it should have the following attributes:

(*a*) ***Flexibility:*** This means that the innovator should have easy access to the equipment, chemicals, raw materials, skilled help and finances so that he may be able to smoothly change over to more effective directions of his other colleagues, is likely to promote innovativeness since one's best ideas are often generated, by causal remarks made by one's colleagues. Moreover, let us recognize that R&D today involves teams of different specialists working on a single set of problems. Team work based on oneness is conducive to cross fertilization of ideas.

(*b*) ***Openness:*** A organization, which encourages open exchange of ideas and finds ways and means to appreciate the work or effort of an individual scientist and/or technologist within a term as well as that of his other colleagues, is likely to promote innovativeness since one's best ideas are often generated, by casual remarks made by one's colleagues. Moreover, let us recognise that R&D today involves teams of different specialists working on a single set of problems. Team work based on oneness is conducive to cross fertilization of ideas.

(c) ***Trust:*** Hierarchic bureaucracy assumes lack of trust of the individual scientists. This has to be replaced by trust. This also means predominance of objectives. A constant battle between the '**administrators**' and the '**intellectuals**' can often demoralize the latter. If the system supports rule boundless, the scientists would soon learn to be bureaucrats themselves. Research would, however, be the real victim.

In view of this, following points must be taken into account:

1. A clear-cut identification of developmental projects for each laboratory based on the national identified technological areas is required. It is one area in which the project teams would have do the self-evaluation as well as evaluation by the research team. Based on such a close periodic scrutiny, the project team could be given either enhanced support for research or disbanded or allowed to continue with the same level of funding.
2. Strengthening of the infrastructure for development whenever such infrastructure is found to be weak for carrying out the developmental stage of the projects.
3. Allow the scientists to take up less expensive appropriate technology relating to R&D projects. These projects should be time bound and should undergo progress evaluation as in (2) above.
4. It is necessary that, while scientists should be asked to be cost and time conscious, there should be an encouragement for greater initiative on their colleagues in the industry. This would be helped if the anomalies (possibilities) regarding travel expenses etc. are removed and made more generous. It is essential to build up an atmosphere of trust in the scientists so that bureaucratic paper work is reduced to the minimum.
5. Market intelligence work needs to be strengthened in each laboratory. Based on the market demand and potential, 'task force' teams of scientists, engineers and workmen could be established to undertake specific technological tasks. They could have part time cons[illegible] m the industry or other private consultancy organizations. It may also be n[illegible] en[illegible]fy entrepreneurs who could take up the task of producing the end result to the [illegible]. Th[illegible]e, entrepreneurs could also be associated with the work of the 'task forces'
6. The personnel rules should be amended to permit, on a long term [illegible] R&D personnel to go along with the entrepreneurs, after the laborato[illegible] less finished, and help the entrepreneurs in the stage of setting up the [illegible]lines. [illegible]is could also facilitate the scientists in taking active interest in the work [illegible]untary organizations interested in the appropriate technology package.
7. Since a large number of technologies are being imported with the intention of indigenizing these technologies, the research institutes could be associated with the task of doing adaptive research and development. This could best be done if the industry and the institutes set up collaborative consultancy and time bound programmes. The scientists and technologists could be given an attractive percentage of the estimated value of the considerable portion of their maintenance costs by participating in such consultancy operations.
8. There is need to streamline the system of collaboration with industry. While unnecessary paper work and the involvement of third parties such as the NRDC needs to be reviewed. There is also a need to ensure that the legal rights of innovators and the institutes are protected against some of the not so honest industrialists.
9. Most of the small farmers and industrialists have problem in adopting new technology due to costly inputs and they are not readily available.

10. Adoption of technology also depends upon a number of factors such as irrigation and credit facilities because modern technology is quite costly.
11. The risk involved is the main cause of not permitting the new technologies in case of small farms. The heavy capital investments and the high variability of yields under unpredictable weather conditions make the farmers not readily accept the new technology.
12. No single attitude, trait, factor, or farming condition explains the patterns of small farm adoption of new technologies.
13. Lack of awareness and knowledge about the technology is also an important reason for the non-adoption of technology.
14. The new technologies which recommend the use of heavy doses of fertilizers are sensitive to rainfall. The technologies developed and tested under adequate moisture conditions did not perform better than local technologies when subjected to rainfall deficiencies.
15. The availability of various inputs and other biological factors such as climate and availability of water particularly to farmers is also a great obstacle in the way of adopting technology.
16. In certain regions of underdeveloped and developing economies a majority of the farmers belong to tribal communities. It was also found that with higher tribal population only about 13 percent of the small farmers adopted new technology with regard to the maize cultivation. Even among those who adopted new technologies, a majority did not follow completely the recommended practices. The problem with the tribal farmers was not only a lack of resources but also a lack of extension efforts directed towards them.
17. There is a also a need for close interaction, association and discussion between the scientists of research institutes and the small farmers, so that the problems of small farmers can be better understood and appropriate technology suiting to their conditions can be developed. The scientists have to interact with small farmers so that they are able to identify and popularize the technology generated by farmers through their (by farmers) informal research and development. As has been said earlier, farmers do not accept the technology passively; they make suitable alterations or modifications to suit their local conditions and environment.

QUESTIONS

1. What is transfer of technology? What are the pitfalls of technology transfer?
2. What prior steps should be taken before transplanting new technology in backward rural areas?
3. Explain the role of technology in assessment and control of physical environment.
4. What do you mean by technological up gradation? Has it got any effect on ecology and environment?
5. Discuss the importance of technology in planning process of a nation.
6. What is the need of technology transfer?
7. What problems are faced by the technocrats during technology transfer?
8. What kinds of organizations are most conducive for innovative research?

10

Social Responsibility of Business and Social Audit

10.1 INTRODUCTION

Public demands on private business have expanded manifolds in recent years. Business enterprise was once expected to denote its efforts to producing and distributing goods and services as efficiently as possible and to making innovations and improvements in products and processes with the ultimate objective of maximization of profits. Now, business is increasingly called upon to promote a variety of social purposes in addition to these long standing economic ones. The demands for "social responsibility" have provoked enthusiastic discussion and debate on what new roles. If any, business firms should play in the social system.

Social responsibility is the personal obligation of every one as he acts for his own interests to assure that the rights and legitimate interests of all others are not impinged. **—Koontz and O'Donnell**

Social responsibility of business refers to the obligation of business to pursue those polices to make those decisions or to follow those lines of action which are desirable, in terms of the objectives and values of the society. **—Howard R. Bowen**

Conceptually social responsibility may be taken to mean intelligent and objective concern for the welfare of the society. **—K. R. Andrews**

In real sense, the assumption of social responsibilities implies recognition and understanding of the aspirations of society and determination to contribute to its achievement. **—George A. Stenier**

10.1.1 Obligation of Business Towards Different Groups

Business is responsible to several social group *e.g,*:

1. Responsibility to shareholders or owners
2. Responsibility to employees
3. Responsibility to customers
4. Responsibility to creditors and suppliers
5. Responsibility to community

6. Responsibility to government
7. Responsibility to society.

1. **Responsibility to Shareholders or Owners :**
   (*a*) Committing funds in the best possible manner
   (*b*) Ensuring a fair rate of return
   (*c*) Fair and honest reporting of business operations from time to time.
   (*d*) To ensure safety of investment
   (*e*) To offer reasonable appreciation of capital
2. **Responsibility to Employees:**
   (*a*) Recognise the social needs of workers and provide adequate participation to employees in matters affecting their life.
   (*b*) Fair and reasonable rates of pay.
   (*c*) Creating a healthy work climate, improving working and living conditions and making the work place safe and pleasant.
   (*d*) Improving the quality of working life of employees.
   (*e*) To provide reasonable opportunities for the advancement of employees in their career.
   (*f*) To protect trade union rights of workers including the right of participation in management, etc.
3. **Responsibility to Customers:**
   (*a*) Providing goods of quality at reasonable prices.
   (*b*) Avoiding deceitful and false, highly exaggerated advertisements and high pressure publicity Gimmicks aimed at wooing the customers.
   (*c*) Management should not indulge in antisocial activities like black marketing, boarding, etc.
   (*d*) To provide products and services this can satisfy the needs and desires of different classes and tastes of consumers and with different purchasing power.
   (*e*) To ensure regular and adequate supply of goods particularly during period of shortages.
   (*f*) To provide goods of proper quality which provide more satisfaction to consumers or match their needs.
   (*g*) To serve as a friend and guide by providing prompt and courteous service.
   (*h*) To ensure a fairly wide distribution of products among all sections of consumers.
   (*i*) To disseminate useful information to consumers about the company, new products and services, new uses of existing products etc.
4. **Responsibility to Creditors and Suppliers:**
   (*a*) Provide accurate information regarding the financial health of the organization
   (*b*) Ensure a reasonable price for the articles supplied, and make prompt repayments (involving interest an borrowings); there should be fairness in transactions
   (*c*) Promote a healthy atmosphere where creditors, suppliers and other interest groups are treated as partners in a co-operative endeavor.
5. **Responsibility to Community:**
   (*a*) Develop constructive relationships with those in the community
   (*b*) Participate in community activities and promote community welfare. Renovating neighbourhood's surrounding the company's head quarters, building public parks with

play ground equipment, and granting employees paid leave of absence to work in social programmers are examples

(c) Provide safe, livable communities with good housing and efficient transportation.

(d) To avoid extravagant living standards and vulgar display of wealth.

(e) To promote social institutions like education, hospitals, sports recreational organizations, religious institutions etc.

(f) To help weaker sections of society and backward regions of the country.

6. **Responsibility to Government:**

(a) Follow fair trade-policies and practices.

(b) Pay taxes to the government honestly.

(c) Obey the laws.

(d) Discourage unhealthy practices like bringing the government officials to carry favours, obtain licences in order to kill the competition, etc.

7. **Responsibility to Society:**

(a) Elimination of poverty and provision of quality health care.

(b) Preservation of the environment by reducing the level of pollution.

(c) Providing sufficient number of jobs and carrier opportunities and facilities for all members of society.

(d) Improving the physical environment and developing human resources fully. Some of the examples in this regard may be : conducting educational programmes to combat drug abuse, provision of seed money to minority suppliers, urban rehabilitation, loans to construct low cost houses, providing managerial training to small entrepreneurs, establishing manufacturing plants in economically backward regions, etc.

In India also a large number of companies like TISCO, DCM, Mafatlal group, Hindustan Steel Ltd., Escorts Ltd., Godrej, Kirloskars, Ambanis etc., have accorded a place of importance to social action programmes. Corporate social responsibility is seen as a relevant constraint to business and a large percentage of managers assign a high place for social responsibility, responsibilities with that of profit. However, the process of social commitment in India is some what different like starting social trusts, anti pollution measures, adopting villages or all around progress and development, starting family planning clinics for the benefit of workers and residents, allocation of a fixed percentage of profit to trained unemployed engineers, technicians and community development activities like provision of drinking water facilities, conducting social audit on a voluntary basis, providing medical, recreational facilities, conducting tournaments to promote sports talent, undertaking consumer education campaigns, avoiding unethical, deceptive advertising and so on.

10.1.2 Social and Economic Objectives of Business

The purpose of business enterprise is to produce quality goods and distribute them and do it ethically so as to serve the general welfare. Since the economic and social objectives of business are not mutually exclusive, but are intertwined, the management has both social and economic responsibilities. In this context the function of management is to see that the business enterprise fulfill its purpose by paying investors a fair return on their capital, by paying workers a fair compensation for their effort, by paying vendors a fair price for the materials and supplies, and by "paying" the consumers in terms of reasonable price, quality and service and above all by helping our society to function in harmony as an organic whole.

This function imposes upon the business and its managers a higher responsibility than that arising out of private property and self interest. Self interest and public good cannot be kept apart. The manager must assume responsibility for public good, because management's responsibility to the enterprise and to society "is inextricably interwoven. This is bases on the fact that the enterprise is an organ of society and that its actions have a decisive impact on the social scene. If a business provide a product or service that is useful or other wise of value to the community, then it is socially desirable that it continues to function.

10.1.3 Arguments for Social Responsibility of Business

1. Public needs have changed, leading to changed expectations. Business, it is suggested, received its charter from society and consequently has to respond to the needs of society.
2. Social involvement discourages additional government regulation and intervention. The result is greater freedom and more flexibility in decision making for business.
3. Business has a great deal of power that, it is reasoned, should be accompanied by an equal amount of responsibility.
4. Social involvement may be in the interest of stockholders.
5. Business try to solve the problems that other institutions have not been able to solve.
6. Business has the resources. Specifically, business should use its talented managers and specialist, as well as its capital resources, to solve some of society's problems.
7. Social involvement creates a favorable public image. Thus, a firm may attract customers, employees and investors.
8. It is better to prevent social problems through business involvement than to cure them. It may be easier to help the hard core unemployed than to cope with social unrest.
9. Problems can become profits. Items that may once have been considered waste (for example, empty soft drink cans) can be profitably used again.
10. The creation of a better social environment benefits both society and business. Society gains through better neighborhoods and employment opportunities: business benefits from a better community, since the community is the source of its work force and the consumer of its products and services.
11. Modern society is an interdependent system, and the internal activities of the enterprise have an impact on the external environment.

10.1.4 Arguments against Social Responsibility of Business

1. The primary task of business is to maximize profit by focusing strictly on economic activities. Social involvement could reduce economic efficiency.
2. Social involvement can create a weakened international balance of payments situation. The cost of social programs, the reasoning goes, would have to be added to the price of the product. Thus, companies selling in international markets would be at a disadvantage when competing with companies in other countries that do not have these social costs to bear.
3. Business has enough power and additional social involvement would further increase its power and influence.
4. There is not complete support for involvement in social actions. Consequently, disagreements among groups within different view points will cause friction.

5. Business people lack the social skills to deal with the problems of society. Their training and experience is with economic matters and their skills may not be pertinent to social problems.
6. There is a lack of accountability of business to society unless accountability can be established, business should not get involved.

Social responsibility practice in Indian organizations presents a mixed picture. There are so many organizations which have contribute a lot for the uplift of the society. On the other hand, many business organizations do not look beyond their immediate profit objective. As the economic conditions of the country has changed considerably, the organizations are not able to take the fullest advantage of the economic conditions like shortages, ignorance of general public and general seller's market. Here we presents the example of ITC and Tisco to suggest how these forward looking companies have taken social responsibility as a basic element of the decision making process.

All persons whether in business, government, a university or any other enterprise are concerned with ethics. Ethics is defined as *"the discipline dealing with what is good and bad and with moral duty and obligation."* The social responsibility of business involves ethics. All individuals whether in business or non business activities are concerned with some standardized form of behaviour, commonly known as ethics. Thus the *personal ethics* has been referred to as "the rules by which an individual lives his or her personal life, accounting ethics pertains to the "code that guides the professional conduct of accountants, Business ethics is concerned with truth and justice and has a variety of aspects such as the expectation of society, fair competition, advertising, public relations, social responsibilities, consumer autonomy etc.

Business ethics relate to the behaviour of the businessmen in a business situation. They are concerned with actions measured by ethical rules. Business ethical behaviour is conduct that is fair and just over and above the various rules and regulations. It is always ethical for a businessmen to obey the laws even though he may personally believe theme to be unjust or immoral.

Sources of Ethical Behaviour : Ethical behaviour on the part of the managers has its roots in numerous aspects of contemporary society. These are value forming activities of various social institution, values and goals of an organization, peers and colleagues in the organization, work and career, and professionalism these affects the ethical behaviour as follows :

1. Values and goals
2. Value forming institutions
3. Work and career
4. Professional code.

1. **Values and Goals :** There is an important relationship between goals and values of an organization. The goals of the organization are influenced by leadership, integrity, knowledge and skills, amity, influence and power, and survival. All these factors change the goals of the organization and consequently the expected behaviour from individuals.
2. **Value Forming Institutions :** The life and development of a society are both based upon and produce values. There are so many organisaions like family, school, state and religion which are the basic sources of values for persons. The basic purpose of value formation by these institutions is that they prescribe what is good or bad for an individual.
3. **Work and Career :** Work consists of the tasks or responsibilities associated with a particular job or position in an organization. In organization the work is organized, directed at the achievement of a desired end. An individual's experience over a period of time constitute his

career. Work and career create special values that give unity, cohesion and meaning to persons and groups.

4. **Professional Codes :** Professional codes are an increasing source of ethical norms for managers in business organizations. These codes are consistant with customs, mores, beliefs, social values, public laws and generally desirable behaviour. Following is described the code of conduct by All India management association for its professional managers.

10.2 ALL INDIA MANAGEMENT ASSOCIATION CODE OF CONDUCT FOR PROFESSIONAL MANAGERS

Guide to Good Management Practices: A professional manager is primarily responsibly to his organization. He is there after responsible to the following six stake holders of his organisaiton:

(*i*) Employees
(*ii*) Investors
(*iii*) Suppliers
(*iv*) Customers
(*v*) Goverment
(*vi*) Community.

1. As regards his ORGANIZATION, a professional manager should:
   - Act both with loyalty and honesty in carrying out the lawful policy and instructions of his organization and not undermine its image or reputation.
   - Plan establish and review objective, tasks and organization structures for himself and his subordinates which are compatible with the codes and practice of other professionals in the enterprise and will help to meet the enterprises overall objectives, and by his own work and through leadership and co-ordination direct all available effort towards the success of enterprise rather than of himself.
   - Fully respect the confidentiality of information which comes to him in the course of his duties, and not use confidential information for personal gain or in a manner which may be detrimental to the organization for which he works or has worked.
   - Provide a working climate in which innovation and free exchange of ideas are encouraged.
   - In his contacts and dealings with other people, demonstrate his personal integrity and humanity and when called upon to given an opinion in his professional capacity shall, to the best of his ability, give an opinion that is objective and reliable.
2. As regards the EMPLOYEES, a professional manager should:
   - Set an example for his subordinates though his own work capability and performance, through his leadership and by taking account of the needs and problems of his subordinates.
   - Pay proper regard to the safety and well being of the personnel for whom he is responsible.
   - Train people as appropriate under to become qualified for higher duties, including his own positions.
3. As regards of the investors in his organisation, a professional manager should :
   - Ensure that the investors are given fairly accurate financial and other important operating information periodically.

- Ensure that surplus generated in the operations of the company is divided fairly amongst investors employees and for further growth opportunities.

4. As regards the CUSTOMER and SUPPLIERS, a professional manager should :
   - In no circumstances supply inherently unsafe goods or services.
   - Ensure that the terms of all contracts and terms of business be stated clearly and unambiguously and honoured.
   - Operate an organisation's policy or require one to be established, which excludes corrupt practice, with regard to the giving or receiving of gifts or benefits and not to tolerate any practice which could lead to commercial or other corruption.
5. As regards GOVERNMENT, a professional manager should :
   - Confirm to the National interest as expressed in Government policy.
6. As regards the community, a professional manager should :
   - Make the most effective use of all Natural resources employed.
   - Be ready to give professional assistance in community affairs.
7. **Peers and Colleagues:** An individual gets clue of behaviour from his peers and colleagues. He develops and applies beliefs altitudes and values derived from the group of peers and colleagues with whom he is associated. An individual in a work group tends to conform to the group's norms as long as he values the friendship and approval of his associates or tears the possibility that they will cut him from rights, privileges and benefits they can offer. Departure from values may invoice no immediate penalty, a sense of estrangement may follow. Therefore, the individual tries to follow that group norm of behaviour.

The ethical codes serve a number of purposes by specifying behaviour that is required for the protection and continuity of the professional groups as a whole. The various groups associated with the functioning of a professional group are served as :

(*a*) Employees of organization benefit by the codes because these provide guidelines for their behaviour. They can know easily what is expected of them.

(*b*) The organization can derive benefits from ethical codes as it requires less control of employees' behaviour because these may serve as regulatory measure for its members.

(*c*) Customers dealing with the organizations following some codes may benefit as they are sure that no undesirable transactions will be conducted.

(*d*) The industry may derive benefits from ethical codes as these provide for fair trade practices resulting into healthy competition among its various constituents.

(*e*) The ultimate benefit of ethical codes goes to society. Ethical environment of managers is a part of their social environment. The social cause is served better if all professional forming its part follow certain specified behaviour.

10.3 SOCIAL AUDIT

Business organizations, whether owned privately, publicly, or co-operatively, operate in a context that is much broader than the immediate marketplace. Their activities can have social, economic, and sometimes political consequences whose impact may be felon a local, national, or even international scale. Increasingly, various interest groups are calling on organizations to recognize the for reaching implications of their decisions and to account for their actions on the grounds of social responsibility. The concept of social auditing is a corporate response to the demand for such accountability.

Developed by the New Economics Foundation and Tradecraft Exchange, social auditing seeks to support, guide and define a process by which on organization identifies, measures, assesses and reports on its social performance.

A social audit is a process of evaluation and reporting which is similar, in some respects, to a financial audit but focuses on social rather than financial performance it also is a tool for evaluating how satisfactorily a company has discharged its social responsibilities. It enables the public as well as the company to evaluate the social performance of the company.

The motivation for change is, of course, not entirely altruistic, any organization will benefit from taking into account the expectations of individuals, groups, and other organizations that success of its operations

Why social audit is required?

The fact that business polices and practices have significant social consequences has been recognized since the industrial revaluation.

As a result of continuing efforts to hold organizations accountable to society for their actions, substantial improvements have been realized in many areas: for example, employment practices, safety in the work environment, job enrichment, disposal of Indus trial waste, management of natural resources, and consumer protection. It is apparent that the potential benefits of a social audit are very substantial. If well facilitated, social auditors from different organizations can become self-supporting for subsequent years. There are of course, numerous problems that remain, and constant monitoring is required to achieve and maintain responsible corporate behaviour.

The collection of information is an on-going process, often done in 12-month cycles and resulting in the organization establishing bookkeeping and the preparation of an annual social audit document/report.

Are there any universal principles, concerns, and values which guide the auditing process?

There are many social issues that can be identified as failing within an organizations sphere of influence. In France, a form of accounting is required by law for companies beyond a certain size, and other counties are considering similar legislation. In addition, the social sequences of any given activity or programs are notoriously difficult to determine, they may indeed be altered by the measurement process itself.

Social auditing information is collected through research methods that include social bookkeeping, surveys and case studies. Indeed for businesses that function in a highly sensitive social environment, social accountability is now a matter of survival, organizations simply cannot afford to ignore it. The majority may be considered to be of concern to organizations of all type.

The areas of social concern noted above constitute only a partial list. Many mangers, understandably, are intimidated by so broad a target. Examples are pollution, discriminatory employment practices, misleading advertising, the production and handing of toxic substances, irresponsible use of non-renewable resources, conflict of interest among corporate directors, business activities that intrude upon neighborhoods or otherwise diminish the quality of community life, and the impact of information technology on the lives of private citizens.

QUESTIONS

1. What are the basic social obligations of management? Do they conflict with the profit objectives of the firm?
2. Define social responsibilities of business towards the state.
3. Discuss the role of professional ethics in the discharge of management functions. What role is played by the professional code of conduct in developing ethical behaviour?
4. What do you mean by the business ethics? What are the different sources of learning ethics?
5. What do you understand by the Social audit? Discuss the nature, objectives and benefits of social audit.
6. Write brief note on following:
   (*a*) Social & economic objectives of firm
   (*b*) Arguments for and against the social responsibilities of firm.

11

INDUSTRIAL SAFETY AND HAZARDS

11.1 INTRODUCTION

The idea of safety is as old as the history of mankind. From ancient times, the man has been directing his efforts to protect himself and his belongings from danger; danger form nature, from animals and other men. Today he is no longer afraid of wild animals but is afraid of bacteria, viruses, diseases, forces of nature and accidents and he has to protect himself and his dependents.

The progress in industrial safety before 1911 was practically non-existent. Before the workmen's Compensation laws, all states handled industrial injuries under the "**common law**". Under the common law the legal defenses available to the management of industry almost ensured that they would not have to pay for any accidents occurring on the job. Under this common law system employees didn't automatically receive payments when injured on the job, as they do today. If employee could show any of four legal defenses he would not have to pay the injured employees for the injuries he suffered:

1. The employee himself contributed to the cause of accident.
2. Another employee contributed to accident cause.
3. The employee knew of hazards involved in the accident before his injury and still agreed to work in the conditions.
4. There was no employer negligence.

In the early years, plant conditions were worse. Factories were little more than shacks. Light, ventilation and sanitation were almost non-existent. Rest rooms were unthought-off. Two third of workers were women and children whose workday was 12–14 hours. Machine guards were unknown. Occupational deaths were frequent.

The state of New York passed the first workmen's Compensation law according to which, regardless of fault, management would pay for injuries occurring on the job. When management found itself in a position, by legislation of having to pay for injuries on the job, it decided that it would be financially better to stop injuries from happening. This decision of industry gave birth to the organized safety movement. In the early years of safety movement, management concentrated heavily, if not entirely, on correcting the hazardous physical conditions that existed. This resulted in reduction of death rate (deaths per million man hours worked). As time went on, legislation shortened working hours, established minimum working age for children and made some improvement in sanitary and working conditions.

11.2 SAFETY MOVEMENT

The safety movement developed in five distinct stages:

Firstly, the improvements of the environment or the removal of physical hazards, **Secondly,** the improvement of personal practices, which is sought through a combination of education and training. It was soon realized that further improvement lay in the better organization of safety in the labor management department of enterprise concerned. The **third** stage could be said to be first step in management's direct involvement. This led to setting up of various departments specifically charged with the responsibility of ensuring health of workers such as personnel, medical, safety, industrial hygiene and training. The **fourth** stage in development of safety movement is of particular importance, with the growing complexity of industry and the recognition that safety is an essential element in the profitability and the effective functioning of the enterprise, safety was accorded first rank status. It was accepted that safety can't be viewed in isolation but has to be considered as in essential element in an industrial approach to the planning, organization and operation of an industrial enterprise. Safety has to be considered right from the project stage and this has to be a continuous process all through design, construction and initial operation of the plant. The **fifth** stage is a distinct stage in the development of safety movement. It is in fact a co-operative effort on the part of industries to prepare codes and standards to pool and disseminate information of technical aspects relating to safety.

11.3 GROWTH OF SAFETY MOVEMENT

There have been continuous efforts from different areas for betterment of working conditions and ensuring safety of workmen. The various factors which have directly or indirectly contributed to this growth are as under:

1. Over these years trade union movement have grown and it is recognized as an integral part of social system. The trade unions have contributed to safety movement in many ways:
   - Their fight for better working conditions.
   - Their fight for better wages to workers.
   - Bringing of fresh legislations due to link in social and political circles and stricter enforcement of existing legal provisions concerning safety and health at work.
   - Involvement in case of compensation to injuries.
2. The safety legislations has undergone revisions during past few decades specifying better conditions of work and increased responsibility for safety on the part of employer.
3. Emergence of new theories in the field of management giving importance to '**man**'. The safety and health of worker has assumed greater importance.
4. The all round consciousness generated and support from management and society have led to the formation of many voluntary bodies to promote safety and health in industry *e.g.*, the National Safety council of USA, Royal Society for prevention of accidents, Consumer Product Safety Commission (CPSC), Environmental Protection Agency (EPA), Food and Drug Administration (FDA), Occupational Safety and Health Administration (OSHA) etc.

5. Improvement of the standard of education of workers, which made them conscious of the hazards around and their right for protection.

11.4 SAFETY MOVEMENT NOW-A-DAYS

One of the main motivating factors for management's effort directed at accident prevention, today, are the statutory provisions, which are to be compiled with. With strict legislations, the government's announced policies in favor of vulnerable sections of community and the trade union involvement, the industrial management find it easier to abide by Safety laws rather than to face legal and other consequences of doing so.

Secondly there are considerations relating to be *prestige or image of the company.* A poor accident record reduces the prestige of the organization in eyes of public. The competitions and incentives provided by the government, national bodies and professional bodies also have helped in this respect.

Then there is *humanitarian angle.* The employer develops relations with his employees-as individuals and as a group. He cannot be immune to sufferings and pain of the persons injured. Taking an injured person to a hospital or meeting him or his family members after the accident is an unpleasant job for any employer, more so when he shares responsibility for the event. The moral responsibility for all accidents rests with the employer and failure to take reasonable precautions adds to it. This in many situations acts as a powerful motivator for management efforts directed at safety.

Material considerations are not least important. Now a days, an accident can be very costly to management-in terms of cost of medical aid, medical facilities, compensation cost, loss of production and so on. Damage to materials, damage to morale, upset delivery schedules can not be ruled out.

Over a period of years the responsibility for safety has shifted from the employee and the legal and moral responsibility for safety today lies on the shoulders of the employer. The current thinking is that the employer has not only to provide safe plant and equipment, but also create a climate where safe operations are possible. His responsibility also includes predicting the chances of human errors and taking reasonable steps to control them. Safety is no longer considered as a social cause, which the management should contribute to or support. It is now recognized as an action area where chances of accidents, creating safe working conditions and controlling human errors in the process of works. The safety function in industry today is not confined to preventing accidents at work only. It also includes:

- Control of occupational health problems
- Creating working conditions which are safe and foolproof
- Design of machines, equipment and work situations to suit the capabilities and limitations of man-both physically and mentally
- Controlling human factors responsible for accidents
- Creating a climate conducive to working.

One of the many achievements of safety movement is that it has generated voluntary efforts towards prevention of accidents in industry. Industries in both public and private sector have been encouraged and motivated to launch accident prevention programs with many of them achieving creditable performance. Creation of awareness about safety at national level is a yet another achievement.

11.5 SAFETY RESPONSIBILITY AND ORGANIZATION

This section deals with safety responsibilities and organization within industrial and commercial firms, public companies and enterprises owned by state and local authorities. They vary in size from very small firms to large multinational corporations.

11.5.1 General Responsibility for Safety

Employees at all levels are involved in accident prevention, but the main responsibility rests with directors and management who alone have the authority to issue orders and direction. The implementation of proper safety measures is sometimes resisted on the grounds that slow down or interface with production. Of course introduction if any new and safe method or piece of machinery is likely at first to cause a temporary drop in production when method is learnt and new habits are being formed. Soon however, production reaches previous level, employees find they are able to work faster, knowing that certain risks has been eliminated and that they no longer have to worry over them.

11.5.2 Management's Responsibility

Top management must accept full and definite responsibility and must apply a good share of attention to the task, just as it does to any other undertaking of vital importance. Every kind of work that men do involves some degree of hazard, and every uncontrolled hazard given time enough, will produce its share of injuries. But proper attention to safety will result in elimination of almost all the injuries that would otherwise occur regardless of industry of type of operation or the occupation. Management must eliminate injuries and make accident prevention a vital part of all activities.

In order to achieve adequate safety performance and prevent accidents management must:

1. Make it clear to all levels of supervision and to all workers that management is behind the safety program and that every supervisor will be held accountable for the success the program in his area of responsibility. The policy must be clearly set forth in write and must be brought to attention of everyone in the plant.
2. Provide a safe and healthy place to work, including:
    - Safe equipment and tools
    - Safeguard machinery
    - Process and operations planned and arranged with careful attention to safety
    - An inspections system to discover and correct all hazards.
3. Provide adequate training and supervision to develop safe work habits in all employee. Train supervisors in safety so they in turn, can train and supervise the people under them in safe methods of performing their jobs.
4. Organizing safety activities as a continuing function which includes :
    - Investigation of all accidents
    - Reporting of accidents to the proper authorities
    - Development of accident to the proper authorities
    - First aid treatment.

11.5.3 Responsibility of Safety Specialist

The safety specialist or engineer is employed by management and is a part of managerial supervisory staff directing the work of employees. The safety specialist shares the responsibility of management. Because of being specially qualified in safety work and often in direct charge it, the safety specialist has the opportunity of advising others. His main duties will be:

- Short term and long term planning of safety activities
- Monitoring safety performance continually
- Advising managers in matters relating to safety
- Supervision of the work of safety committees
- Conducting safety inspections and accident investigations
- Overall direction of activities aimed at prevention of accidents.

11.5.4 Responsibility of Supervisory Staff

The responsibilities of management also apply to supervisors and foreman. The person who gives direct instructions to employees is in a key position in reducing accidents. He is closely associated with the work force, knows them personally and is acquainted with their habits, grievances, attitudes and personal qualities. The foreman controls the workforce by example. This control is a most important factor in safety performance. The intelligent support of able supervisors and foreman is essential to any safety program. Proper training of supervisors in accident prevention method is essential.

11.5.5 Responsibility of Employees

The main responsibility of non-supervisory employee is to oneself and ones dependents, to avoid being injured and thus impairing the ability to support them. Employees have no authority to instruct fellow employees nor are they responsible for inherent hazards in the equipment, which they operate. But they have responsibilities to their employer for the safe conduct of their work and the maintenance of equipment:

1. Every employee should report unsafe working conditions, which are outside his own control to the supervisor.
2. They should not interface in the work of fellow employees in a way liable to create hazards.
3. They should not allow other employees to interface with their own work in such a way.

Despite their lack of authority, employees may have opportunities to assist their fellows by passing their knowledge and experience is safety matters and they should be encouraged to take active parts in safety meetings. Accident prevention is thus a co-operative task in which management has the main responsibility, employees must do their share, and the foreman or the supervisor bears most of the burden of detail.

11.6 OCCUPATIONAL HEALTH AND SAFETY PROGRAM

An occupational health and safety program is a specific plan of action to prevent workplace accidents, injuries and occupational diseases. Its objective is to integrate health and safety into the design of all workplace practices and all workplace conditions. There are many good reasons to make safety and health a top priority. The primary reason is to improve the protection of worker health and safety and dramatically reduce the cost of accidents, injuries and occupational illness. Employers in high-hazard workplaces must develop and put in place an occupational health and safety program.

11.6.1 Safety Program

Developing a safety Program does not have to be an overwhelming project. A safe working environment does not just happen; it has to be created. The degree of sophistication depends on your size and type of business. For example, if you only have a few employees and are in a service related business, not as many serious safety related issues would apply to your company as would to a manufacturing or construction company. The organizations with the best reputations for safety have developed well-planed, thorough safety programs. Concern for safety should begin at the highest level within the organization, and the managers and the supervisors at all levels should be charged with demonstrating safety awareness, held responsible for safety training, and rewarded for maintaining a safe workplace. ***Any safety program could be composed of the following elements:***

1. ***Written Safety Policy Statement.*** Your statement should define your company's commitment to your safety program.
   - It needs to be signed by the company's owner, president or other key senior management official(s).
   - Is should be put on company letterhead and posted where all employees can read it.
   - It should also be reviewed with all new employees at the time they are hired so they know that safety is an important concern to management.
   - Do not fall into the trap of outlining a list of safety rules that your employees will be expected to follow and use this as your Safety Policy Statement. Safety rules should be implemented separately. Your Safety Policy Statement should show your company's concern for your employees and commitment for protecting their safety.
2. ***Support by Management.*** For a safety program to be truly effective, it must start at the top. Safety needs to be considered a core organizational value that directly contributes to the overall success, growth and profitability of the company. Top management must lend a safety program its active support in order to survive and be effective. Safety is an investment in the continued success of the company. Top management should be:
   - Promoting safety as a key organizational value. This needs to be made clear to everyone in the company.
   - Authorizing needed resources to ensure the safety program is successful.
   - Attending safety meetings.
   - Doing periodic personal inspections.
   - Establishing clear management safety responsibilities for all levels of management.
   - Setting high goals and standards for performance.
   - Holding everyone accountable for the success of the program.
   - Measuring results and striving for continuous improvement.
   - Staying actively involved in the process.
   - Accepting no excuse for failure.
   - Constantly promoting the value of the safety program.
   - Enforcing safety policies.

3. ***A safety Director***. To get any program go all round, some other person must be given primary responsibility for its installation and maintenance. If the company is too small to appoint a safety officer, it is still important to assign some one the additional duty of promoting the safety effort. In the larger firm, a staff safety director, often called as *"safety engineer"* is usually appointed. He is responsible for overseeing and enforcing the company's safety procedures on all the company's jobsites, interfacing with governmental agencies like OSHA, updating the safety policy, educating clients about the firm's safety program, developing the training program, and handling the record keeping responsibilities.
4. ***Employee Involvement and Recognition***. The more employees are involved, the easier it is to change long-standing unsafe behaviors. If employees buy into the safety program, the entire company's safety effort will be easier. Remember that a genuine effort by management is necessary on a continuous basis.

   How to get employees involved?
   - Set up a safety suggestion program to solicit good ideas.
   - Review accidents with them and ask them for their recommendations.
   - Ask your employees for suggestions on improving the safety program.
   - Set up a safety committee or safety improvement team.
   - If your company is large enough, a safety committee or safety improvement team may prove an excellent way to get employees involved in the safety process. Select employees who want to be a part of the team, but make sure all employees have the opportunities for input.

   Activities of The committee:
   - Helping write up new safety policies.
   - Conducting new employee or ongoing safety training.
   - Conducting plant safety inspections.
   - Counseling other employees on safe work practices.
   - Helping the company conduct accident investigations.
   - Listening to employee complaints and working with management to resolve them in a positive manner.

   Conducting safety observations to ensure safe work practices are followed.

   In addition to involving employees in the safety program, it is important to recognize and reward employees for their efforts. Everyone likes to win and be associated with a winner. Make being safe a winning goal for your company.

   Other things you can do include:
   - Establish a program to identify, and formally recognize employees for excellence in accident prevention.
   - Recognize good suggestions.
   - Set company safety performance goals and reward employees upon successful attainment.
   - Comment positively to employees who are making the extra effort to be safe.
   - Put articles in the company newsletter about good safety performance.

   A pat on the back, a kind word, public recognition or rewards makes us feel better about what we are doing. If safety is important to the company, make it important to the employees. The money you spend will be returned many times in safer, more productive and happier employees.

5. ***Education and Training for all Employees.*** A large part of safety program must be devoted to the process of educating the employee to act, think, and work safely. Proper safety training is one of the most important aspects of an effective safety program. New employees need to be taught the company's safety policies and specific job safety requirements. The goal of safety training is not only to convey knowledge, but also to develop the proper attitude towards safety and establish safe work practices from the beginning. To ensure that new employees are put through the necessary safety training, and that this training is documented properly, New Employee Safety Orientation Guidelines should be developed.

   In addition, your training program should include:
   - Ongoing refresher safety training.
   - Job specific training when employees are transferred to different jobs.
   - Follow-up training when behavior indicates retraining is needed.
   - Training anytime a new substance, process or procedure or equipment is introduced.
   - Follow-up training about any accidents that could be repeated.
   - All training should be documented including the date, subject, instructor's name and names of employees receiving the training.

6. ***Written and Communicated Safe Work Practices.*** Safe work practices are essential to a safe work environment. Written procedures should be developed and used to guide your company's safety effort. These in turn need to be communicated to all management and hourly employees. The safety manual or handbook should include company wide work practices and job or task specific safe operating procedures. The manual should also include a Safety Policy Statement, Management and Employee Responsibilities and General Work Rules. Examples of some of the information that should be included:
   - Safety policy statement
   - Management and employee safety responsibilities
   - Safety coordinator responsibilities
   - General work rules
   - Housekeeping
   - Proper ventilation
   - Personal protective equipment
   - Proper lifting
   - Hazard communication
   - Fire extinguisher operation
   - First aid procedures
   - Safe use of tools and equipment
   - Machine guarding and safe operating procedures
   - Job specific safe work practices
   - Emergency evacuation if applicable
   - Safety inspection guidelines.

7. ***Safety and Health Program Coordinator.*** Each company is required by the new rules to appoint an individual as the company Safety Coordinator. This individual should be given the necessary time, authority and resources to facilitate developing the company's safety program. The Safety Coordinator should demonstrate an interest in employee safety and have a

high level of trust from the employees. All levels of management should be willing to work with the Safety Coordinator and support the company safety efforts. The Safety Coordinator is responsible for overseeing the company safety program and working with management and employees, but he/she cannot assume duties and responsibilities that need to be carried out by management.

The following is a list of some of the responsibilities for the Safety Coordinator:

- Overall coordination of the safety and health program
- Training of management and employees in safety responsibilities and specific safety related topics.
- Maintaining an updated knowledge of safety and health regulations.
- Tracing the overall progress of the safety program and communicating this to management and employees.
- Providing assistance and advice to management and employees on safety related issues
- Reviewing of all accidents and assisting supervisors in conducting accident investigations.
- Conducting regular safety inspections.
- Compiling injury and illness statistics and related records.
- Following up on employee concerns and complaints
- Working with the company safety committee
- Developing new safety policies and procedures as necessary to help improve the safety program.

8. ***Medical Treatment and Return to Wok Practices.*** Proper medical care for employees and early return to work procedures can greatly reduce claim cost and reduce unnecessary legal involvement and expenses.

   The following steps can be helpful in ensuring quality medical care and quick return to work for employees:

   - Establish a close working relationship with the Managed Care Organization selected by your company.
   - Communicate with the physicians, clinics and hospitals anytime an employee is injured.
   - Educate medical providers about the nature of your business.
   - Educate your employees about your procedures for obtaining medical treatment.
   - Have employees report all injuries immediately.
   - If employees are off work, set up a program of regular and frequent communication with the injured employee.
   - Provide any assistance the employee needs during his/her recovery.
   - Keep the employee advised of company functions and activities.
   - Establish a modified duty program that will allow injured workers, when possible, to return to work in a productive capacity during the recuperation period.

9. ***Verification of Program Success.*** Like any successful program, what is measured gets done. This applies to effective safety management, as well. One of the major problems in getting support for a safety program is lack of communication about the effectiveness of the program. Results oriented data such as workers' compensation cost, injury frequency, rates and trends

need to be reviewed, evaluated and communicated regularly. Performance targets should be established and communicate to all management and hourly employees.

The following are examples of areas that should be measured to reflect the success of the safety program :

- Days without either a recordable accident or lost time accident. The days should be posted in the work area and updated daily.
- Workers compensations claim cost compared yearly.
- Cost per hour or item produced.
- Supervisory tasks completed during the year against established targets.
- Number of safe miles driven without an incident or accident.
- Safety program evaluation compared yearly.

11.6.2 Health Program for Employees

The prevention of accidents is a major part of the function of employee maintenance and constitutes only one segment of a comprehensive program. The employee's physical condition can be harmed through disease, stress, and strain and as well as through accidents. It is important for the firm to be concerned with the general health, both physical and mental, of its employees for both economic and humanitarian reasons.

For employees whose job performance is suffering because of physical, mental, or emotional problems, Employees Assistance Programs (EAPs) are designed. EAPs address a variety of employee problems ranging from drug abuse to martial problems. Moreover, poor employee health leads to a high level of absenteeism and a low level of productivity.

Physical Health

There is considerably more industrial activity in the field of safety then there is in the field of health. When health and medical problems are provided, it has been customary in too many firms to attach this function to the safety unit. Physical health program should include the following elements:

- A stated health and medical policy
- A competent medical consulting staff.
- The availability of facilities for physical examinations of all employees.
- A well-equipped and staffed medical dispensary.
- Systematic attention to sanitation, safety precautions, and industrial hygiene.
- A chief medical officer who reports to a responsible member of management.

Mental Health

In recent years, an increasing amount of attention has been devoted to industrial mental health. It is difficult to determine how far a business firm should go in such as this. Mental difficulties of various types are reflected in the firm's productivity and profit records. Mental illness takes its toll through alcoholism high accident rates, high tardiness rates, high turnover, and poor human relationships. Mental health programs should include the following elements:

- Availability of psychiatric counseling.
- Cooperation with outside psychiatric specialists and institutions.

- Educations of company personnel concerning the nature and importance of the mental health problem.
- Development and maintenance of a proper human relations programs.

11.7 SAFETY PRINCIPLES

Managing safety at work should involve following principles:

1. ***Policy Decision.*** The company should first decide the level of safety it is going to achieve. The attitude of top management, accidents liability, profits, relations with employees, government policy, previous accident records, funds available etc. will influence this decision. Policy decision should include:
    - Targets for performance-targets could be spelt out in terms of reduction from earlier results, or maximum limits in terms of number of accidents.
    - A statement as to what the company proposes to do, to achieve the targets.
    - What it expects from its employees-right from the manager to the worker?
    - Action required at executive/supervisory level.
    - How it proposes to monitor and evaluate the performance?
2. ***Assignment of Responsibility for Safety.*** The management should assign the responsibility for safety to its executives in clear terms. Some companies employ a full time manager. If there is no safety manager, the job of coordinating safety activities is given to an officer as an additional charge. There are also safety committees constituted in some of the industrial units. The different department heads and other line executives are also involved in the safety program. The safety manager and the safety committees should help the supervisors and the line executives to achieve safety. The responsibilities, duties and authorities in different levels should be clearly spelt out by company.
3. ***Staff Safety Assistance.*** Every industry should have one executive who is assigned the duty of coordinating the safety activities within the plant. In small industries this job can be assigned to one of the executives as an additional charge if it can be ensured that the person will have sufficient time at his disposal to look after safety function without affecting his main functions. For the safety program in the plant to be successful, the manager should:
    - Have clearly defined responsibilities and targets of performance.
    - Have full support and backing from the chief executive/have sufficient facilities to perform the tasks assigned to him, and
    - Have the ability to think, come to logical conclusions and put his ideas across to others.
4. ***Communications with Employees.*** Even in cases where the top management is genuinely interested in the safety of the workmen, it is often seen that the employees are unaware of the management's intentions. As such it is essential that the company's policy on safety should be made known to each and every employee in clear terms. Employees, should be told:
    - That safety of workmen is of importance to the company
    - That the management has assigned the job of looking after the safety of workmen to the executives
    - That it expects the employees to work safely and co-operate with the management in ensuring safety
    - An outline of proposed safety program and safety organization in the plant.

5. ***Participation of Employees.*** One of the weak links in the safety programs in industry is employee himself. The success of safety program depends to a great extent on the interest created in the employees and so the emphasis on the participation of the employees should be more in safety program than in any other area of management. Since safety is a matter of direct interest to the employees, naturally it can be expected that they would like to participate. Employees' participation may be in the form of:
   - Safety committees
   - Safety campaigns
6. ***Overall Management Direction.*** Managing Safety should have the four basic elements of direction and control :

Planning : Decisions for safety, targets for performance, setting standards, working out detailed activities, budgeting, assignment of responsibility and procedures, data collection and analysis.

Activities : Carrying out activities, which are planned. These include :

- Providing safe conditions of work
- Directing people to be safe.

Measurements : To evaluate actions taken, to compare performance with standards and targets set, to identify weak areas.

Control : Using the information from measurements to bring about changes in the next cycle/ future plans.

11.8 SAFETY POLICY

Managing the safety function must start with a written safety policy-a definition of management's desires on safety. Policy is defined as "a settled course adopted and followed by a body" when top executives determine and announce a "settled course" they are affirming a shared purpose or one in which they want to enlist the voluntary cooperation of every member.

"A health, safety and environment policy is a plan, unique to your business, which sets out how you are going to mange health, safety and environment issues."

The policy document should:

- Set out your business' commitment to manage your risks and meet your legal duties for safety
- Tell people in your business what their duties are for health, safety and the environment
- Explain the steps that they need to take to meet their duties.

11.8.1 Need of Safety Policy

Written safety policy is important in every organization. It is a starting point for all our various activities in a safety program. If the principle *"safety is line responsibility"* is true, then it is important not only that we as safety people believe it, but also that the line organization believes it. The line organization will believe that safety is their responsibility only when it is definitely assigned to them by management.

Another principle is that *"management should direct the safety effort by setting safety goals and by planning, organizing and controlling to achieve them"*. A safety policy is management's expression of direction to be followed. It is management's first step in organizing to accomplish its desires. It is important that management's safety policy be in writing in almost all cases, to ensure that there would be no confusion concerning direction and assignment of responsibility. It is only through policy that the worker on the line, or on the construction job, will know management's desires on the subject of safety. It is for the reason we say that regardless of how most company policy is handled, safety policy must be stated on paper and published at least to the point where each and every employee is aware of it.

11.8.2 What should be Included?

The Health and Safety at Work Act says you need to set your policy out in writing and then bring it into effect. Most businesses set out their policy in three parts:

- The *statement of intent* section sets out your commitment to managing health, safety and the environment effectively, and what you want to achieve.
- The *organization* section states who is responsible for what.
- The *arrangements* sections are the detail of what you are going to do in practice to achieve the aims set out in your statement of intent.
- What is included in the safety policy may vary from company to company?

The Statement of Intent

The statement says what your commitment to health, safety and the environment is and how you are going to manage it. There are no set rules on what you should include in your statement, but statements of intent are often only one page long. The most senior person in the business must sign and date the statement.

Most statements of safety policy will say:

- You are committed to ensuring the safety of your employees, customers and anyone else affected by your business' activities.
- Who has ultimate responsibility for health, safety and environment in the business?
- Which staffs have special responsibilities - give the name of the person incharge of health, safety and environment?
- That all employees are responsible for taking care of their own health and safety and that of the people they work with.
- That you recognize the legal duties your business is bounded by and that you will provide a safe working environment, safe work equipment and safe methods of work.
- What organization and arrangements will be made to support the policy?

The Organization Section of your Policy

The organization section of your policy clearly says who is responsible for what. The overall responsibility for safety rests on the employer, but day-to-day responsibility can be delegated to others within the business.

In smaller businesses, it may be very easy to decide who needs to do what for health, safety and environment.

You can identify for example :

- Who will undertake risk assessments.

- Who will make workplace inspections.
- Who will ensure the safety of specific tasks or work activities or areas of the workplace.

As businesses get larger, the organization section of the policy may need to set out the responsibilities of staff at different levels in the business.

The organization section may include a diagram or chart showing the **business structure** and the responsibilities of :

- The managing director
- Other managers
- Supervisors or team leaders
- All employees.

Your business must designate a ***"competent person"*** who will take responsibility for health, safety and the environment and advise you on your **legal duties** and how to comply with them. The duties of your competent person should be mentioned in this part of the policy, along with their contact details.

Before you Write your Policy

Before writing your health, safety and environment policy you should consider the following.

Your Legal Duties

You should ensure that your systems for managing health, safety and the environment comply with the law.

Your "competent person" for health, safety and environment should be able to advise you on what your legal duties are, and whether you are doing enough. They are appointed by the company to take responsibility for health, safety and the environment and advise you on your legal duties and how to comply with them.

Risks

You have a legal duty to carry out a risk assessment. The outcome of your risk assessment may be:

- Providing training for employees
- Providing a written safe working method
- Safety rules for particular areas
- Providing special tools to enable a task to be carried out in safety.

There will form part of the arrangements section of your policy. Your risk assessment can help you to ensure that you cover the main risks within your policy.

Involve Your Employees

You have a legal duty to consult with your employees on matters to do with their safety at work. In practice it can be useful to involve your employees in preparing and implementing the safety policy, because:

- Where employees are involved in making safety rules, or devising safe working methods, they may be more likely to adhere them, because they will understand the reasoning behind them.

- The person who is actually doing a job can be best placed to advise on whether the proposed safe working method will work in practice.

You have a legal duty to bring the *statement of policy* to the attention of your employees. Some of the ways that businesses bring the policy statement to their employees' attention are by:

- Including it in any employees' handbooks
- Providing it at induction
- Including a copy with the contract of employment
- Posting it on their intranet sites
- Posting it on notice boards.

11.8.3 Monitoring the Effectiveness of Policy

Your health, safety and environment policy should be a practical guide to how your manage health, safety and environment within your business. It sets out the actions you are taking to:

- Manage risks
- Prevent accidents and ill health
- Meet your legal duties

There are a number of ways you can check your policy :

- Where you have set out clear responsibilities for various employees, you can check that they have actually carried them out.
- You can monitor whether people are working in accordance with any rules, or safe methods set out in the arrangements.

You can check whether any records or registers – part of your arrangements, are being used. For example, it is good practice to record workplace safety inspections and training given to staff.

11.9 SAFETY INSPECTION

Plant safety inspection, which is a method of discovering potential accident risks, is one of the oldest of accident prevention techniques. When an employees was caught between a pair of gear wheels and mangled, naturally preventive action was confirmed not only to the particular dangerous unit, which caused the unfortunate accident, but to other units also. Thus plant safety inspection became an important tool in accident prevention right from the early days. Development of the statutory regulations covering safety and the inspections by factory inspectors has given stimulation to in plant inspections and guarding program.

Just as inspection is an important function in quality control of the manufactured product, plant safety inspection plays an important role in accident prevention. The objective of safety inspections is to identify the unsafe working practices and unsafe conditions with a view to take corrective action before the accident itself can happen. It may not be possible to remove and spot all the unsafe conditions or unsafe practices in the entire plant. But the ultimate aim is to make all the operations safe and efficient. An unsafe action is a causative factor an all accidents and as it may not be possible to make the physical conditions safe and fool proof in all circumstances, it is very important that unsafe actions or work practices are also carefully noted during plant safety inspections.

Safety inspections help the management in many ways:

The purpose and function of workplace inspections must be seen within the context of the whole safety and health program. It is not an isolated function, but relates to the major objectives of the program, namely:

- Identifying and recording potential and actual hazards associated with buildings, equipment, environment, processes and practices.
- Identifying any hazards which require immediate attention, whether they are unsafe conditions or unsafe acts.
- Ensuring that existing hazard controls are functioning adequately; and
- Where appropriate, recommending corrective action.

Effectively carried out inspections are used to assist and improve other elements of the program:

1. Regular inspections carried out on shop floor cannot escape the notice of the workmen. This in fact, is an indirect way of demonstrating management's interest in the safety and welfare of employees and thus contributes to better understanding and better shop floor relations.
2. Safety inspections bring to light areas where waste can be decreased, processes can be improved and productivity increased. This results in better management of resources.
3. Safety inspections result in contacts with employees and discussions with them on their difficulties, in ensuring safety.

Who should conduct inspections?

The following points should be considered:

- Regular planned inspections could be conducted by a team consisting of the plant manager and the floor supervisor, along with an employee familiar with the work process and a member of the safety and health committee or the safety and health representative;
- Rotating teams could be used taking into account different shifts at the workplace and other factors.
- When inspecting special equipment or processes, it may be useful to have an appropriate specialist on the inspection team, *e.g.*, an engineer, electrician or other.
- The floor or area supervisor should be included for the part of the inspection covering his or her department, or at least be made aware of the inspection.

What should be inspected?

No workplace can be considered entirely free from hazards. If follows, therefore, that all workplace within an establishment should be inspected including, for example, the office, storage areas, and the maintenance areas. Also included should be areas where normally no work is performed, such as the parking lot, the canteen, and locker rooms. However, in deciding how many separate inspections are necessary and the timing and frequency of each inspection, the following should be considered:

- The number of different processes or operations and their scale, since different processes involving different machinery or employees may warrant separate inspections.
- Certain hazardous equipment requires inspections at set intervals.
- Work processes with a high hazard potential may require separate and more frequent inspection.

- Number of shifts (inspections should be conducted on every shift, since the nature of the activity may vary from one shift to another).
- Special inspections are necessary when a new process or piece of machinery is introduced into the workplace.

11.9.1 Responsibility for Inspections

Obviously, safety inspection cannot be done effectively by any one individual or agency. The responsibility should be shared by all these concerned. Within the plant, it is important that all personnel at various levels should clearly understand their responsibilities and functions so that management's objectives in this respect are implemented through their coordinated efforts.

1. *Safety Officer*. Industries of a fairly large size normally have a safety department with one or more qualified safety officers but in smaller units, safety is normally under the charge of a senior person, the personnel or production department. Though safety inspection will have to be carried out by all levels of personnel, the safety officer and his assistance have a special responsibility in this respect. The inspection made by the members of the safety development could bring out many unsafe conditions, methods and actions which are not reported by those in charge of the departments concerned.
2. *Line Management*. Safety being principally a line management function, persons from the line management should themselves take the main responsibility for plant inspection.

Senior Plant Management

Interest at all levels in safety can be stimulated and maintained if the senior management makes it obvious that they are eager to maintain safety. For example, while going round the department, the action on the part of a plant superintendent who notices an unsafe condition or unsafe action which is contrary to the approved practice, to stop and point it out to the supervisor, will go a long way in educating all those under him.

First-Line Supervisor

The duties of supervisor require that he should continuously look out for hazardous conditions and unsafe practice on the part of the men in his section. In addition to these inspections of continuous type, he should make periodic inspections at regular intervals to see that all hazards have been properly safeguarded and that the safe procedures are being followed. He should also make periodic and special inspections of equipment in his department.

Maintenance Engineers

Supervisory personnel both of senior and first line levels of the above department could make effective contribution to safety by making frequent safety inspection and taking necessary steps for maintaining the plant and equipment in safe conditions.

Workers

It is workers who ultimately carry out the various operations using the machinery and equipment. Since they came into close contact with the machinery, equipment, tools and other part of the plant during working they can easily detect the defects associated with these. They should be encouraged to report defects or hazardous condition, which can cause accidents to them or others.

11.9.2 Information Requirements

The extent to which anyone can carry out an effective inspections depends on his or her ability to identity hazards. This involves knowledge and understanding of:

- The nature of the industrial process including an understanding of working procedures;
- The applicable safety standards and requirements
- The range of potential, hazards associated with the equipment, the machinery, the process, or the environment
- Previous accidents and problem areas.

Basically, what one needs to know to conduct an effective safety inspection is the following:

Plant layout:

- Buildings
- Basic layout plans showing equipment and machinery used
- Process flow, start up and shut down times
- Information on possible hazardous substances used
- Storage areas
- Exists
- Other specific requirements of the workplace.

Legal Requirements and Standards:

- Company rules/regulations
- Job procedures and safe working practices
- Manufactures' specifications
- Personal protective equipment
- Engineering controls.

Other Background Information:

- Accident data
- Investigation reports
- First-aid reports
- Employee reports/complaints with regard to particular hazards in the workplace
- Recommendations made by safety and health committee or safety and health representatives
- Results of previous inspections
- Maintenance reports
- Other specific problems.

11.9.3 Steps for Safety Inspection

The steps involved in establishing a sound workplace inspection system are:

- Planning
- Inspection

- Reporting
- Monitoring.

1. ***Planning***. The first step in preparing for effective inspections is planning. This involves considering and establishing appropriate policy and procedures.

   As for any other element of the safety program, it is important that senior management show their commitment to the activity and to its objectives. This can be done by establishing and making known a policy related to the overall safety and health program. The form of the policy and its content will vary depending on the company, but the following general points should be considered in developing it:

   - Commitment of senior management
   - The role of inspections in attaining overall workplace safety and health objectives
   - Who is responsible and accountable for carrying out an effective inspection system?
   - What the employers and employees must do to comply with the spirit and intent of the occupational safety and health legislation?

2. ***Inspection***. To ensure consistent and comprehensive coverage of all areas in the workplace, it is useful to develop checklists. Such lists have to be continually reviewed and revised to reflect changed in equipment, process, and accident records.

Sample check list for plant safety inspection	
Machines Safety in design Standards of maintenance Provision of guards Safe operation Housekeeping around machines	**House Keeping** Layout/passages Condition of flooring Orderly arrangement Cleanliness Water logging
Electricity Approach to installation Condition of installation Protection Standards of maintenance	**Hand Tools** Condition Use Maintenance standards
Fire Protection Approach to fire points Equipment in position Condition of equipment Ignition sources Storage of flammable materials	**Environment** Lighting Noise Dust/Fumes/Gases Leakage of chemicals Spillage of chemicals Functioning of exhaust systems
Potential for Major Hazards Toxic Release Chances of fire, explosion High pressure areas High temperature areas	**Actions of People** Posture Methods Clothing Use of PPE

Each organization, however, should devise its own checklist appropriate to its own processes and experience. In conducting inspections the following basic principles bear consideration:

- While it may be necessary to ask questions, the person inspecting should not unnecessarily disrupt work activities nor start attaching blame for hazards observed:
- Attention must be drawn to the presence of any immediate danger; other items can await the final report.

3. ***Reporting.*** If action is to be taken to control or eliminate hazard, management needs to be made aware of the problems in a concise, factual way. Good reports help to gain support from management for the findings of inspections. A sample form is shown in fig. 11.1.

SAFETY INSPECTION REPORT

Company name

SAFETY INSPECTION REPORT

Date ....

Department ....

Name of supervisor

Section ....

....

No. of employees.

S. No.	Machine No.	Observations,	Corrective action	* Priority rating

*Priority Rating

Class A: Major Hazard, immediate action called for

Class B: Hazard exists, action to be taken soon

Class C: Relatively minor hazards, action can wait

Fig. 11.1. Safety Inspection Report

An inspection will be effective only if the results are reported to the right persons. These might include all or some of the following:

- Plant manager
- Department managers
- Supervisors
- Safety and health committee or representative
- Safety coordinator
- Maintenance manger.

4. ***Monitoring.*** The information obtained from regular workplace inspections must be considered and used if inspections are to be a valid part of the safety and health program. For this to be achieved, it should be clear who has ultimate responsibility for considering and making decisions on action, which is to be taken. Equally important is the need to ensure that there is timely feedback to those persons responsible for undertaking inspections. The information

obtained from regular inspections should be reviewed carefully to identity trends as part of overall monitoring of program effectiveness. Analysis of inspection reports over a period of time may, for example:

- Highlight the need for training in certain areas:
- Provide insight as to why accidents are occurring in particular areas
- Establish priorities for corrective action
- Assists in establishing or improving safe work practices.

11.9.4 Types of Inspection

Safety inspections can be of following types:

- Periodic inspections
- Intermittent inspections
- Continuous inspections
- Special inspections.

1. ***Periodic Inspections***. These inspections are well planned and made at regular intervals. Machinery equipment tools and all parts of the plant should be inspected properly. Their periodicity should be decided according to the requirements, while making periodical general inspections, efforts should be made to cover the entire plant. Fro periodic inspections of pressure vessels, cranes, hoists, elevators, chains, cables, ropes etc. a prearranged schedule as required by law should be followed and these should be carried out by competent persons. All concerned persons should be notifies well in advance of such inspections.
2. ***Intermittent Inspection***. These are announced inspections made at irregular intervals. These may be made by safety department personnel, supervisor, safety committees, or individual workmen and may cover a particular department, place of equipment or small work area.
3. ***Continuous Inspections***. In a system of continuous inspection, some selected employees spend all their time on serving certain equipment or operation. Maintenance men, electricians, etc. fall under this category. Continuous inspection of tools are done by store staff in some factories.
4. ***Special Inspections***. Special inspections are those, which are occasionally made to locate hazards which are suspected to be present in certain situations or processes. Examples of such inspections are :

- Inspections during special campaigns such as fore prevention week, waste elimination.
- Inspections of new buildings, installations and processes.
- Health surveys to determine the extent of the suspended hazard and the necessary preventive measures.
- Inspections of hand tools, personal protective equipment, lighting facilities, general ventilation equipment, construction work etc.

11.9.5 How to Make Safety Inspection Useful?

1. The inspection should be slow and the inspector should give sufficient time to study each job situation thoroughly.

2. Co-operation of workers is important and consulting them is useful. The purpose of the inspections should on no account be treated as a secret or held back from the workmen.
3. Once the inspections is complete, it will be advisable for the inspector to go through the points and whenever in doubt, collect more information. If necessary, a second round can be made.
4. Efforts should be made to find out contributory causes for unsafe acts/conditions.
5. Inspection reports (Fig. 1.2) should be presented in such a way that the reader gets the idea quickly. Vague and general statements are best avoided.
6. The recommendations should be made known to persons who are responsible for corrective action.
7. The inspections should be regularly conducted, at least once in a month.
8. Senior executives in the company should go through the inspection reports and follow up corrective action.

11.10 SAFETY PLANNING

Planning is the process of deciding in advance the future course of action. Planning is must for achieving industrial safety. Planning forms the backbone of any management activity including industrial safety management. Planning for safety includes considerations like role of employees/ management, participation of workers, experience of workers, age and other factors of workers, actual working procedures and conditions etc.

Purpose of Safety Planning

Safety program must be a planned continuous effort to promote plant safety. An effective safety program recognize that accidents are results o unsafe acts and unsafe conditions. Hence the safety program should begin with the safety planning. The underlying objectives of safety planning are:

- Humanitarian
- Reduction of production and operating costs
- Intangible benefits
- Performance standards.

The above said objectives can also be extended as benefits of safety planning which are:

- To promote standards of safety and health which comply fully with requirements of Factories Act, 1948.
- To maintain safe and healthy workplaces and safe systems and methods of work.
- To protect all employees from unforeseen hazards.
- To develop safety awareness.
- To encourage effective participation from employees on safety matters.

Procedure of Safety Planning

Industrial safety management always starts by having a plan. This furnishes not only some specific steps mapped out for reaching the established aim, but also enables estimates of strong points and weak ones. If there is no structured plan, signs of success or failure cannot be evaluated.

The various steps involved in safety planning are:

- Write and announce safety plan.
- Appoint a safety director.
- Analyze the operational record.
- Appraise the scope and seriousness of hazards.
- Select, organize and schedule common methods.
- Establish a periodic review.
- Determine long term and short term goals.

Variety Types of Safety Plans

A safety plan consists of safety objectives, policies rules and regulations and the method chosen for their implementation. Based on time span planning is classified as :

- Long Range Planning (Strategic Planning) ≥ 5 years
- Medium Range Planning (Tactical Planning) between 1–2 years
- Short Range Planning (Operational Planning) < 1 year.

11.11 SAFETY MEASURES IN MANUFACTURING INDUSTRY

The term measures means the provision of devices, equipment or other modes which ensure safety e.g. presence of guard enclosing a rotating part of machine, fire extinguishers, emergency exits etc. Safety procedures relate to methods to be followed to minimize accidents.

Safety measures for design of factory premise

These includes:

- Proper ventilation and lighting.
- Adequate layout.
- Proper control of overhead cranes.
- Sharp bends and undue steps.
- Emergency exits.
- Adequate controls on electric installation.
- First-aid arrangements.

Safety Measures Against Machine Accidents

These includes:

- Enclose dangerous parts with suitable casing.
- Dangerous parts should be out of reach during operation.
- All belts, pulleys, gears etc. should be covered with adequate guards.
- Provision of interlock guards.
- Don't oil/grease during running of machines.
- Don't allow trainees on complicated machines.

Safety Measures Against Fire

These includes:

- Adequate exits
- Doors should be open outwards.
- Proper warning signals should be provide.
- Interior doors shouldn't be locked.
- Fire fighting equipments.
- Ensure proper electrical contacts.
- Proper storage of explosives.

Safety Measures Against Electrical Hazards

These includes:

- Insulate all electrical, writing etc.
- All metallic cover over electrical machinery should be properly guarded.
- No hanging of loose wires.
- Warning signals.
- Underground cables.
- Safety fuses should be there.

11.12 EMPLOYEES PARTICIPATION IN SAFETY

Employee participation is an important part of developing and implementing health and safety practices in a place of work. Employers must provide reasonable opportunities for employees to participate in ongoing processes for the improvement of health and safety, and employees are expected to actively participate in health and safety matters. Employers, employees who wish to be involved, and unions representing them are required to co-operate in good faith and develop, agree, implement and maintain a participation system.

To promote safety, one of the best alternatives is to involve the employee. They usually know better than anyone where hazard exists. Each worker should have a safety responsibility, because it is the worker and his family who suffers a lot due to accidents. It must be made clear to all the employees that all the safety regulations and instructions are expected to be followed just as seriously as any other directions. It is always better to lead than to loss and safety should be more cooperative than enforced. After all employees should be willing to act in the interests of their own safety.

The term 'employees participation' involves encouraging the people to participate in safety function of an organization to accomplish improved results and satisfaction. Employee's participation in safety can be achieved by providing employees:

- Means to report signs, symptoms and hazards
- Means to make control recommendations
- Prompt responses to their reports and recommendations
- Access to information about ergonomic programs and
- Ways to become involved in program development, implementation and evaluation.

Employee's participation has several benefits such as:

- Enhanced worker motivation/job satisfaction.
- Adding problem solving capabilities.
- Greater acceptance of change.
- Greater knowledge of the work and organization.

In general, the greater the employee participation in the safety program the more effective it will be. The organized safety programs should include as many activities in which workmen can take part. It should give them the major roles whenever and wherever possible. ***Such activities include;***

- Safety campaigns
- Safety contests
- Safety suggestion schemes
- Safety committees
- Safety cartoons.

Safety Campaigns

Safety campaigns are of undoubted value if properly timed and properly conducted. They are timely when a special situation arises, for example, general laxity in wearing goggles, a general staleness in the safety work etc. Safety campaigns helps a lot in making employees aware of the safety rules and regulations.

Safety Contests

Safety contests are used to develop a spirit of rivalry in the interest of safety between individuals in the department, departments in a plant, or between establishments in a given industry. Many organizations have found safety contests between their members to be very effective in improving the injury experience of the participants. The success of safety contest should not be judged mainly by the excellence of the winning performance or the attendance, much more important are factors that indicate the benefits occurring to the non-winners.

Safety Suggestion Schemes

Under this system, workers are invited and encouraged to offer suggestions for improving the safety in an organization. A suggestion box is installed. Any worker can write his suggestions and put it into the box. Periodically all the suggestions are scrutinized by the suggestion committee. Good suggestions are suggested for implementation and suitable awards are given to the concerned workers. Suggestion schemes encourage worker's interest in safety activities of an organization.

Safety Committees

One of the most successful means of promoting safety work is through safety committees. A safety committee consists of representatives of both the employer and the workers. The employer's representatives include the staff members who have direct responsibility for safety such as manager, the safety engineer, plant engineer etc. The worker representatives are usually rotated so that an increasing number have the experience of safety work. The worker members may be elected by the workers themselves or themselves or nominated by the employer. Its functions can be broadly classified as follows :

- To promote safety through co-operation between management and workers.
- Planning for the control of physical hazards concerning standards to be followed guarding machinery, the design of equipment, layout of process, etc.
- Arousing and maintaining interest in safety of all workers and supervisors and to help them to understand safety in their business.
- Investigation of accidents.
- Study of accidents and accident record.
- Safety training.
- To provide an opportunity for free discussion of accident problems and preventive measures, and
- To make safety activities an integral part of operating policies and methods.

Safety Cartoons

Cartoons showing bad practice and bad conditions are widely used because of their effectiveness in promoting interest in safety.

11.13 SAFETY AND PRODUCTIVITY

Industrial safety is conducive to improve productivity. It affects productivity as:

- If a worker is mentally ensured that he is working in safe working conditions his morale will be high and he will work with more concentration and thus productivity will increase.
- Safe working environment helps in motivation employees to perform their jobs in an efficient and effective manner and thus the productivity will increase.
- Safe working environment helps in reducing absenteeism, which will in turn lead to increase in productivity.
- Safe workplace environment, leads to reduction of accidents and hence reduction in direct and indirect cost of accidents. This leads to increase in productivity.

11.14 RELATIONSHIP OF SAFETY WITH PLANT DESIGN, EQUIPMENT DESIGN AND WORK ENVIRONMENT

A good layout and working conditions play a major role in preventing many accidents, which would have otherwise occurred. The following aspects must be considered for the safe working conditions:

- Illumination
- Ventilation
- Plant layout
- Housekeeping and maintenance
- Color schemes and safety color codes
- Equipment design.

Plant Layout

It should be designed in such a way that it will provide safety and security to all employees at all costs. Optimum plant layout calls for high operating efficiency coupled with proper safety provisions. For preventing accidents, the layout should be such that:

- Every employee has enough space to move and operate.
- Passageways between working places, roads, tracks etc. must never be obstructed.
- It should prevent the inrush of hot/colt air to the working place.
- Floors must be non-skid type and must be capable of absorbing sound.
- Workers operating on machines should have easy access to safety switches provided or machines.

Work Environment

It work environment housekeeping and maintenance are important. Good housekeeping I one of the most important factor in preventing injury and accidents. Everything should be orderly placed in its suitable position. A well-designed, well-built factory is easier to keep, clean and free from disorder and hazards. The science of Ergonomics made the workplace safe and healthier place.

Equipment Design

Equipment should be designed by keeping in mind safety measures.

11.15 SAFETY AND ECONOMICS

Safety and Economics are closely related. The economics occupational safety suggests for many little more than how can better working conditions be made profitable for business. It focuses primarily on the ways in which individuals and communities produce, distribute and consume goods and services. The general purposes that economics can serve for occupational safety.

- First, identifying and measuring the economic costs of occupational injury and disease can motivate the public to take these problems more seriously.
- Second, understanding the connections between the way firms and markets function and types, of occupational safety problems that arise is crucial for the success of public polity.

11.16 SAFETY LEGISLATION

Indian government has made following legislation to ensure safety:

- Factory Act, 1948
- Indian Boilers Act, 1923
- Mines Act, 1952
- Indian Electricity Act, 1910
- Petroleum Act, 1934
- Bihar Rules
- Indian Boiler Act, 1923
- The Minimum Wages Act, 1948

- The Payment of Wages Act, 1936.

Purpose: To consolidate and amend the law relating to steam boilers

Provisions:

- Prohibition of unregistered/uncertified boiler.
- Prohibition of using any boiler by the owner, which has been transferred from one state to another until the transfer has been reported in prescribed manner.
- Prohibition of using any boiler by the owner at a pressure higher than maximum pressure recorded in the certificate.

The Minimum Wages Act, 1948

Objectives

To determine the minimum wages in industry and trade where labour organizations are non-existent or ineffective.

Scope and coverage:

- Applicable to all employees engaged to do any work, skilled, unskilled manual or clerical, in a scheduled employment, including out-workers.
- Fixation of minimum wages.

Main provisions:

- Fixation of minimum wage of employees.
- Procedure for fixing and revising minimum wages.
- Obligation of employees.
- Rights of workers.

The Payment of Wages Act, 1936

Objectives

To ensure regular and prompt payment of wages and to prevent the exploitation of a wage earner by prohibiting arbitrary fines and deductions from his wages.

Scope and coverage:

- Application for payment of wages to persons employed in any factory.
- Not applicable to wages which average Rs. 1600/-per month or more.
- Wages include all remuneration, bonus, or sums payable for termination of service, but do not include house rent reimbursement, light vehicle charges, medical expenses, TA, etc.

Main provisions:

- Responsibility of the employer for payment of wages and fixing the wage period.
- Procedures and time period in wage payment.
- Payment of wages to discharged workers.

- Permissible deductions from wages.
- Nominations to be made by employees.
- Penalties for contravention of the Act.
- Equal remuneration for men and women.
- Obligations and rights of employers.
- Obligations and rights of employees.

11.17 REGULATORY AGENCIES AND STATUTORY BODIES DEALING WITH SAFETY IN INDIA AND ABROAD

1. ***Director General of Employment and Training (DGE & T).*** This Office is responsible for laying down the policies, standards, norms and guidelines in the area of vocational training throughout the country and also for coordinating employment services.
2. ***Office of the Chief Labour Commissioner (Central).*** The office is responsible for:
    - Prevention, investigation and settlement of industrial disputes in the Central sphere
    - Enforcement of awards and settlements
    - Implementation of labour laws in industries
    - Verification of membership of Unions affiliated to the Central Organizations of workers for giving them representation in national and international conferences and committees and
    - Fixation and revision of dearness allowance component of minimum wages by notification under the Minimum Wages Act, 1948 in scheduled employments.
3. ***Directorate General of Factory Advice Service and Labour Institutes (DGFASLI).*** This Office is concerned with formulation of policy relating to the safety, health and welfare of workers in factories and docks. It is responsible for coordinating the implementation of the measures of the Factories Act, 1948 by the State Governments and formulation of Model Rules there under. It undertakes research in industrial safety, occupational health, industrial hygiene, industrial psychology and industrial physiology. It provides training mainly in the field of industrial psychology and industrial safety and health including a diploma course of one-year duration in industrial safety. The Diploma is an essential qualification for appointment of Safety Officers in factories.
4. ***Directorate General of Mines Safety (DGMS).*** This Office is entrusted with enforcement of provisions of the Mines Act, 1952 and the Rules and Regulations framed there under. The provisions of the Indian Electricity Act, 1910 as applicable to mines and oil fields are also enforced by it.
5. ***Employees' State Insurance Corporation (ESIC).*** The Corporation is responsible for implementation of the Employees' State Insurance Act, 1948 which provides for medical care and treatment to insured persons and their families. Assistance is given in terms of benefits during sickness and maternity, compensation for employment injury, pensions for dependents on the death of workers due to employment injury, etc.
6. ***Employees' Provident Fund Organization (EPFO).*** This Organization is responsible for administration of the Employees' Provident Funds and Miscellaneous Provisions Act, 1952. Scheme for Provident Fund, Family Pension and Deposit Linked Insurance are implemented by the Organization for the benefit of workers covered under the scheme.

7. ***Central Labour Service (CLS).*** The Central Labour Service (CLS) was constituted with effect from 3rd February 1987 to ensure better industrial relations, labour law enforcement and labour welfare. Factories and mines employing 500 or more workers and plantations employee 300 or more workers are required to appoint a prescribed number of welfare officers under the relevant statutes. Assistant Labour Welfare Commissioners (Central) and Deputy Labour Welfare Commissioner (Central) under the supervision of Welfare Commissioners discharge these statutory functions and they also advise and assist the management of the concerned establishments in maintaining harmonious industrial relation in the areas of safety, health and welfare of workers etc.
8. ***Occupational Safety and Health Administration (OSHA).*** OSHA stands for the Occupational Safety and Health Administration, an agency of the U.S. Department of Labor, "to assure so far as possible every working man and woman in the nation safe and healthful working conditions and to preserve our human resources."

   For business, protecting workers' safety and health is the right thing to do. It saves money and adds value to the organization. Safety add value to workplaces by increasing moral, improving productivity, and reducing turnover. Establishing a safe and healthful working environment requires every employer and every worker to make safety and health a top priority. OSHA uses three basic strategies, authorized by the *Occupational Safety and Health Act*, to help employers and employees reduce injuries, illnesses, and deaths on the job:

   - Strong, fair, and effective enforcement
   - Outreach, education, and compliance assistance and
   - Partnerships and other cooperative programs.

   Based on these strategies, OSHA conducts a wide range of programs and activities to promote workplace safety and health. The agency:

   - Encourages employers and employers and employees to reduce workplace hazards and to implement new safety and health management systems or improve existing programs
   - Develops mandatory job safety and health standards and enforces them through worksite inspections, employer assistance, and sometimes by imposing citations, penalties, or both
   - Promotes safe and healthful work environments through cooperative programs, partnerships, and alliances
   - Establishes responsibilities and rights for employers and employees to achieve better safety and health conditions
   - Supports the development of innovative ways of dealing with workplace hazards
   - Establishes training programs to increase the competence of occupational safety and health personnel and
   - Provides technical and compliance assistance and training and education to help employers reduce worker accidents and injuries.
9. ***Consumer Product Safety Commission (CPSC).*** The U.S. Consumer Product Safety Commission (CPSC) is an independent federal regulatory agency that was created in 1972 by Congress in the Consumer Product Safety Act. In that law, Congress directed the Commission to "protect the public against unreasonable risks of injuries and deaths associated with consumer products".

CPSC woks to reduces the risk of injuries and deaths from consumer products by:

- Developing voluntary standards with industry.
- Issuing and enforcing mandatory standards, banning consumer products if no feasible standard would adequately protect the public
- Obtaining the recall of products or arranging for their repair
- Conducting research on potential product hazards and
- Informing and educating consumers through the media, state and local governments, private organizations, and by responding to consumer inquiries.

10. ***Institution of Occupational Safety and Health (IOSH).*** IOSH (Institution of Occupational Institution of Occupational Safety and Health) is an independent, not-for-profit organization with over 27,000 members. It is Europe's leading body for individuals with a professional involvement in occupational safety and health.

Clearly communicated expert advice from competent safety and health practitioners is an essential component in defining the safety, health and welfare policies of employers, ensuring that high standards of occupational safety and health are achieved and maintained. IOSH promotes the important function of the safety and health professional while also working to encourage employees to protect themselves and their colleagues from workplace hazards.

11.18 INDUSTRIAL ACCIDENTS

Industrial accidents are one of the problems of industrialization. One such is Industrial Accident. Industrial Accident has become a part of our set up in spite of continuing efforts by individuals and organizations to put a check on their numbers. With rapid advances in industrial processes, new type of dangers to life, limb and health are being increasingly introduced. Every year several lakhs of employees are injured in factories due to accidents. These accidents may be caused as a result of any unsafe activity, or act in their part or chance occurrences, or as a result of some unsafe working conditions or unsafe act of employees themselves, or defective plant or shop layout. These accidents represent a social loss of great magnitude in the form of pain, loss of earning capacity and cost due to disturbance of economic efficiency. The pain and suffering of the injured as well as the emotional loss to the victims of the accidents causing permanent disabilities are impossible to, be summed up and evaluated.

"It is an unplanned and uncontrolled event which causes or is likely to cause an injury".

According to Factory Act of 1948 is "an occurrence in an industrial establishment causing bodily injury to a person which makes him unfit to resume his duties in the next 48 hours".

However injuries inflicted with the consent of person cannot be regarded as accidents. An industrial injury can be defined as "a personal injury to an employee which has been caused by an accident or an occupational disease, and which arises out of, or in the course of employment, and which could entitle such employee to compensation under the Workmen's Compensation Act, 1923".

11.19 INDUSTRIAL HAZARDS

A hazard is something that has the potential to cause harm. This harm can affect people, property and processes as follows;

- People-Injury, illness, death, psychological trauma

- Processes-Work disruption and interruption to production.
- Property-Damage, contamination, theft and wastage.

In an industrial environment Hazard is defined as *any existing or potential condition in the workplace which by itself or by interacting with other variables, can result in unwanted effects of deaths, injuries, property damage and other losses.*

or

A **hazard** *is a any practice, behaviour or condition or combination of these that can cause injury or illness in people or damage to property.*

or

"An unsafe condition or practice that could cause an injury or illness to an employee and it is preventable".

The prime objective of any accident prevention program is to control hazards at work, so as to reduce the chances of occurrence of accidents.

It takes a ***hazard*** and someone ***exposed*** to the hazard to produce an ***accident.***

Hazard + Exposure → Accident

Hazard is the potential to cause harm; **risk** on the other hand is the likelihood of harm (in defined circumstances, and usually qualified by some statement of the severity of the harm). The relationship between hazard and risk must be treated very cautiously. If all other factors are equal – especially the exposures and the people subject tc them, then the risk is proportional to the hazard. However all other factors are very rarely equal.

11.20 CLASSIFICATION OF HAZARDS

Hazards are generally classified into two broad categories:

- Those dealing with safety *i.e.,* resulting in injuries.
- Those dealing with health *i.e.,* resulting in illness.

11.20.1 Safety Hazard Categories

1. ***Material Handling Hazards.*** Manual material handling can involve lifting, carrying lowering, pushing, and pulling. All of these activities can lead to muscle strains, tears and pulls of the back, shoulders, arms and abdomen. Mechanical material handling can involve such devices as forklift trucks, conveyors, cranes, and handcarts and trucks. These devices can introduce many hazards including accidental contact with moving equipment or parts, loads, or electricity. Handling of hazardous materials, such as corrosives, flammables and reactive is another key area. Exposure can cause serious harm to people and extensive damage to property.
2. ***Machine Hazards.*** Any machine can be a hazard, especially those with moving parts that can get tangled in a worker's clothes or come into contact with a worker's body. Here are some examples: worker may be crushed if they get caught in rotating shafts, belts or pulleys body parts may be injured or severed by presses, blades and saws workers may be struck by flying projectiles from machines.
3. ***Energy Hazards.*** Workers can be seriously injured by the sudden movement of machine components, electrical shock or other releases of energy when they are adjusting or maintaining

equipment. Energy sources include: electricity, steam, heat, pneumatic or hydraulic pressure and gravity as well as mechanical and chemical energy.

4. *Work Practice Hazards.* Failure to have or to follow safe work practices is a significant cause of injuries. Performing work safely in accordance with established safe work procedures is a fundamental element in the control of safety hazards. Employers, supervisor and workers all share a responsibility in this.
5. *Confined Space Hazards.* Confined spaces are workspaces where hazardous gases, vapours, dusts or fumes may build up or where on oxygen-deficient atmosphere may be created. Examples include: storage tanks, vaults, pits, vats, silos, pipelines, ducts and tunnels. Other hazards of confined spaces include: difficulty of entry and exit and working in awkward spaces, poor walking surfaces, poor visibility, and extremes of temperature and noise.

11.20.2 Categories of Health Hazard

1. *Physical Hazards.* Physical hazards are forms of energy that can harm the body if exposed. Examples include: noise, vibration, temperature extremes (hot or cold), and radiation. The effects of exposure can respectively include: temporary or permanent hearing loss; damage to the small blood vessels and nerves; heat cramps, exhaustion and stroke; frostbite and hypothermia; cancer and eye damage.
2. *Chemical Hazards.* Chemical hazards can take the form of solids, liquids, vapours, gases, dusts, fumes or mists. They can be inhaled, ingested or absorbed into the body. Examples include: paints, solvents, cleaners, degreasers, acids, and cutting oils. Exposure to chemical hazards can cause irritation, allergic reactions, depression of the nervous system, asphyxia, lung disease and cancer. Some chemicals can also have harmful effects on the reproductive system.
3. *Biological Hazards.* Biological Hazards are living things or substances produced by living things that can cause illness in humans. These hazards enter the body by inhalation, ingestion or absorption. Examples of biological hazards include: bacteria, viruses, fungi, parasites and plants. Effects of exposure include: tuberculosis, tetanus, food poisoning, boils, blood poisoning, ringworm, thrush, hepatitis, measles, and rabies.
4. *Ergonomic or Work Design Hazards.* Ergonomic hazards arise from the design and organization of work. They can harm the body by placing strain on the musculoskeletal system and overloading the muscles, tendons, joints, ligaments, nerves and blood vessels. Look for ergonomic hazards in: workstation layout and design, tool and equipment design, the work environment, and general work organization.
5. *Stress or Psychosocial Hazards.* Workplace stressors can lead to excess stress or distress and have been identified as important factors in many types of illness, including heart disease and high blood pressure. There are two main types of stressors: physical (e.g., noise and vibration) and organizational stressors (*e.g.*, lack of job control, work overload, role uncertainty and conflict, isolation and workplace violence).

11.20.3 Hazard Management Program

It can be aid that accidents results from inadequate hazard management.

The steps in the management of hazards are:

- Identification or Recognition

- Evaluation
- Control.

11.21 HAZARD IDENTIFICATION

There is no single ideal system of hazard identification procedure but there are several ways to identify hazards ranging from very simple to highly sophisticated depending upon type of industry, process and various stages like pre design, design, commissioning of the project etc. Some of the commonly used hazard identification techniques are:

1. Past Accident Analysis
2. Safety Survey/Study
3. Safety Audit
4. Job Hazard Analysis
5. Fault Tree Analysis (FTA)
6. Failure Mode and Effect Analysis (FMEA)
7. Fire and Explosion Index or Dow Indies
8. Hazard and Operability Studies (HAZOP).

There hazard identification techniques plays vital role in evaluating the qualitative risk and also in assessment of quantitative risk i.e., estimation of probability of an unexpected event and its consequences in quantitative terms of damage to property, environment or personnel.

1. ***Past Accident Analysis.*** Keeping a record of accidents help in identifying hazards based upon past experience. Though non-occurrence of accidents in previous years does not necessarily means that plant is totally safe. An accident history definitely indicates the need for identifying and controlling some inherent hazards.

   In order to properly utilize past accident records for hazard identification purpose, it is very important that such record is suitable analyzed department wise, section wise, shift wise, accident causing agency wise, reportable/non- reportable wise. Regular workers/contract workers wise etc. A careful study of such an analysis is very useful in identifying hazards in technical systems as well as management systems. Past accident analysis also provides guidance about the quantity of material involved, failure mode, energy released, damage and injury, type of event, time factor for development of hazard from initial accident, warning time available to the emergency team for taking emergency measures and emergency preparedness.
2. ***Safety Survey/Study.*** A safety survey is a detailed examination in narrow field to activity relating to industrial safety. Such activities may involve identifying hazards due to air-borne chemical contaminants, identifying hazards due to excessive noise levels. Identifying hazards due to heat stress or poor illumination etc. This technique can identify hazards even before occurrence of a loss-producing event.
3. ***Safety Audit.*** A safety audit differs from safety survey in a sense that a safety survey is a detailed examination limited to a narrow field whereas safety audit covers a plant in its entirety. It covers various management and technical aspects to identify the hazards due to ineffective use of various safety systems. Safety audit is carried out thoroughly by a team of experts in various fields. This study is based on information collected through questionnaire, checking documents/records, interviewing people at various levels and plant visits. The report highlights the strength and weakness of various systems and suggests recommendations

for improvements. Safety audit gives only a qualitative assessment of hazards. Safety audit has five elements:

- Identification (of less producing situations)
- Assessment (of losses by then situations)
- Selection (of measures to minimize)
- Implementation (of measures in the organization)
- Monitoring (the changes).

4. ***Job Hazard Analysis.*** One way to increase the knowledge of hazards in the workplace is to conduct a job hazard analysis on individual tasks. A job hazard analysis (JHA) is a technique that focuses on job tasks as a way to identify hazards before they occur. It focuses on the relationship between the worker, the task, the tools, and the work environment. In a JHA, each basic step of the job is examined to identify potential hazards and to determine the safest way to do the job. Other terms used to describe this procedure are Job Safety Analysis (JSA) and hazard breakdown.

   A job hazard analysis is one component of the larger commitment of a safety and health management systems. Supervisors can use the findings of a job hazard analysis to eliminate and prevent hazards in their workplaces. This is likely to result in fewer worker injuries and illnesses; safer, more effective work methods; reduced workers' compensation costs; and increased worker productivity. The analysis also can be a valuable tool for training new employees in the steps required to perform their jobs safety.

   Steps in conducting a JHA are:

   - Selecting the job to be analyzed
   - Breaking the job down into a sequence of steps
   - Identifying potential hazards
   - Determining preventive measures to overcome these hazards.

   Selection of job

   A job hazard analysis can be conducted on many jobs in our workplace. Priority should go to the following topics of jobs:

   - Jobs with the highest injury or illness rates
   - Jobs with the potential to cause severe or disabling injuries or illness, even if there is no history of previous accidents
   - Jobs in which one simple human error could lead to a severe accident or injury
   - Jobs that are new to your operation or have undergone changes in processes and procedures and
   - Jobs complex enough to require written instructions.

   Breaking jobs into sequence of steps

   Nearly every job can be broken down into job tasks or steps. When beginning a job hazard analysis, watch the employee perform the job and list each step as the worker takes it. Be sure to record enough information to describe each job action without getting overly detailed. Avoid making the breakdown of steps so detailed that it becomes unnecessarily long or so broad that it does not include basic steps. You may find it valuable to get input from other workers who have performed the same job.

Identifying potential hazards

To help identity potential hazards, the job analyst may use questions such as:

- Can any body part get caught in or between objects?
- Do tools, machines, or equipment present any hazards?
- Can the worker make harmful contract with objects?
- Can the worker suffer strain from lifting, pushing, or pulling?
- Is the worker exposed to extreme heat or cold?
- Is excessive noise or vibration a problem?
- Is there a danger from failing objects? Is lighting a problem?
- Can contact be made with hot, toxic, or caustic substances?
- Are there dusts, fumes, mists, or vapours in the air?

Determine preventive measures

The final stage in a JHA is to determine ways to eliminate or control the hazards identified. The generally accepted measures, in order of preference, are:

Eliminate the hazard: This can be done by

- Choose a different process
- Modify an existing process
- Substitute with less hazardous substance
- Improve environment (ventilation)
- Modify or change equipment or tools.

Contain the hazard

If the hazard cannot be eliminated, contact might be prevented by using enclosures, machine guards, worker booths or similar devices.

Revise work procedures

Consideration might be given to modifying steps which are hazardous, changing the sequence of steps, or adding additional steps (such as locking out energy sources).

Reduce the exposure

These measures are the least effective and should only be used if no other solutions are possible. One way of minimizing exposure is to reduce the number of times the hazards is encountered. An example would be modifying machinery so that less maintenance is necessary. The use of appropriate personal protective equipment may be required. To reduce the severity of an accident, emergency facilities, such as eyewash stations, may need to be provide.

5. ***Fault Tree Analysis (FTA).*** Fault Tree Analysis (FTA) is a deductive technique that focuses one particular accident event and provides a method for determining the causes of that accident event. Negative analytical trees or fault trees are excellent troubleshooting tools. They can be used to prevent or identify failures prior to their occurrence, but are more frequently used to analyze accidents or as investigative tools to pinpoint failures. When an accident or failure occurs, the root cause of the negative event can be identified.

   Each event is analyzed by asking, "How could this happen?" In answering these questions, the primary causes and how they interact to produce an undesired event are identified. This logic process continues until all potential causes have been identified. Throughout this process, a tree diagram is used to record the events as they are identified. Tree branches stop when all events leading to the negative event are complete.

A typical example of this method of analysis is seen in fault locating charts supplied with household appliances or automobiles, where the possible failure (*e.g.*, engine not starting) and the probable reasons for the same (*e.g.*, engine cold, defective plug, poor mixture) are told to customer in logical order.

On the basis of analysis, a block diagram is constructed which depicts:

- The possible ways of failure,
- The events which can logically be responsible for such failures,
- The sequence of events leading to each failure, and
- Relation between failures of different components

FTA involves the following steps:

1. Define the top event.
2. Know the system.
3. Construct the tree.
4. Validate the tree.
5. Evaluate the tree.
6. Study tradeoffs.
7. Consider alternatives and recommend action.

6. ***Failure Mode and Effect Analysis (FMEA).*** Failure Mode and Effect Analysis is defined as a procedure and tools that help to identify every possible failure mode of a process or product, to determine its effect on other sub-items and on the required function of the product or process. Here the entire system is divided into sub systems, sub-sub systems and then to individual components. Each component is then studied in detail as to:

   - The function it is performing
   - The ways in which it can fail
   - The effect of failure of the component on the sub-systems and the system.

   For example: A worn out bearing can overheat. Overheating of bearing leads to temperature rise of the motor and the pump, temperature rise can damage the motor which in turn can stop the pump. The pump failure affects the flow rate and so on.

7. ***Fire and Explosion Index or Dow Index.*** This technique is used to rate a chemical process unit numerically for hazards to suggest the particular hazards involved in process plant and to indicate measures that may be taken to neutralize or minimize them. Dow Index system of risk analysis is developed by Dow chemical company of USA. It involve simplified ways of calculating.

   - Area of exposure in case of fire and explosion.
   - Percentage of damage expected in the above area.
   - Maximum probable property damage.
   - Total business loss in terms of money value.

   The above approach involves general process hazards (increased hazard potential due to process itself *e.g.*, physical changes taking place, reactions during transfer of chemicals) an Special process hazards (like low/high pressure, low/high temperature, corrosion, vibration which are difficult to control).

Thus the technique attempts to evaluate the maximum loss that can occur due to fire, explosion and reactivity potential of process equipments by identifying them, qualifying the expected damage and presenting to the management a criterion for acceptable risk.

8. ***Hazard and Operability Studies (HAZOP).*** Hazard and Operability Studies (HAZOP's) are structured critical examinations of plant or processes undertaken by an experienced team of company staff in order to identify all possible deviations from an intended design, along with the consequent undesirable effects concerning safety, operability and the environment. The possible deviations are generated by rigorous questioning, prompted by a series of standard 'guidewords' applied to the intended design.

Definitions

Hazard any operation that could possible causes a catastrophic release of toxic, flammable or explosive chemicals or any action that could result in injury to personnel.

Operability-any operation inside the design envelope that would cause a shutdown that could possibly lead to a violation of environmental, health or safety regulations or negatively impact profitability.

Standard guidewords and their generic meanings

Guideword	*Meaning*
No (not, none)	None of the design intent is achieved
More (more of, higher)	Quantitative increase in a parameter
Less (less of, lower)	Quantitative increase in a parameter
As well as (more than)	An additional activity occurs
Part of	Only some of the design intention is achieved
Reverse	Logical opposite of the design intention occurs

The deviations from the intended design are generated by coupling the guideword with a variable parameter or characteristic of the plant or process, such as reactants, reaction sequence, temperature, pressure, flow, phase, etc.

In other words: **Guideword + Parameter = Deviation**

Systematic examinations are made of each part of a plant or process using these guidewords. Having examined one part of the design and recorded any potential hazards and operability problems associated with it, the study progresses to focus on the next part of the design or the next step in the operation. The examination is repeated until the whole plant has been studied for all major modes of operation. Recommendations for changes in design, operating procedure, materials or for referral outside for further consideration can then be made to overcome the problems which have been identified.

The success or failure of study depends on four aspects:

- The accuracy of the design drawings and other data used as the basis for the study
- The technical skills and expertise of the team
- The ability of the team to use the approach as an aid to their imagination in visualizing
- Possible deviations, causes and consequences and
- The ability of the team to maintain a sense of proportion, particularly when assessing the seriousness of the hazards, which are identified.

11.22 HAZARD ASSESSMENT AND EVALUATION

Once a list of hazards within a company has been developed, each hazard on the list should be evaluated in order to assess what its impact on the company would be if control action was not taken. This assessment should take account of legal, humanitarian, and economic considerations in addition to frequency (number of times spotted), the potential severity (number of persons at risks) and the maximum or worse possible injury, damage or loss that might occur. Considerations should also be given to the probability that the hazard will result in loss. From these assessments a list of priorities for control hazard can be established and used on the basis of control action.

Relative Consequences (severity) of Various Hazards

Catastrophic Hazard	Imminent danger exists, the hazards is capable of causing death, possible multi-deaths, wide spread occupational illness and loss of facilities.
Critical Hazard	The hazard can result in severe injury, serious illness and property and equipment damage.
Marginal Hazard	The hazards can cause injury, illness and equipment damage, but it would not be serious.
Negligible Hazard	The hazard will not result in a serious injury or illness. It would be a very minor first aid case.

11.23 HAZARD CONTROL

The control of hazards within the organization requires careful planning

A hierarchy of controls-preferred order

Select controls from the highest level you can.

Elimination-removing the hazard or hazardous work practice from the workplace. This is the most effective control measure. The best way to eliminate the risk is to completely remove the hazard. For example, buy pre-sawn timber instead of using a circular saw on site; or if the power cord of an electric drill is in danger of being cut, creating an electrocution hazard, use a cordless drill instead. If a hazard cannot be eliminated there are a number of control options that can be used alone, or in combination, to minimize the risk.

Substitution-substitute the plant or the hazardous aspects of the plant with safe options, such as replacing a pedestal fan with a ceiling fan in a restaurant kitchen, substituting or replacing a hazard or hazardous work practice with a less hazardous one.

Isolation-isolating or separating the hazard or hazardous work practice from people involved in the work or people in the general work areas from the hazard. This can be done by installing screens or barriers or marking off hazardous areas or isolating the plant. For example, placing an air conditioning plant on the roof, building a booth from which a machine can be operated remotely, moving a photocopier to a dedicated room with its own ventilation system.

Engineering control-if the hazard cannot be eliminated, substituted or isolated, an engineering control is the next preferred measure. This may include modifications to tools or equipment, providing guarding to machinery or equipment. Rather than plant, modify its design by fitting a frame to a tractor for rollover protection. These include such things as cutout switches, screens and guards.

Administrative control-includes introducing work practices that reduce the risk. This could include limiting the amount of time a person is exposed to a particular hazard. Safe work practices should be used at all time to minimize exposure to a hazard and hence minimize the risk. For example, signs should warn of any hazards. A system might be used to ensure plant is isolated from its power source during maintenance. Work programs might be arranged to minimize the time spent near noisy machinery. Rest breaks should be prescribed for computer operators.

Personal protective equipment-should be considered only when other control measures are not practicable or to increase protection.

Techniques Available to Control Hazards within the Workplace

Mechanical hazard may be engineered out of the process, or effectively enclosed by means of fixed guarding. Alternative forms of guarding involve the use of interlocked guards light sensitive barriers.

Hazards from working environment may be controlled by effective ventilation systems, maintaining suitable temperatures, illumination and the general provisions of the good working conditions.

Chemical hazards may also be controlled by effective ventilation, regular monitoring, substitution with less hazardous material, change of process, and use of personal protective equipment.

Hazard Reduction Requires Team Efforts by Employees and Management.

The engineering departments can design facilities free of uncontrolled hazards and provide technical services to other departments to aid in hazard identification and analysis. They must be sure that designs comply with standards.

Manufacturing departments can reduce hazards through such efforts as effective tool design, change in processes, job hazard analysis and control, and co-ordinating and scheduling production.

Quality control can test and inspect all materials and finished products and conduct studies to determine whether alternate design material and methods of manufacture could improve quality and safety of the product.

Maintenance can perform construction and installation work in conformance with good engineering practices and provide planned preventive maintenance on electrical systems, machinery and other equipment to prevent abnormal deterioration, or safety and health hazards.

11.24 MAJOR INDUSTRIAL HAZARDS

The phrase '**major hazard**' is used in the context of possible accidents in chemical or other industrial installations handling hazardous substances which might cause extensive damage and kill or injure a person in or outside the boundary.

A major hazard can be defined as an industrial accident which causes injury or death to large number of persons. It could be an industrial fire, explosion, a toxic release or even a structural collapse.

The following are considered major hazards:

- Industrial fires
- Explosions
- Toxic releases.

Industrial Fires

Major Industrial fires arise due to leakage of flammable material from the system and their getting ignited from some source. In case of large discharge of flammable materials, formation of vapour clouds, the cloud traveling to other location and causing a fire or explosion is possible. This poses danger to communities nearby or even at some distance. The damage in both cases could be extensive.

In case of industrial fires the quantities of flammable materials involved are large. Hence there is possibility of generating toxic materials in large quantities, which are released into atmosphere. This poses another major hazard from toxic material. In case of fires the extent of damage will depend upon the quantity of material in flame, the properties of the material (flash point, products of combustion), atmospheric factors such as wind, velocity, fog, temperature, humidity and also the site situation.

Explosions

Flammable material released can mix with air and form explosive mixture at certain concentrations. When this mixture is ignites, the result is an explosion. Explosion can result in:

- Structural damage to buildings
- Flying objects
- Persons getting buried under collapsed structures.

Examples:

- Gases (*e.g.*, LPG)
- Dust
- Confined vapour clouds
- Boiling liquid expanding vapour explosion.

Toxic Releases

Large quantities of toxic chemicals released into the atmosphere due to accidents can cause serious injury or death in large numbers. The tragedy in Bhopal is vivid example of this. In case of toxic releases the extent of damage depend upon several factors. The higher the quantity released, naturally higher will be the damage. The toxicity of the material is important. Physical properties of material released (lighter or heavier than air, soluble in water or not) also plays its role. Wind can spread toxic clouds over a large area. Hence factors like wind direction, wind velocity, fog and humidity needs consideration in estimating the extent of damag.

11.25 LIST OF INDUSTRIES INVOLVING HAZARDOUS PROCESSES

- Ferrous metallurgical industries

- Non-ferrous metallurgical industries
- Foundries
- Coal (including coke) industries.
- Power generating industries
- Pulp and paper industries
- Fertilizer industries
- Cement industries
- Petroleum industries
- Drugs and pharmaceutical industrial
- Fermentation industries (Distilleries and breweries)
- Rubber industries
- Paints and pigment industries
- Leather tanning industries
- Electro-plating industries
- Chemical industries
- Insecticides, fungicides, herbicides and other pesticides industries
- Man-made fiber industry.

11.26 FACTORY ACT, 1948

Objectives:

- To ensure adequate safety measures and to promote the health and welfare of the workers employed in factories.
- To prevent haphazard growth of factories through the provisions related to the approval of plans before the creation of a factory.

Scope and Coverage:

- Regulates working condition in factories.
- Basic minimum requirements for ensuring safety, health and welfare of workers.
- Applicable to all workers.
- Applicable to all factories using power and employing 10 or more workers, and if not using power, employing 20 or more workers on any day of the preceding 12 months.

Main Provisions of Factory Act 1948:

1. Compulsory approval, licensing and registration of factories.
2. Health measures.
3. Safety measures.
4. Welfare measures.
5. Working hours.
6. Employment of women and young persons.
7. Annual leave provision.

8. Accident and occupational diseases.
9. Dangerous operations.
10. Penalties.
11. Obligations and rights of employees.
12. *Compulsory Approval, Licensing and Registration of Factories.* The State Goverment may make rules
    - Requiring, the submission of plans of any class or description of factories to the Chief Inspector or the State Government; requiring, the previous permission in writing of the State Goverment or the chief Inspector to chief.

QUESTIONS

1. Define safety policy. What should be included in safety policy? Give some examples of safety policies.
2. What are the various factors that lead to growth of safety movement?
3. Discuss the occupational health and safety program in an industry.
4. Define safety inspection in an industry. How is it carried out?
5. Explain the procedure of safety planning in an industry.
6. Discuss the various safety measures in an industry.
7. How can employees participate in safety measures of an organization?
8. Explain the relationship of safety with plant design and equipment design.
9. How is safety related to productivity?
10. Name the statutory bodies, which are viewing safety seriously in India and abroad.
11. Define major industrial hazard.
12. Define industrial hazard. What are its types?
13. Briefly explain HAZOP technique of hazard identification.
14. How hazards can be controlled in an industry?
15. Name some industries involved in hazardous processes.
16. Write short notes on:
    (*i*) FTA
    (*ii*) FMEA
    (*iii*) Hazard management program in an industry.

12

Code of Professional Ethics for Engineers

12.1 INTRODUCTION

Ethics refers to most important beliefs and values of an individual or a society. *Honesty, Integrity, fulfilling commitments, abiding by agreements, being fair and open minded and willing to admit mistakes, being compassionate, having respect for human dignity, pursuit of excellence and being accountable for one's decisions are the universally accepted ethical principles.*

Moral is concerned with the principles of right and wrong behaviour. Very often the terms morality and ethics often used interchangeably but it is essential to identify the difference between these two. Morality tends to be more general and prescriptive.

Code of ethics is a framework for ethical judgment of any profession. It does not develop new moral principles but it guides an engineer, how to solve a specific problem by applying moral and ethical principles stated in it. Several engineering societies have framed the code of ethics of engineers by considering the fundamental principles of safety, health and welfare of the public in the performance of their professional duties.

The different professional engineering societies and their framed codes of ethics are detailed below

12.2 ASME—THE AMERICAN SOCIETY OF MECHANICAL ENGINEERS

ASME requires ethical practice by each of its members and has adopted the following code of ethics of engineers as referenced in the ASME constitution. Article C-211

12.2.1 Code of Ethics of Engineers

The Fundamental Principles

Engineers uphold and advance the integrity, honour and dignity of the engineering profession by :

- Using their knowledge and skill for the enhancement of human welfare.

- Being honest and impartial, and serving with fidelity the public, their employers and clients, and
- Striving to increase the competence and prestige of the engineering profession

The Fundamental Canons

1. Engineers shall hold paramount the safety, health and welfare of the public in the performance of their professional duties.
2. Engineers shall perform services only in the areas of their competence.
3. Engineers shall continue their professional development throughout their *careers* and shall provide opportunities for the professional and ethical development of those engineers under their supervision.
4. Engineers shall act in professional matters for each employer or client as faithful agents or trustees, and shall avoid conflicts of interest or the appearance of conflicts of interest.
5. Engineers shall built their professional reputation on the merit of their services and shall not compete unfairly with others.
6. Engineers shall associate only with reputable persons or organizations.
7. Engineers shall issue public statements only in an objective and truthful manner.

12.3 THE ASME CRITERIA FOR INTERPRETATION OF THE CANONS

The ASME criteria for interpretation of the canons are guidelines and represent the objectives toward which members of the engineering profession should strive. They are principles which an engineer can reference in specific situations. In addition, they provide interpretive guidance to the ASME board on professional practice and ethics on the code of ethics of engineers.

1. Engineers shall hold paramount the safety, health and welfare of the public in the performance of their professional duties.
   (*a*) Engineers shall recognize that the lives, safety, health and welfare of the general public are dependent upon engineer's judgments, decisions and practices incorporated into structures, machines, products, processes and devices.
   (*b*) Engineers shall not approve or seal plans and/or specifications that are not of a design safe to the public health and welfare and in conformity with accepted engineering standards.
   (*c*) Whenever the engineer's professional judgments are overruled under circumstances where the safety, health and welfare of the public are endangered, the engineers shall inform their clients and/or employers of the possible consequences.
      - Engineers shall endeavor to provide data such as published standards, test codes, and quality control procedures that will enable the users to understand safe use during life expectancy associated with the designs, products, or systems for which they are responsible.
      - Engineers shall conduct reviews of the safety and reliability of the designs, products, or systems for which they are responsible before giving their approval to the plans for the design.
      - Whenever engineers observe conditions, directly related to their employment, which they believe will endanger public safety or health; they shall inform the proper authority of the situation.

(*d*) If engineers have knowledge of or reason to believe that another person or firm may be in violation of any of the provisions of these canons, they shall present such information to the proper authority in writing and shall cooperate with the proper authority in furnishing such further information or assistance as may be required.

2. Engineers shall perform services only in areas of their competence :

(*a*) Engineers shall undertake to perform engineering assignments only when qualified by education and / or experience in the specific technical field of engineering involved.

(*b*) Engineers may accept an assignment requiring education and / or experience outside of their own fields of competence, but their services shall be restricted to other phases of the project in which they are qualified. All other phases of such project shall be performed by qualified associates, consultants, or employees.

3. Engineers shall continue their professional development throughout their careers, and should provide opportunities for the professional and ethical development of those engineers under their supervision.

4. Engineers shall act in professional matters for each employer or client as faithful agents or trustees and shall avoid conflicts of interest or the appearance of conflicts of interest.

(*a*) Engineers shall avoid all known conflicts of interest with their employers or clients and shall promptly inform their employers or clients of any business association, interests, or circumstances which could influence their judgment or the quality of their services.

(*b*) Engineers shall not undertake any assignments which would knowingly create a potential conflict of interest between themselves and their clients or their employer.

(*c*) Engineers shall not accept compensation, financial or otherwise, from more than one party for services on the same project, or for services pertaining to the same project, unless the circumstances are fully disclosed to, and agreed to by all interested parties.

(*d*) Engineers shall not solicit or accept financial or other valuable considerations, for specifying products or material or equipment suppliers, without disclosure to their clients or employers.

(*e*) Engineers shall not solicit or accept gratuities, directly or indirectly, from contractors their agents, or other parties dealing with their clients or employers in connection with work for which they are responsible. Where official public policy or employers policies tolerate acceptance of modest gratuities or gifts, engineers shall avoid a conflict of interest by complying with appropriate policies and shall avoid the appearance of a conflict of interest.

(*f*) When in public service as members, advisors, or employees of a governmental body or department, engineers shall not participate in considerations or actions with respect to services provided by them or their organization(s) in private or product engineering practice.

(*g*) Engineers shall not solicit an engineering contract from a governmental body or other entity on which a principal, officer, or employee of their organization serves as a member without disclosing their relationship and removing themselves from any activity of the body which concerns their organization.

(*h*) Engineers working on codes, standards or governmental sanctioned rules and specifications shall exercise careful judgment in their determinations to ensure a balanced viewpoint, and avoid a conflict of interest.

(*i*) When, as a result of their studies, engineers believe a project(s) will not be successful, they shall so advise their employer or client.

(*j*) Engineers shall treat information coming to them in the course of their assignments as confidential, and shall not use such information as a means of making personal profit if such action is adverse to the interests of their clients, their employers or the public.

- They will not disclose confidential information concerning the business affairs or technical process of any present or former employer or client or bidder under evaluation, without his consent, unless required by law or court order.
- They not reveal confidential information or finding of any commission or board of which they are members unless required by law or court order.
- Designs supplied to engineers by clients shall not be duplicated by the engineers for others without the express permission of the client(s).

(*k*) Engineers shall act with fairness and justice to all parties when administering a construction (or other) contract.

(*l*) Before undertaking work for others in which engineers may make improvements, plans, designs, inventions, or other records which may justify seeking copyrights, patents, or proprietary rights, engineers shall enter into positive agreements regarding the rights or respective parties.

(*m*) Engineers shall admit their own errors when proven wrong and refrain from distorting or altering the facts to justify their mistakes or decisions.

(*n*) Engineers shall not accept professional employment or assignments outside f their regular work without the knowledge of their employers.

(*o*) Engineers shall not attempt to attract an employee from other employers or from the market place by false or misleading representations.

5. Engineers shall built their professional reputation on the merit of their services and shall not compete unfairly with others.

(*a*) Engineers shall negotiate contracts for professional services on the basis of demonstrated competence and qualifications for the type of professional service required.

(*b*) Engineers shall not request, propose, or accept professional commissions on a contingent basis if, under the circumstances, their professional judgments may be compromised.

(*c*) Engineers shall not falsify or permit misrepresentation of their, or their associates, academic or professional qualification. They shall not misrepresent or exaggerate their degrees of responsibility in or for the subject matter of prior assignments. Brochures or other presentations used to solicit personal employment shall not misrepresent pertinent facts concerning employers, employees, associates, joint ventures, or their accomplishments.

(*d*) Engineers shall prepare articles for the lay or technical press which are only factual. Technical Communications for publication (theses, articles, papers, reports, etc.) which are based on research involving more than one individual (including students and supervising faculty, industrial supervisor/researcher or other co-workers) must recognize all significant contributors. Plagiarism, the act of substantially using another's ideas or written materials without due credit, is unethical.

(*e*) Engineers shall not maliciously or falsely, directly or indirectly, injure the professional reputation, prospects, practice or employment of another engineer, nor shall they indiscriminately criticize another's work.

(*f*) Engineers shall not use equipment, supplies, laboratory or office facilities of their employers to carry on outside private practice without consent.

6. Engineers shall associate only with reputable persons or organizations.
   (*a*) Engineers shall not knowingly associate with or permit the use of their names or firm names in business ventures by any person or firm which they know, or have reason to believe, are engaging in business or professional practices of a fraudulent or dishonest nature.
   (*b*) Engineers shall not use association with non-engineers, corporations, or partnerships to disquise unethical acts.
7. Engineers shall issue public statements only in an objective and truthful manner.
   (*a*) Engineers shall endeavor to extend public knowledge, and to prevent misunderstandings of the achievements of engineering.
   (*b*) Engineers shall be completely objective and truthful in all professional reports, statements or testimony. They shall include all relevant and pertinent information in such reports, statements or testimony.
   (*c*) Engineers, when serving as expert or technical witnesses before any court, commission, or other tribunal, shall express an engineering opinion only when it is founded on their adequate knowledge of the facts in issue, their background of technical competence in the subject matter, and their belief in the accuracy and propriety of their testimony.
   (*d*) Engineers shall issue no statements, criticisms, or arguments on engineering matters which are inspired or paid for by an interested party, or parties, unless they preface their comments by identifying themselves, by disclosing the identities of the party or parties on whose behalf they are speaking, and by revealing the existence of any financial interest they may have in matters under discussion.
   (*e*) Engineers shall be truthful in explaining their work and merit, and shall avoid any act tending to promote their own interest at the expense of the integrity and honor of the profession or another individual.
8. Engineers accepting membership in The American Society of Mechanical Engineers by this action agree to abide by this society policy on ethics and procedures for its implementation.

12.4 THE AMERICAN SOCIETY OF CIVIL ENGINEERS (ASCE)—CODE OF ETHICS

Fundamental Principles

Engineers uphold and advance the integrity, honor and dignity of the engineering profession by

1. Using their knowledge and skill for the enhancement of human welfare and the environment
2. Being honest and impartial and serving with fidelity the public, their employers and clients
3. Striving to increase the competence and prestige of the engineering profession and
4. Supporting the professional and technical societies of their disciplines.

Fundamental Canons

1. Engineers shall hold paramount the safety, health and welfare of the public and shall strive to comply with the principles of sustainable development in the performance of their professional duties.

2. Engineers shall perform services only in areas of their competence.
3. Engineers shall issue public statements only in an objective and truthful manner.
4. Engineers shall act in professional matters for each employer or client as faithful agents or trustees, and shall avoid conflicts of interest.
5. Engineers shall build their professional reputation on the merit of their services and shall not compete unfairly with others.
6. Engineers shall act in such a manner as to uphold and enhance the honour, integrity, and dignity of the engineering profession.
7. Engineers shall continue their professional development throughout their careers, and shall provide opportunities for the professional development of those engineers under their supervision.

12.5 IEEE—CODE OF ETHICS

We, the members of the IEEE, in recognition of the importance of our technologies in affecting the quality of life throughout the world, and in accepting a personal obligation to our profession, its members and the communities we serve, do hereby commit ourselves to the highest ethical and professional conduct and agree :

1. To accept responsibility in making engineering decisions consistent with the safety, health and welfare of the public, and to disclose promptly factors that might endanger the public or the environment
2. To avoid real or perceived conflicts of interest whenever possible, and to disclose them to affected parties when they do exits
3. To be honest and realistic in stating claims or estimates based on available data
4. To reject bribery in all its forms
5. To improve the understanding of technology, its appropriate application, and potential consequences
6. To maintain and improve our technical competence and to undertake technological tasks for others only if qualified by training or experience, or after full disclosure of pertinent limitations
7. To seek, accept, and offer honest criticism of technical work, to acknowledge and correct errors, and to credit properly the contributions of others
8. To treat fairly all persons regardless of such factors as race, religion, gender, disability, age, or national origin
9. To avoid injuring others, their property, reputation, or employment by false or malicious action
10. To assist colleagues and co-workers in their professional development and to support them in following this code of ethics.

12.6 INSTITUTION OF ENGINEERS (INDIA)—CODE OF ETHICS

Code of Ethics for Corporate Members

Preamble

The Corporate Members of The Institution of Engineers (India) are committed to promote and practice the profession of engineering for the common good of the community bearing in mind the following concerns:

- Concern for ethical standard
- Concern for social justice, social order and human right
- Concern for protection of the environment
- Concern for sustainable development
- Public safety and tranquility.

The Code

- A Corporate Member shall utilize his knowledge and expertise for the welfare, health and safety of the community without any discrimination for sectional or private interests;
- A Corporate Member shall maintain the honour, integrity and dignity in all is professional actions to be worthy of the trust of the community and the profession.
- A Corporate Member shall act only in the domains of his competence and with diligence, care, sincerity and honestly.
- A Corporate Member shall apply his knowledge and expertise in the interest of his employer or the clients for whom he shall work without compromising with other obligations to these tenets.
- A Corporate Member shall not falsify or misrepresent his own or his associates qualifications, experience, etc.
- A Corporate Member, wherever necessary and relevant, shall take all reasonable steps to inform himself, his employer or clients, of the environmental, economic, social and other possible consequences, which may arise out of his action.
- A Corporate Member shall maintain utmost honesty and fairness in making statements or giving witness and shall do so on the basis of adequate knowledge.
- A Corporate Member shall not directly or indirectly injure the professional reputation of another member.
- A Corporate Member shall reject any kind of offer that may involve unfair practice or may cause avoidable damage to the ecosystem.
- A Corporate Member shall be concerned about and shall act in the best of his abilities for maintenance of sustainability of the process of development.
- A Corporate Member shall not act in any manner which may injure the reputation of the Institution or which may cause any damage to the Institution financially or otherwise.

12.7 IIMM—INDIAN INSTITUTE OF MATERIALS MANAGEMENT

Ethics

- To consider first the total interest of one's organization in all transactions without impairing the dignity and responsibility to one's office.
- To buy without prejudice seeking to obtain the maximum ultimate value for each rupee or expenditure.
- To subscribe and work for honesty and truth in buying and selling, to denounce all forms and manifestations of commercial bribery and to eschew anti-social practices.
- To respect one's obligations and those of one's organization consistent with good business practice.

12.8 ENVIRONMENTAL ETHICS

The birth of environmental ethics as a force is partly a result of concern for our long-term survival, as well as our realization that humans are but one form of life, and that we share our earth with forms of life.

Ethos, the Greek word from where **"ethic"** is derived, means the character of a person as described by his or her actions. This character was developed during the evolutionary process and was influenced by the need for adapting top the natural environment. Our ethics is our way of doing thing. Our ethics is a direct result of our natural environment. Doing the later stage of the evolutionary process, Homosapiens began to modify the environment rather then submit to what, millennia later, became known as Darwanian natural selection. As an example consider the cave-dweller who, in the chilly dawn of prehistory, realized the value of the saber-toothed tiger's coat and appropriated it for personal use. Inevitably a pattern of appropriation developed, and our ethics became more self modified then environmentally adopted. Thus, we are no longer adapted to our natural environment but rather to our self made environment. In the ecological context, such maladaptation results in one of two consequences:

a. the organism (Homo sapiens) dies out or

b. the organism evolve to a form and character where it once again is compatible with the natural environment.

Assuming that we choose the latter course, how can this change in character (ethic) be brought about? Each individual must change his or her character or ethic, and the social system must change to become compatible with the global ecology.

The acceptable system is one in which we learn to share our exhaustible resources – to regain a balance. This requires we reduce our needs and that the materials we use must be replenish able. We must treat all of the earth as a sacred trust to be used so that its content is neither diminished nor permanently changed we must release no substances that cannot be reincorporated without damage to the natural system. The recognition of the need for such adaptation (as a means of survival) has developed into what we now call the environmental ethics.

Technology that people use today are quite often difficult to use and do not fit to our way of doing things. We have to learn many types of user interfaces, new paradigms and tools, and understand many logical schemes before we are able to use technology.

As human we have different aims and goals, culture, habits, skills and knowledge. We can ask, what is human centered technology? Different users see and experience technology differently. Technology should be adaptable. It should adapt to the user skills and level of knowledge and to the way the user uses the technology. When discovering new ways to use technologies, developers should always bear in mind the determines of different users. From global perspective people are using more and more technology. Technology should be functional in different cultures, in different age groups and overall in different contexts.

Technology should be considered as a tool that helps and simplifies normal human everyday processes. The whole usage context including the factors relating to the user, should define the characteristics of the technology solutions. The use of technology is not a value itself. The most important value in new technologies is the fact that user is able to accomplish his/her tasks effectively. Too often the design of technologies is dictated by the technologies themselves. Human centered technology can be understood as being technology that supports human actions in various usage contexts.

Engineering Ethics Definition

It is defined as a systematic study of the rules and ideals of the engineering profession. It is study of the moral issues and decisions confronting individuals and organizations engaged in engineering / profession. It is the study of related questions about the moral ideals, character, policies and relationships of people and corporation involved in technological activity. The set of behavioral standards that all engineers are expected to follow.

Dimensions of Engineering Ethics

The word 'ethics' has many distinct although related meaning. Corresponding to them are various additional senses of expression "engineering ethics". Ethics is an activity or area of enquiry it is the activity of understanding moral value, resolving moral issues, and justifying moral judgments, as well as the discipline or area of study resulting from the activity. Like ethics, engineering ethics also aims at finding out moral values related to engineering, knowledge accurate solutions to the moral problems in the engineering and justifying moral judgment of engineering. The term "ethical" indicates a contrast between moral questions and questions of a legal, political and artist nature. Engineering ethics in this sense refers to the set of especially moral problems moral problems, and issues related to engineering.

Ethics is used as a means of describing particulars set of beliefs, attitudes and habits related to individual's group's morality. Where as in engineering ethics, accepted codes and standard which are to be followed by the group of engineering or engineering societies will be used.

Ethics and the grammatical variants of ethics can also be described as "normally correct". In this connection, engineering ethics is also concerns with discovering moral principals, such as obligation, rights and ideals in engineering and by applying them to make a correct decision.

Variety of Moral Issues

An engineer is consistently moving with the environment. And he is required to make continuous decisions regarding technical, financial, managerial and ethical matters. The areas of possible disagreements on moral factors are as follows:

Organization Related Issues

It is obvious that all the engineers are not self-employed; they are employee of some organization or other. As an employee, an engineer should utilize his skills to the betterment of the organization and should take decisions accordingly. He is not expected to take decision for his own benefits at the cost of organization.

Environment Related Issues

It is important to use the resources carefully without depleting them. An engineer should take almost care to allocate resources in an effective manner and not to spoil the natural resources like air, land and water.

1. Society Related Issues

   The engineers should have a certain amount of social responsibility in addition to his core activities. This, his motive is not only to earn profit, but also to maintain the welfare of the society to a large extent.

2. Product relate issues

   In general the engineers are required to find out solutions to the following question with regard to the product:

   - Whether the product is safe
   - Does it match the requirement?
   - Will it perform the required functions?
   - Does it have concealed defects which will emerge during the later stages?

3. Finance/Cost Related Issues

   If the price of the product is priced higher than the optimum, it denotes that the engineers are charging his efficiencies upon the innocent customers. For a long lasting business, a right pricing strategy should be adopted.

4. Customer Related Issues

   Following are the customer related issues:

   The product is made available to the market (existing customers), the manufactures should not create artificial scarcity- because some organizations board the product and practice black marketing techniques to boost up price and reap unjustified profits.

5. Supplier Related Issues

   As engineers get his raw materials and other necessary inputs from certain sources and he is duty-bound to given right prices for them. He is duty-bound to given right price for them. He also has a task of developing ancillary industries and sub contractors around him. He must not give false hopes to ancillaries to expand their production/finance capacity and desert them after that.

6. Employees Related Issues

   As an employer, an engineer must ensure safe working condition for his workers.

12.9 WHISTLE BLOWING

It is the act by an employee informing the public or higher management of unethical or illegal activity by an employer or supervisor.

Whistle blowing is 'bringing an activity to a sharp conclusion as if by the blast of a whistle'

(Oxford English Dictionary)

Raising a concern about wrong doing within an organization or through an independent structure associated with it

(U K Committee on Standards in Public Life)

Giving information (usually to the authorities) about illegal or underhand practices (Chambers' Dictionary)

It is defined as bringing to the knowledge of relevant persons acts of moral or legal corruption with an expectation that some co-rrective measures will be initiated by the relevant persons.

In general, there are many groups/forums/people who may not belong to an organization but gather information on corruption and whistle-blow by writing articles for newspapers, magazines etc., or by informing suitable agencies to take action.

Example: Consumer forums, journalists, politicians.

There are four main features of whistle blowing that are given below:

1. Act of disclosure: Information is intentionally conveyed to the outside approved organization channels or in situations the person conveying it under pressure from supervisors or others.
2. Topic: Information is an important moral problem of the organization. For example, corruption, unethical policies, injustice to employees, threats to the employees and public safety and so on.
3. Agents: The person disclosing information is an employee or previous employee of the organization.
4. Recipient: Information conveyed to a person or organization in a position to act on the problem.

12.10 TYPES OF WHISTLE BLOWING

- **External whistle blowing**: Information is passed outside the organization.
- **Internal whistle blowing:** Information is conveyed to someone within the organization.
- **Open whistle blowing:** Individuals openly reveal their identity as they convey the information.
- **Anonymous whistle blowing**: Revealing information without revealing oneself.

There are also overlapping cases that are partly open and partly anonymous, such as when employees acknowledge their identities to a journalist but insist their names be withheld from anyone else.

12.10.1 Moral Guidelines (Whistle Blowing on Safety Issues)

Engineers are morally allowed to whistle blow when the following conditions are met :

- If the product can harm to the public to a considerable extent.
- If the engineers make their concerns about the harmful effects of the product to their superior.
- If all organizational channels have been informed about the harmful effects of the product to the society.
- Reasonable documented evidence should be present with the engineer.
- Engineers should have convincing evidence that making the information public will in fact prevent the threatened serious harm.

12.10.2 Protecting Whistle-Blowers

- Various laws had been made to protect whistle blowers.
- Unions also provide the required protection.
- Professional societies should also help in protecting whistle blowers. They can take steps to publish about companies that adversely treat whistle blowers.

12.10.3 Commonsense Procedures and Practical Advice

There are several rules of practical advice and common sense that should be needed before taking the action of whistle blowing.

1. Try to work through the normal organization channels.
2. Be prompt in expressing objections.
3. Proceed carefully and respect the feelings of others. Avoid personal criticisms.
4. Let the superiors to know the action through formal informally measures.
5. To be accurate in observations and claims and keep formal records documenting relevant events.
6. Consult colleagues for advice – avoid isolation.
7. Consult the ethics of committee of the concerned professional society.
8. Consult a lawyer concerning potential legal liabilities.

12.10.4 Right to Whistle Blow

Both the engineers working in government sector and private sector do have the professional moral right to whistle blow provided their motive is good and is done especially to safeguard the interests of the public with respect to health, safety and welfare measures.

12.10.5 Beyond Whistle Blowing

Internal whistle blowing should be permitted with greater freedom and open communication within the organization. These will be good steps to create an ombudsperson or an ethics review committee to investigate complaints and make recommendations to the top management. In order to take care of external whistle blowing, ethics committees of professional bodies can be approached to mediate (in confidence) between the employee (who whistle blows) and employee.

QUESTIONS

1. Define Ethics and Code of ethics.
2. Explain the Code of Conduct developed by the Institution of Engineers (India).
3. Explain the Code of Conduct developed by the American Society of Civil Engineers.
4. What is whistle blowing? State types of whistle blowing and also explain the guidelines on whistle blowing.

13

Politics and Technology

13.1 INTRODUCTION

Strong tendencies exist to apply technology and particularly information technology to support centralist, authoritarian world views. It is argued that alternative architectures can be readily created, which are more attuned to the openness and freedoms and which are supposed to be the hallmarks of democratic government. It is questioned whether authoritarianism will be capable of surviving the complexities, dynamism and widely distributed power which are features of the emergent information societies.

13.2 AUTHORITARIANISM'S UNDERLYING ASSUMPTIONS AND VALUES

An authoritarian society favours obedience to Authority over individual freedoms, to the extent of demanding subservience of the individual to the State. The notion clusters with tyranny (the cruel exercise of power), despotism and dictatorship (the exercise of absolute power), totalitarianism (single-party government) and fascism (a usually savage blend of authoritarianism with nationalism).

Authoritarianism is associated with logical positivist and utilitarian philosophies. These perspectives place very high value on rational social engineering, law and order, and resource efficiency. The populace is perceived as unsophisticated, uneducated, unreliable, chaotic, and/or incorrigibly venal and immoral. For their own good, the organised State must impose control on the unruly people.

A further assumption of the authoritarian perspective is that there exist humans with a level of both intelligence and morality superior to the common herd. In different ideologies, their innate superiority derives from different sources, such as the divine right of kings, wealth, force of arms, mystical power, what Machiavelli called virtú, wisdom, intellectual merit, technical capability, political cunning, demagogery, and/or public popularity. These superior humans are accepted as being the appropriate ones to make judgements on behalf of their society, with a minimum of checks and balances. They do this through social engineering; that is to say by organising and re-organising society in what they consider the rational way of achieving order and efficiency, and hence of delivering material well-being, and therefore spiritual happiness, for all.

13.3 THE AUTHORITARIAN MODEL OF I.T. APPLICATIONS

Under an authoritarian regime, the populace must be managed. Tools and techniques that have proven effective in managing raw materials, manufactured goods and animals, can be applied to humans too. A unique identifier for each person, and its general use by government agencies and other organisations which conduct transactions with people, are highly desirable tools for efficient social administration. Public administration systems must be designed to exercise control over people, in all of their various roles. There may be scope for at least some semblance of choice by individuals, but employees need to operate within a corporate culture, consumer demand needs to be statistically predictable, and citizens' freedom of choice needs to be constrained, lest unworkable parliaments eventuate, with too many splinter parties, independents and conscience votes.

It is only logical that an authoritarian society should recognise the benefits of a unary executive branch, in which the boundaries between agencies are porous. In this way, data can flow freely (such that transaction data and client histories can be cross-verified, and changes of address and status cross-notified), and systems can be tightly integrated and efficient (and hence misdeameanours by people in one arena, such as traffic fines, can be readily punished through another, such as denial of a marriage licence, permission to move apartments, or approval for travel).

Authoritarian IT-based systems use a centralised architecture. Elements may be physically dispersed, however, to achieve efficiency in data transmission, and to provide resilience against localised threats such as natural disasters and sabotage by dissidents. The general shape of such systems is that provided by cyberneticians: a cascade of control loops, culminating in a master-controller. In authoritarian regimes, information privacy and data security play important roles. These have little to do with the protection of people, however, but rather serve to protect the integrity of data, and of the system, and to legitimate the repressive system through the provision of nominal rights for data subjects.

13.4 INSTANCES OF AUTHORITARIAN APPLICATION OF I.T.

The reader is by now (hopefully) annoyed by the extent to which the foregoing description has been a caricature, hyperbole, a 'straw man' designed to be easily criticised. However there are manifold instances of just these features in IT-based public administration systems, both those in operation and being conceived, in countries throughout the world. In North America, whose use of IT has been well ahead of that in most other countries, a 'national data center' was proposed as early as 1966. Elements of it have emerged, such as the widespread use of the Social Security Number (SSN) as a unique identifier, proposals for a health id card, and the all-but uncontrolled use of computer matching and profiling. Some protagonists in the current debates surrounding the national information infrastructure (NII) are seeking a network consistent with authoritarian control; for example, by insisting on use only of those cryptographic techniques which are 'crackable' by national security agencies.

Australia has followed the North American tendency. It flirted with a national identification scheme in the late 1980s (Greenleaf & Nolan 1986, Clarke 1987, Graham 1990). When that was overwhelmingly rejected by the populace, senior executives in public sector agencies 'went underground'. They have variously gained Parliamentary support for, and smuggled through, a series of measures whose cumulative impact is in some ways already more emphatic than the 'Australia Card' would have been (Clarke 1992).

The cultures of many Asian nations are well-suited to authoritarian regimes. There are elements of high-social-efficiency applications of IT in such nations as Singapore. Busy Asian countries have shown especial interest in vehicle monitoring systems. Thailand and the Phillipines appear eager to act as laboratories for United States corporations developing identification and surveillance technologies. Under China's strongly authoritarian political system, it is unlikely that IT will be applied in any way other than to bolster existing relationships between its citizens and the State.

In Western Europe, Scandinavian countries lead the way with their social welfare systems and the heavy dependence of their citizens on the State. Denmark's citizen register is a model for authoritarian regimes everywhere, and a looming export success. Other countries are keenly adopting proposals to use IT to constrain the populace, by such means as identification cards (variously for football fans, patients, and the public in general), and the integration of data systems between government agencies, and between countries within and beyond the European Community.

In Central and Eastern Europe, there was an expectation that democratic, free enterprise systems would arise to replace the authoritarianism of the collapsed communist regimes. In practice, few of those countries have ever known freedom of choice, and genuine democracy (as distinct from variants of authoritarianism referred to in local lexicons as 'democracy') is not on the agenda of many of these countries. Their focus is on economic growth, rationalist solutions to economic and social problems, and centralism. IT is seen as a tool of authoritarianism, not of democracy, of centralised power, not of a pluralist body politic; and of control, not of freedom.

It can come as no surprise that public administration systems are being conceived in these ways. Applications of all kinds are developed by well-trained and self-confident engineers, using unequivocally rationalist techniques. System design comprises the expression of relevant parts of the present and near-future world in a formal model which has the important characteristic of being 'mappable' onto a machine. The application's users and 'usees' (*i.e.* the people affected by it) are treated as objects to be modelled, not as stakeholders with interests in the process and its outcomes. Human language is treated as though it were an (imprecise) formalism, rather than a means of interaction among people. The designers fail to notice that their formalisms cannot reflect the complexities, ambiguities and dynamism inherent in social systems, and the negotiation and accommodation processes which take place among humans and social groups (Clarke 1992b, Ciborra 1992, Agre 1994, Gronlund 1994).

Hence the problems highlighted in this paper are to a considerable degree inherent in the techniques currently used to develop IT applications generally. Nonetheless, their greatest impact on people's freedom is by way of public administration systems.

13.5 THE DEMOCRATIC ALTERNATIVE

The technological determinism notion has been applied to IT. In particular, IT has been accused of being inherently de-humanising, centralist and authoritarian (*e.g.*, Roszak 1986). The standpoint adopted by this author is that technology is essentially morally 'ambivalent' (*i.e.*, it has potential applications and potential impacts variously supportive of, and inimical to, any particular social value - Ellul, 1990). IT may make some paths easier than others, but the choice is made not by blind fate, but by politicians, government executives, and, not least, IT professionals.

The alternative political philosophy to authoritarianism is democracy, popularly expressed as 'government of the people by the people for the people', and commonly implemented through representatives chosen regularly and frequently by the combined and equal vote of all competent adults. The democratic ideal derives from the assumption that no class of people has the right to dominate other classes. It reflects the renaissance conception of mankind, whereby each individual should have the opportunity to access and interpret for themselves the ideas of other people and of Gods and, in more modern terms, should have the scope for self-determination and self-fulfilment.

Early computer technology may indeed have encouraged centralisation, but since the market-place debut of integrated circuitry and the mini-computer about 1970, modern IT has been readily applied in the service of democracy. Open IT-based systems involve nodes which are 'peers', with equal authority in respect of particular functions. For example, in a national health network, each node might take responsibility for all processing and storage relating to a particular aspect of the system's functionality (*e.g.*, support of a particular regional clinic, or epidemiological research into a particular class of diseases), and have special rights recognised by all other nodes in that regard (*e.g.*, the right of access, respectively, to identified data relating to specific patients, and to identifiable data relating to particular diseases and procedures). Similarly, particular kinds of data held at each node (*e.g.*, data identifying a patient) might be recognised as being controlled by that node and require special authority before it could be released to any other node.

One form of democratic topology is the unconstrained network, with maximum inter-connectivity, and dominion by each node over the services it provides. Another model is a variant on simple-minded cybernetics, a cascade of controllers which folds around, such that the ultimately controlled (the populace) are also the ultimate controllers (the voters). Before modern communications became available, the only practicable democratic mechanism for geographically large countries was periodic (typically, 3- or 4-yearly) election of representatives. In information societies of the very near future, however, major policy decisions can be instigated, formulated, and decided by direct democracy. Voters may choose to delegate the articulation of broad policies to their elected representatives, but even this can be subject to the over-riding of unpopular decisions, and the removal of representatives the electorate considers are not performing their functions.

Hierarchical topologies serve authoritarianism, whereas non-hierarchical ones are consistent with a free society. Access to data under the control of each node must be restricted, until and unless, via due process, disclosure is justified in fulfilment of some higher interest. Such topologies provide not only robustness and adaptability, but also integrity, because clients can trust them, and there is a lower risk of loss of quality (through suspicion and uncooperativeness), and of sabotage (through active attempts to mislead, and direct, destructive action).

13.6 INSTANCES OF DEMOCRATIC APPLICATION OF I.T.

Is this image of democratic computing just a caricature too? Possibly, but examples exist. Local Area Network architectures are inverting the old notion of centralist processors accessed by terminals. The now-conventional names reflect the fact that 'client' workstations demand data and processing from 'servers' the user's device is in control, and the central facility performs at its bidding. In wide-area networking also, peer-to-peer protocols are rivalling and may be progressively replacing the older, hierarchical or 'star' configurations. At the level of inter-networking, the topology of the world-wide TCP/IP-based Internet is essentially flat, the systems software is highly distributed, the redundancy is very high, and its robustness, its resilience and its capacity to resist authoritarian governments are therefore all of a high order.

The Internet's technical features have resulted in a culture very different from that on hierarchical nets. It provides a space in which imaginations have substantial freedoms. Some people use those freedoms to create new services and products, others to experiment with self-expression and group-experiences, some as a 'cybernetic' analogue to psychotropic drugs, and some just to distribute pornography or racist materials. Nor are the boundaries between these activities always clear-cut.

It seems ironical that the Internet was sponsored by the United States military complex, but the irony is more apparent than real. Systems which support military operations cannot risk the fragility of centralisation, but rather demand robustness and resilience, and therefore redundancy. Moreover, aero-space-defense R. & D. is dispersed across vast numbers of universities and private sector research laboratories. It then seeks to complement competition by collaborative interaction among individual researchers and among potential research partners. To retain its technological and intellectual leadership, it was essential that the U.S.A. avoid the temptation to sustain centralised, authoritarian topologies, and to its credit it knowingly spawned a dynamic, world-wide, democratic network laboratory.

In the above paragraphs it has been considered the extremes of authoritarianism and democracy. Clearly, any society will demand not only freedoms, but also protections against those who use those freedoms to harm others. Naive authoritarian models are doomed to fail, because they deny freedoms; and naive democratic models are doomed to fail too, because they deny protections. Ronfeldt concluded that IT-based public administration (which he calls 'cyberocracy') "far from favoring democacy or totalitarianism ... may facilitate more advanced forms of both" (1990, p.283). How should new 'cyberocracies' be designed, and how should existing public administration systems be adapted to exploit the new opportunities, while balancing the needs for control and freedom?

Authoritarian aspects of schemes could be justifiable in some societies as interim measures. Lenin and then Stalin judged that the country's large peasant population, and its institutions, were insufficiently mature for immediate implementation of the full Communist platform. Unfortunately the repression inherent in their interim arrangements became ingrained, and was only relieved by counter-revolution. Authoritarian elements in public administration should therefore be not only justified, but also demonstrably interim, *i.e.*, the means must be shown whereby they will be replaced, by evolutionary processes, with alternative mechanisms consistent with democratic principles.

In any case, the feasibility of grafting democratic features onto an essentially hierarchical model must be regarded as very slim. All power vests in the centre, and any softening of the system's features is by gift of the powerful. Moreover, the system can be manipulated by the powerful (for example, by monitoring nominally confidential communications), and privileges can be withdrawn by the powerful. No freedom-loving populace could regard such a system as credible, and would therefore only submit to it as a result of coercion.

Is the alternative feasible, to graft control mechanisms onto an essentially open model? Communication channels can still be tapped and storage devices searched (under warrant). Evidence arising from such interceptions and searches can still be presented in a court of law. Certain actions and uses of IT can be expressly made illegal. The ex post facto controls can therefore still function within open, democratically conceived public administration. Toffler distinguished this form of IT application by coining the term 'practopia' (1980, p.368).

What is not so simple to contrive within open systems is effective real-time monitoring and control: Foucault's 'prison' is readily implemented using hierarchical topologies, but if the nodes and arcs of networks are not all under the control of Authority, then preventive controls become much harder to bring to fruition. That, then, is the essential battleground between authoritarian and democratic models of IT, should someone or some class of people, and in particular politicians and senior public sector executives, be permitted to have the power to prevent transgressions? Because it is that kind of control over the public which is at the very heart of the anti-utopian nightmare.

13.7 CONCLUSIONS

Power does not need to be explicitly and consciously granted to public administrators by the voting public, or by their elected representatives. It can accrue, slowly and gently, through developments in IT, through new applications of established techniques, through the gradual 'creep' of existing schemes into new functions, and through seemingly harmless refinements to statutes. As frogs are reputed to do, a society might resist being put into boiling water, yet be lulled to sleep in warm water slowly brought to the boil.

Little echoes of these patterns are evident in contemporary societies. The use of the Internet for anti-social purposes is proving much harder to control than similar behaviour using the telephone network. IT contributed significantly to the breakdown of the Soviet Union because, in addition to improving production effectiveness and efficiency, PCs delivered 'samizdat' - the means for cheap reproduction of dissident newsletters. Lies that had been lived for seven decades could not withstand the heat generated by eager users of a potentially democratising technology. And that was before inter-networking and computer-mediated communications had achieved any degree of sophistication.

IT may be applied to public administration in ways consistent with authoritarianism or with democracy. Proponents of hierarchical structures and social engineering, chief amongst them senior public sector executives, must at the very least appreciate the limits of tolerance of authoritarian measures within their society. Preferably, governments should ensure that social administration schemes are not emphatically centralised and incapable of adaptation towards more liberal patterns. And most desirably, public servants, governments and voters themselves, should be exploiting the opportunities for more effective democracy which are being created by information technology.

13.8 DEMOCRATIC POLITICS OF TECHNOLOGY

Nations such as the United States have evolved elaborate checks and balances, enshrined in formal constitutions, for ensuring that we never enact a new law that would subvert our basic constitutional principles and cherished political values. By what perverse logic can we justify holding technologies — which plainly produce repercussions on social structure as sweeping and profound as any law — to a lesser standard?

No nation on earth has an effective system for taking into account the profound effects that technologies exert on basic social and political structures, including on democratic values and institutions. The absence of such a capability stunts fulfillment of the democratic promise, thwarting people's opportunities to establish the lives, communities, and societies they wish.

13.9 DEMOCRATIZING TECHNOLOGY: HISTORICAL AND THEORETICAL BACKGROUND

No nation on earth has an effective system for taking into account the profound effects that technologies exert on social and political structures, including on fundamental democratic values and institutions.

Historic Example — Improving Public Health/Compromising Democracy: *In the late 19th century, U.S. cities experienced epidemics of typhoid fever and other diseases. Over time, public health experts identified the culprit as sewage-contaminated drinking water supplies. Either local or state governments could have taken charge of the needed clean-up, but political centralization won out. State governments began appointing new public authorities to manage water and sewage on a translocal, regional scale. The result: public health improved dramatically, but local autonomy and municipal democracy suffered.*

Moreover, this case set a precedent emulated in other areas of infrastructural improvement: roads, ports, energy sources, and telephone services. In each case, civic decisions were shifted from municipalities (where decision-making forums were often accessible to workers and everyday citizens) to larger, more remote state and national political arenas (where generally only wealthy businesses and individuals possessed the resources needed to exert influence). By neglecting more local means that were available at the time for addressing urban needs, the U.S. underwent a fundamental change in political structure — yet without any of the political deliberation or due process that would normally be considered appropriate to transformations of this character and import.

In a more contemporary vein, I have written elsewhere of a "Cybernetic Wal-Mart Effect" — the danger that unregulated electronic commerce, by draining revenue from local economies, could erode community vibrancy, public spaces, and the buffering that a robust local economy affords against the vagaries of impersonal global market forces. On all three counts, this would impair conditions vital to healthy democratic self-governance.

Interest in the relationship between technology and democracy has mushroomed over the past decade, both among scholars and in practice. The vast majority of this constructive effort has involved investigating the privileged role of technical experts in technological and environmental decisions, or else is concerned with new mechanisms for broadening participation in such decisions.

But from another perspective, all of this recent work is covering only half of the necessary terrain. Consider a simple distinction between democratic *procedures* versus their *substantive results.* Imagine, for example, a public referendum on whether to abolish the Bill of Rights of the U.S. Constitution (that is, freedom of the press, the right of assembly, and so on). In this case, the public referendum process — that is, citizens voting for their policy preferences — is the democratic procedure, while the decision ensuing from that process is the substantive result. Notice that democratic procedures do not invariably result in democratic outcomes (for instance, suppose the preceding hypothetical referendum were actually to abolish the Bill of Rights).

This distinction can illuminate the sense in which recent work on the relationship between democracy and technology, while immensely valuable, is also seriously incomplete: These recent efforts are concerned exclusively with *democratic procedure* in making decisions about technologies. But they omit the equally important question of whether technologies are *substantively* democratic — that is, whether a technology's design and use is compatible with perpetuating democratic social relations.

This omission is non-trivial. Just recall the preceding examples of U.S. sewage systems that compromised municipal democracy or the democratic hazards associated with a Cybernetic Wal-Mart Effect. As Langdon Winner wrote in a widely cited 1978 essay, "Do Artifacts Have Politics?":

Technological innovations are similar to legislative acts or political foundings that establish a framework for public order that will endure over many generations. For that reason the same careful attention one would give to the rules, roles, and relationships of politics must also be given to such things as the building of highways . . . .

The dearth of practical, systematic effort to ensure that democratic procedures issue in technologies that are *substantively* democratic — *i.e.*, that help, both individually and collectively, and both directly and indirectly, to sustain democratic values, procedures and institutions — is what called "the missing half" of a democratic politics of technology.

For instance, despite aspiring to conceptual comprehensiveness, the massive *Handbook of Science and Technology Studies* — compiled by the international Society for Social Studies of Science — does not even mention the issue. Neither does *Fairness and Competence in Citizen Participation*, an otherwise outstanding comparative evaluation of alternative methods in participatory environmental decision-making prepared by Ortwin Renn and colleagues.

In a world of pervasive technology, the recurring failure to consider the impact of technologies on basic social and political structure is as astonishing as it is ominous. In the absence of such effort, *democratic institutions will continue to be distorted, thwarted, and further jeopardized by decision-making procedures* — ranging from market interactions and technocratic policy analyses to even the most highly evolved democratic procedures — *that fail to assist participants in considering the structural social consequences of their decisions.*

Now, were it true that lay participants normally inquire deeply into the implications of technologies for their society's basic structure, then this concern could be addressed simply by continuing the political struggle for "more democratic participation." However, case studies reveal that while lay people *do* routinely raise significant social issues about technologies that technical experts are prone to slight, they don't do so sufficiently deeply and systematically to prevent unintended sociotechnological results, including adverse effects on a society's basic democratic structure.

Thus while contemporary industrial and postindustrial societies have grown attentive to economic, health, environmental and selected ethical repercussions of technologies, apparently our cultures still lack the vocabulary, concepts, illustrative examples, and so on that would make it feel natural and legitimate for lay participants to consider broader social, cultural and political impacts in depth and systematically. That is where new assistive tools or participatory procedures would be valuable.

Seek technologies that can enable disadvantaged individuals and groups to participate fully in social, economic, and political life. Avoid technologies that support illegitimately hierarchical power relations between groups, organizations, or polities.

The complete list of criteria developed in *Democracy and Technology* is posted on the Web at A Provisional System of Design Criteria for Democratic Technologies. Inasmuch as democracy is a precondition for setting other social priorities fairly and effectively, the book argues that use of such design criteria should, in principle, be a first-order consideration in technological decisions. Against this standard, it is striking that, in practice, so far no nation on earth has evolved a systematic capability to take technologies' impacts on democracy into account even as a last-order consideration. (Granting first-order consideration to technologies' democratic impacts is an unfamiliar concept, but it is directly analogous to the practice of ensuring that proposed legislation is consistent with the first-order principles embodied in a nation's constitution.)

Thus one approach to an adequate "democratic politics of technology" would be to incorporate democratic processes for debating, refining, customizing to local circumstances, and then applying such criteria in order to evaluate technologies' effects on democratic social relations. My stress on the need to debate and refine criteria underscores the notion that I intend these criteria not as an endpoint, but merely as a conversational stimulant and starting point — albeit for a democratically vital conversation that, in the absence of such stimulation, rarely occurs.

However, *Democracy and Technology* has at least one core limitation that has inhibited practical applications of its provisional democratic design criteria. The book observes that crucial social impacts often emerge from the interaction among seemingly unrelated technologies. Consider in the United States the role of automobiles in driving pedestrians away from streets — a result now reflected and reinforced materially by the elimination of front porches from homes and the creation of sprawled suburbs that don't even have sidewalks — and the complementary role of air conditioning, central heating, television, and other home entertainment devices in drawing people indoors. All of these technologies interact to weaken face-to-face social interaction at the local level. A conventional evaluation of a single technology — say, air conditioning — is unlikely to detect such combined or synergistic effects.

On this basis, *Democracy and Technology* reasons that — beyond debating and refining democratic design criteria — democratic processes are needed for applying these criteria not only to one technology at a time, but also comprehensively across a society's technological order.

That's a pretty tall order: lay participants are expected to debate and refine *multiple democratic criteria*, and then use them to evaluate the *combined effects of multiple technologies*. Can human beings really do that? Thus a central limitation of *Democracy and Technology* is that it does not describe an institution or democratic procedure that would make this complex task practicable.

QUESTIONS

1. What is authoritarian versus democratic control of technology?
2. Define authoritarian model of I T applications.
3. State democratic politics of technology.
4. What are the underlying assumptions and values of Authoritarian model?

14

MAN MACHINE SYSTEM AND AUTOMATION

14.1 INTRODUCTION

Facilities layout provides overall integration of design of production-systems. However, while doing facilities planning an extremely important decision about the basic organization of the work has usually been left out, *i.e.*, the system is organized to have compatibility with the work to be done. In the present context, work means the operations involved in manufacturing the work. So, there must be compatibility between the physical configuration, the materials and facilities available. Within each of these physical configurations, there may be a number of possible ways to divide up the work to be done as a basis for the design of jobs. There must be a continuous effort to consider alternatives of overall organization of work job content which will be determined by the process, existing machine designs, the physical layout, time requirements, etc.

In an entire complex assembly process, we break it down into a series of operations so that the product can be produced on an assembly-line and the line is designed to meet capacity requirements.

The basic requirement is to achieve higher quality levels in produce, lower the production cost, employee's satisfaction, reduce the risks of hazards and finally job and methods design. Meaning thereby the 'work design and method design must economically match'.

Exhibit 14.1 shows the relationship of constraints, and other forces acting in deciding the work/job content.

Constraints	**Criteria and forces**
(*i*) Product design	(*i*) Economy
(*ii*) Processes	(*ii*) Job specialization
(*iii*) Production targets	(*iii*) Job satisfaction
(*iv*) Layout	(*iv*) Job risk factors
(*v*) Pacing of equipment.	(*v*) Make skills.

JOB CONTENT

JOB METHODS
- (i) Physiological data
- (ii) Work flow principles
- (iii) Data on fatigue
- (iv) Work Schedule.

Exhibit 14.1. *Typical relationship between various forces.*

Therefore, in discussing work design, we shall concentrate on Man-machine system focusing on job methods component of work design. Man's great potential is not to be used in repetitive mechanical activities, but rather in situations requiring improvisation, reasoning and judgment. In the era of highly organized and expensive labour, it is becoming obvious that we may also be misusing man's capabilities from an economic point of view.

14.2 WORK AND MAN-MACHINE INTERACTION

In performing work, man's functions fall into three categories:

(i) Receiving information through various sense organs.

(ii) Making decisions based on information received and information stored in the memory of individual.

(iii) Taking action based on deciɹion.

There are few objectives guidelines to the allocation of tasks to man and machine i.e. there are tasks that are appropriate for man only while there is a list of tasks appropriate for machine only:

(i) Human being appear to surpass existing machines in their ability to :

- (a) Detect small amount of light and sound.
- (b) Improvise and use flexible procedures.
- (c) Store large amount of information for long periods and recall facts at appropriate time reason inductively and excessive judgment.
- (d) Develop concepts and create methods.

(ii) Existing machines appear to surpass human in their ability to :

- (a) Respond quickly to control signals.
- (b) Apply great force smoothly and precisely.
- (c) Perform repetitive and routine tasks.
- (d) Store information and erase it completely.
- (e) Perform rapid computations.
- (f) Perform many different functions simultaneously.

The following schemes show man-machine system functions in three different ways :

Man as power source and controller

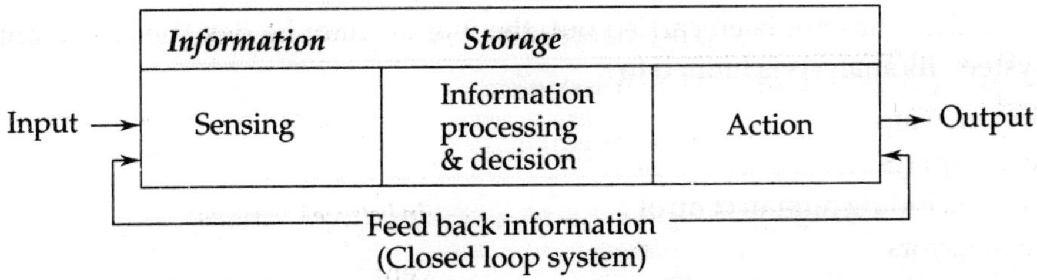

(*a*) Manual system where in man directly transforms input to output. The system may be operating in some working environment which may have an impact on the man and the output.

Man as power source and controller

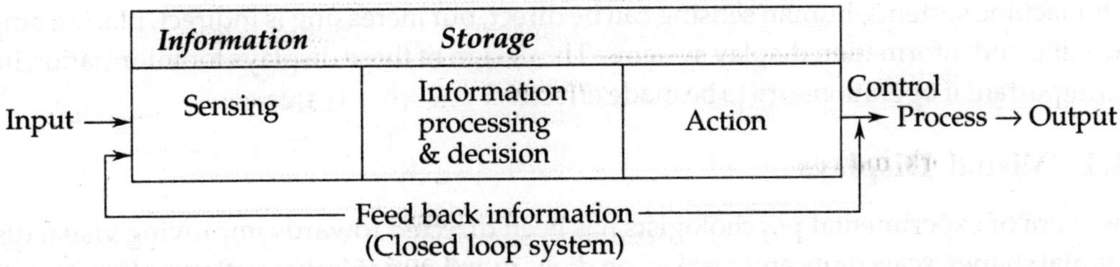

(*b*) Semi-automatic system

Machine

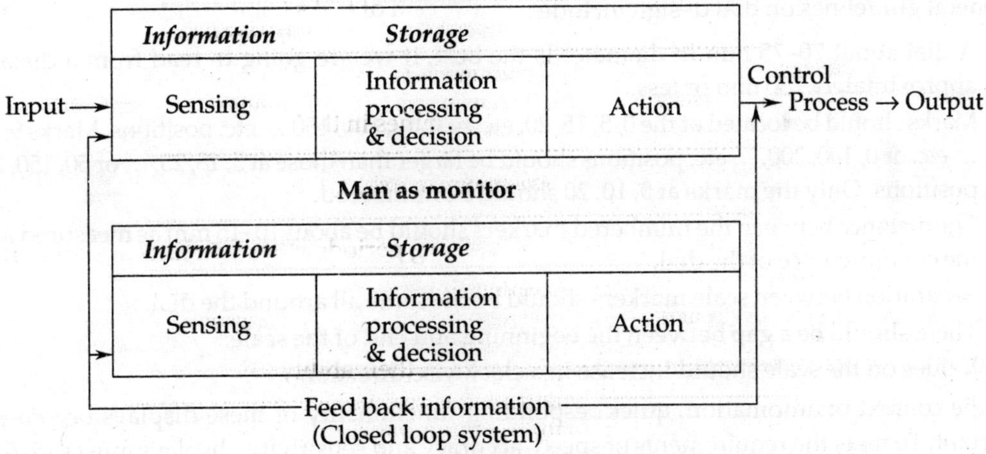

(*c*) Automatic system

(*c*) Semi-automatic system wherein man functions as a controller of the process and interacts with machine by sensing information about the process, interpreting it and using a set of controls to carryout possible adjustment as the work requires. In includes the off and on of the machine system. Most of the machine tools fall into this category.

(*d*) Automatic system which may not need man. However, the role of man is to monitor the control of the process. "Man periodically or continuously maintains surveillance over the process through displays which indicates the present status of crucial parameters of the process.

Therefore, in the context of designing the work system, compatible to man-machine system, the analysis of the entire man-machine system is essential. The analysis needs to discuss the following aspects of man-machine system.

14.3 ANALYSIS OF MAN-MACHINE SYSTEM

Unless, the analysis has not been carried out, the work cannot be designed compatible to man-machine system. Its analysis is limited to :

(*i*) Information input
(*ii*) Visual displays
(*iii*) Control including human control
(*iv*) Human factors
(*v*) Working environment.

14.4 INFORMATION INPUT

In man-machine systems, human sensing can be direct, but increasing is indirect, placing emphasis on encoding and information display systems. The design of these displays for information input to man is important if operations are to be made effective.

14.4.1 Visual Displays

Post war-era of experimental psychologists has been directed towards improving visual displays such as dial shapes, scale units and marked on dials, number preference patterns affecting interpretation of dial readings, characteristics of numbers and letters to make them most legible, size of letters and numerals and line thickness, height, width, arrangement of dials, etc.

General guidelines on dial design include :

(*i*) A dial about 70–75 mm in diameter is the best, if we are going to read from a distance of approximately 750 mm or less.
(*ii*) Marks should be located at the 0, 5, 15, 20, etc. or 0, 50, 100, 150 ... etc. positions. Marks 9, 10, 20, ... etc. or 0, 100, 200, ... etc. positions should be larger than those at 5, 15, 25, ... or 50, 150, 250, ... positions. Only the marks at 0, 10, 20 should be numbered.
(*iii*) The distance between the numbered markers should be about 10–15 mm as measured around the circumference of the dial.
(*iv*) Separation between scale markers should be the same all around the dial.
(*v*) There should be a gap between the beginning and end of the scale.
(*vi*) Values on the scale should increase in a clockwise direction.

In the context of automation, quick responding functionality of these displays becomes very important. To meet the requirements of speed, accuracy and sensitivity, displays must satisfy three criteria – Detection, Recognition and Understanding. The pre-requisite information desired for the design of displays are as follows :

(*i*) The type of technology to be the most suitable to present the information visual, auditory, mechanical, electronic.
(*ii*) The necessary accuracy and sensitivity required from the point of view of communication.
(*iii*) The total range of information to be presented and to be displayed.
(*iv*) The necessary speed requirement for communication.
(*v*) The permissible equipment error.
(*vi*) The normal and maximum distance for user to use display.

(*vii*) The environmental conditions under which the display will function.

14.4.2 Controls

Displays and controls are usually associated with every man-machine system. Displays work to communicate information, while controls monitor the state of process. Wide variety of control devices (Knobs, levers, pedals, cranks, wheels etc.) is available for the application. Controls are basically developed as extensions of the human body. From these simple controls, larger and complicated machines were developed in order to extend human ability and to relieve people from the need to spend larger amounts of energy. The human ability to control the human body must be transferred to control the movement and operation of the machine and its components. The control is no longer a tool itself, but is an intermediary between the user and the machine. So, the control has also got to fit the body of the user, but must also be able to indicate results with minimum loss of time. Control, being the direct link of communication between the two (user and machine) must be designed to meet the human capabilities and the system as a whole, if the system has to operate at maximum efficiency. A control must be accessible, identifiable and usable. Depending upon the application of control, the factors like force, speed and range and accuracy are important.

The design of controls and control systems has an important impact on the effectiveness of the man-machine system. Knowledge of forces that man exerts may be of importance in some systems so that these capabilities are not exceeded in the design of controls.

14.4.3 Human Controls

Having information as input by direct or indirect means the human operator of a man-machine system responds by performing work in the physical sense. To accomplish this, analysis of the hand and body motions and how they contribute to effective operations is important.

Manipulative activity in handling controls has been studies with considerable care and this knowledge can be used to design effective systems. Workplace layout can be based on a knowledge of anthropometry so that manual motions can take place within a prescribed area, and chair and table heights can be set at levels appropriate to human body sizes. The following table is a summary of the areas of application of different manual activity methods of analysis.

TABLE 14.1. Summary of areas of application of different man and machine analysis methods

Nature of activity	*Analysis of method*
1. Repetitive short cycle task, low to moderate production volume	Operation chart or operation chart Supplemented by motion time standard data
2. Repetitive short cycle task, high production volume.	Micromotion analysis chart, motion time standard data
3. Repetitive long cycle tasks	Man flow process charts, activity chart
4. Repetitive tasks involving areas and/or a machine	Activity Chart
5. Jobs involving tasks occurring at irregular intervals	Activity classification

While employing human controls in an effective way, the 'principles of motion economy' are also very important. Normally, these motion economy principles are used to reduce fatigue and they have wide applications in work system design.

14.5 PRINCIPLES OF MOTION ECONOMY

A Check Sheet for Motion Economy and Fatigue Reduction

These twenty-two rules or principles of motion economy may be profitably applied to shop and office work alike. Although not all are applicable to every operation, they do form a basis or a code for improving the efficiency and reducing fatigue in manual work.

Use of the Human Body:

1. The two hands should begin as well as complete their motions at the same time.
2. The two hands should not be idle at the same time except during rest periods.
3. Motions of the arms should be made in opposite and symmetrical directions and should be made simultaneously.
4. Hand motions should be confined to the lowest classification with which it is possible to perform the work satisfactorily.
5. Momentum should be employed to assist the worker wherever possible, and it should be reduced to a minimum if it must be overcome by muscular effort.
6. Smooth continuous motions of the hands are preferable to zigzag motions or straight-line motions involving sudden and sharp changes in direction.
7. Ballistic movements are faster, easier, and more accurate than restricted (fixation) or "controlled" movements.
8. Rhythm is essential to the smooth and automatic performance of an operation, and the work should be arranged to permit easy and natural rhythm wherever possible.

Arrangement of Work Place:

9. There should be a definite and fixed-place for all tools and materials.
10. Tools, materials, and controls should be located close in and directly in front of the operator.
11. Gravity feed bins and containers should be used to deliver materials close to the point of use.
12. Drop deliveries should be used wherever possible.
13. Materials and tools should be located to permit the best sequence of motions.
14. Provisions should be made for adequate conditions for seeing. Good illumination is the first requirement for satisfactory visual perception.
15. The height of the workplace and the chair should preferably be arranged so that alternate sitting and standing at work are easily possible.
16. A chair of the type and height to permit good posture should be provided for every worker.

Design of Tools and Equipment:

17. The hands should be relieved of all work that can be done more advantageously by a jig, a fixture, or a foot-operated device.
18. Two or more tools should be combined whenever possible.
19. Tools and materials should be propositioned whenever possible.
20. Where each finger performs some specific movement, such as in typewriting, the load should be distributed in accordance with the inherent capacities of the fingers.

21. Handles, such as those used on cranks and large screwdrivers, should be designed to permit as much of the surface of the hand to come in contact with the handle as possible. This is particularly true when considerable force is exerted in using the handle. For light assembly work the screwdriver handle should be so shaped that it is smaller at the bottom than at the top.
22. Levers, crossbars, and hand wheels should be located in such positions that the operator can manipulate them with the least change in body position and with the greatest mechanical advantage.

14.6 HUMAN FACTORS

As stated previously, Engineering Design is concerned with the satisfaction of human needs. With this in mind, it is quite obvious that there are many areas in which human factors must be taken into account in the design analysis. The basic consideration in determining whether or not human factors are important is the possibility of a direct man-machine relationship. Having determined the possibility and extent of involvement in is then necessary to consider the form of this relationship, be it :

(*i*) Anthropometric
(*ii*) Ergonomic
(*iii*) Physiological
(*iv*) Psychological.

From this general appraisal, the Design Engineer will be in a position to determine the relative importance of human factors in his design and the possibility of trade-offs between these factors and others, such as cost and performance. Here we shall be concerned with developing an understanding and appreciation of the variety of human factors, their relative importance and the limitations they imposed on the design situations.

14.6.1 Anthropometric Factors

Anthropometric human factors are those which are related to the physical size of humans. They are thus mostly related to the accommodation of the person but they do also relate to operation and control of the machine in so far as space requirements are concerned. One of the most common problems of this nature with which most people are familiar is that of driver accommodation in automobiles. Some of you will have experienced difficulties in either getting in or outreaching the controls or adequate visibility. These may have been due to inadequate attention to the nature of the physical dimensions of humans.

We all know that not two people are alike, and this concept of difference obviously applies to the physical characteristics of size, shape and weight. No matter what physical characteristics are, we consider, there will be a range of measured values depending on age, sex, ethnic origin, diet and health. Even when we consider a selected population-persons over 18 years old having in India we will find a considerable scatter in height.

This information can be conveniently presented in a statistical distribution diagram which shows the relative frequency of occurrence of various heights. An alternative method of presentation is by the use of the cumulative distribution diagram, which indicates the percentage of the population whose height is less than the value indicated.

However, an average height can be taken for the population. It would seem on first consideration that one should design for the average person, because we are presumably designing for the largest group in the community. (The peak in the relative frequency diagram is often close to the average value), *i.e.*, by designing for the average person we often preclude 50% of the population.

An example which illustrates this concept is that of the forward visibility of a car driver. Consider the requirement that a car driver must be able to see an obstacle of height 'h' up to a minimum distance 'L' from the front of the car (the stop lights of the car in front, or a child crawling across the road).

Now it is also recognized that one cannot design for the minimum height person because:

(*i*) The tail of the distribution diagram is very drawn out and the numbers of people involved are very small.

(*ii*) It is virtually impossible to determine the minimum unless one conducts a 100% survey of all potential drivers.

For these reasons, and also for economy it is usual to restrict the minimum for design purposes to the 5th percentile level *i.e.*, 5% of the population will not be adequately accommodated using the standard product. These disadvantaged people will be required to make their own special arrangements. It would be a useful exercise for you to determine what percentile level should be considered in defining the minimum roof height.

The designer should therefore need to define his criteria for satisfaction and which side of the demarcation is acceptable, depending on feature considered.

But there is still more to the problem then being merely based on physical size of the driver. We must realize he will be seated on a flexible seat. The deflection of the seat will depend on the driver's weight and postural dimensions. This now becomes a more complicated procedure which is difficult to save in the drawing office. For this reason, automobile manufacturers carry out many design studies using "manikin" which represents the essential physical characteristics of the human being.

There are many other situations in which anthropometric considerations need to be taken into account in the design of equipment to be used by human operators. Some typical cases may be observed in the following list :

(*i*) Crane operator's cabin

(*ii*) Furniture

(*iii*) Hand held implements

(*iv*) Typewriters

(*v*) Domestic appliances.

It would be a valuable exercise for you to examine some object which you make use of Identify the nature of the anthropometric considerations and determine the relevance of the 5th or 95th percentile levels (or even in some cases the 50th percentile level).

14.6.2 Ergonomic Factors

These are of concern when the human operator is involved with the machine in the dynamic sense. He is required to exert a force perhaps supply work to the machine. Once again it should be obvious that the effective operation of a machine over a long periods of time will depend upon the matching of requirements to human capability. One simple example is that when the brake operating force is higher than can be normally tolerated. The driver will cope with the situation at first, but if the brake operation is frequent, muscle fatigue will eventually occur and the dangerous situation will develop. Other situations where similar types of failure may occur are found when moments rates are excessive. Again muscle fatigue will eventually develop and inadequate operation will result.

The capability for performing mechanical tasks of this type depends upon:

(*i*) The physical ability of the operator
(*ii*) The range of moment required
(*iii*) The speed of moment
(*iv*) The duration of activity
(*v*) The position of operator
(*vi*) The environmental conditions.

A variety of information is available for the design engineer concerned with satisfying the ergonomic factors involved. As an example the static strength of arm movements have been determined in one survey on seated subjects. The subjects were to apply either a horizontal pull or push a vertical up or down force or a horizontal inward or outward directed force. The upper arm was positioned at various angles with the forearm horizontal straight ahead

Ergonomic factors are of great importance in the design of workplaces for human operators. The arrangement of components for assembly is a very common application of ergonomic principles which is also known as motion and time study. The usual situated is that several components must be assembled together to form a product (electric power plug, carburetor, domestic appliance etc.). The product may be either small or large so allowing the assembler to sit or require him to stand. If seated, there is a limited area of reach available in which all of the components must be located. Given the maximum extent of the workspace available it is necessary to consider the location of specific components and the manner in which these are presented (in a bin or in a dispenser). The selection of the location of individual components is essentially an exercise in a systems analysis of the operation. That is, it is necessary to determine the relationship between the components, the significance of this relationship and the order in which events must happen. The success of an assembly workplace layout will depend on how well the operation was examined from this systems point of view.

The design of the system from an ergonomic point of view must commence with the establishment of the system representation showing the inter-relationships and their relative value. A useful measure of the relative value of a relationship is the product of the importance of the particular event by the frequency of occurrence. If these can be established, the designer has a logic available to assist in the planning of the layout. An aircraft instrument panel can be considered as an example of this analysis. During a climbing manoeuvre with constant heading it was found that the pilot devoted varying lengths of time to looking at a variety of instruments and at different frequencies of observation.

14.6.3 Physiological Factors

Physiological factors are of great concern when the physic-chemical characteristics of the body are significant. As far as body functions are concerned these involve the neurological, muscular, respiratory, vascular and sensory systems. However, from the point of view of the design we only need to consider the exterior manifestations of these systems. The exterior manifestations can be grouped according to the response to various inputs such as :

(*i*) Visual

(*ii*) Auditory

(*iii*) Tactile, kinesthetic and taste senses

(*iv*) Environment.

Human operators receive a great deal of information visually. The visual processes enable us to perceive form, colour, brightness and motion and so read printed instructions and instruments, observe moving objects and react emotionally to combinations of shape and colour. The light energy collected on the retina of the eye sets off a chain of chemical, neural and memory processes which leads to certain interpretations of the meaning of the signal received. In order to achieve the discrimination necessary for correct interpretation it is necessary to achieve satisfactory intensity and colour discrimination and resolution. The retina consists of two types of sensitive areas known as rods and cones. The 130×10^6 rods which predominate in the sides of the eyeball can only perceive black, white and shades of grey. The cones are fewer in number at the back of the eyeball and are responsible for colour discrimination. The cone system is used when the illumination exceeds $10^{-2}\,1x$. On the other hand the rod system is used for levels of illumination less than $10^{-3}\,1x$. Thus colour discrimination is impaired when illumination levels are low, and this can lead to unexpected difficulties in comprehension. Some of the problems of dark adaptation are also due to this change-over. In fact adaptation to a darkened room takes 15.30 minutes although light adaptation is much faster taking only a few seconds.

Actuity is the ability to perceive black and white detail at various distances. It is measured as the reciprocal of the visual angle subtended by the smallest discriminable visual detail. Because of the structure of the eye there is a marked relationship between brightness, contrast, and actuity. Thus when fine discrimination is required the level of illumination required for any given task can be obtained from a number of reference sources.

An excessive level of noise pollution is, in fact, undesirable for a number of reason :

(*i*) Leads to degradation of speech intelligibility.

(*ii*) Will lead to physical damage to the human auditory system.

(*iii*) Hinders mental activity due to distracting influences.

(*iv*) Can lead to psychological and mental disorders if sustained.

The effect of noise environments on speech intelligibility has been widely studied because of the obvious importance. The Speech Interference Level (SIL) is in certain frequency bands.

The psychological disorders which can be induced by noise pollution depend largely on the largely on the nature of the noise and the persons involved. Intermittent noises have greater effect, particularly when they are sharp sounds as in hammer blows. In addition, if the person has no control over making the sound (*e.g.*, he is not wielding the hammer) he is more likely to be disturbed than if he had caused it.

As far as the design is concerned, the following steps are available for the acoustic treatment of working environments:

1. Control the noise at its source by changing the dynamic behavior of the machine, modifying fluid jet flow, etc.
2. Create barriers between the source and the listener.
3. Provide personal protective device.
4. Modify operating procedures so that the exposure of personnel to noise is reduced.

Acoustic design can also take on a positive aspect when we are concerned with the quality of sound. This is important in the design of concert halls, recording studios, amplification equipment, to name a few areas. In these cases it is necessary to consider the harmonic analysis of the sounds and the reflection and absorption characteristics of surfaces over the appropriate frequency range. The geometric design is also of considerable importance since this determines the reflection of sound waves and the possible interferences.

A vibrating environment can also cause distress and impairment of function visual acuity and depth perception are affected by vibrations in the frequency range 1–70 Hz with amplitudes of more than 0.05 mm. Other functions such as tracking, aiming, attention are mostly affected by vibrations in the 1.5Hz range with amplitudes of 0.5–12 mm. The greatest damage comes, however, when the vibration occurs at the resonant frequency of the body. In the pelvic and abdominal regions resonance occurs at 4–8 Hz whilst the head shoulders resonance occurs at about 30 Hz. At these frequencies the person finds the vibration intolerable and even painful.

There are many sensory inputs to which the body responds and which must be taken into account, or made use of in the man-machine relationship. These senses are all due to the fine nerve fibres which terminates on the surface of the skin. The sense of touch is one which is of great value in various recognition situations. Braille printing of coded impressions is an example of the recognition process by the sense of touch. The same nervous system will detect temperature differences and contact pressure. The skin senses can adapt to certain ranges of stimuli under continuous non-changing stimulation. Under these conditions they cease to give their characteristics sensation. The perception of change in form leads to an important practical consideration. The coding of knobs and handles for unconfused recognition is one of very important applications.

The atmospheric environment in which the human performs his tasks may considerably affect his working efficiency and accuracy. When the temperature is below 10^{0}C, physical conditions. Above 25^{0}C physical fatigue begins and above 30^{0}C the mental processes begin to slow down. A temperature of 50^{0}C is tolerable for a short time but mental or physical effort is almost impossible.

The human body is continually generating heat. At rest, an adult generates about 80 W. As physical activity increases in intensity the heat generated rises to about 1600 W for severe exertion. On the other hand, the temperature regulation system aims to maintain body temperature at about 36.8°C. As a consequence there is a need for heat transfer to, or from, the atmosphere to maintain this control. Heat transfer to and from the body takes place by radiation, convection and evaporation. Convective heat transfer depends on the temperature of the air and the speed of its movement past the body. Of considerable importance also is the nature of the clothing being worn. The following data illustrates this aspect. Humidity has little effect on heat exchange for normal temperatures. At high temperatures, however, humidity has an important effect on heat transfer, comfort and physiological tolerance. As all are aware, there is a relationship between temperature and humidity which leads to similar degrees of comfort.

There is a wide range of environmental atmospheric requirements for human comfort. Aesthetic considerations are largely concerned with the emotional response which the product develops. We must realize at the outset that aesthetic design does not mean the covering of unattractive parts or the addition of adornment for the sake of 'beauty". Aesthetics can only be honest when the technically functional and human factors (ergonomics etc.) have been properly integrated together. However, the designer often has the freedom to manipulate the shape and arrangement of components of the machine to look artistic.

When the product is intended to be open to public gaze, then it is important to make sure that it will look right and be visually exciting. The designers of buildings and bridges obviously have a great responsibility in this regard because of the size and mass visual impact of their creations. It is not satisfactory to think of aesthetics after the design has been completed, but rather to integrate it with the initial design development. It is only in this way that "dressing-up" can be avoided.

14.7 ASSEMBLY LINES

An assembly line is a type of flow production. It was developed in USA for automobile industry. A large scale manufacturing unit prefers to develop and employ assembly line because it helps to improve the efficiency of production. This method results in cost reduction. It is particularly useful when a limited variety of similar products are to be manufactured on a mass scale or in fairly large batches or on regular or continuous basis.

The most vital decision is the proper layout of assembly line. The design involve proper balancing of technology and other manufacturing facilities so as to develop a rational approach for optimization of results. The design largely depends on product design and location of production. The machines are positioned keeping in view the following considerations;

- The rate of flow of work
- The direction of manufacturing operations
- The convenience and comfort of workers
- The availability of service facilities like water, electricity, compressed air etc
- The supply and demand of material.

In an assembly each machine must directly receive material from the previous machine and pass it on to the next machine. Therefore the location of machine is automatically regulated by the sequence of operations.

Assembly line process is employed in;

- Automobiles
- Computers
- Radios Television
- Videos
- Electrical products
- Electronic products.

Steps in Assembly Line Manufacturing of Cars

Consider the assembly of a car, assume that certain steps in the assembly line are to install the engine, install the hood, and install the wheels (in that order, with arbitrary interstitial steps). A car on the assembly line can have only one of the three steps done at once. After the moves on to wheel installation, the second car to hood installation, and a third car begins to have its engine installed. If engine installation takes 20 minutes, hood installation takes five minutes, and wheel installation takes 10 minutes, then finishing all three cars when only one car can be operated at once would take 105 minutes.

On the other hand, using the assembly line, the total time to complete all three cars is 75 minutes. This is possible because there are three different installation stations, an engine station, a hood station and a wheels station. By having three stations, a total of three different cars can be operated on at the same time, albeit involving different steps. After finishing on the first car, the engine installation crew can move to the second car. While the engine installation crew works on the second car, the first car can be moved to the hood station and then the wheels station, to be fitted with the hood and the wheels respectively. After the engine has been installed on the second car, the second car moves to the hood and then the wheels assembly (5 minutes after the wheels assembly is done with the first car). At the same time, the third car moves to the engine assembly. When the third car's engine has been mounted, it then can be moved to the hood and then the wheels installation facilities; meanwhile, subsequent cars (if any) can be moved to the engine installation station. Because it takes 20 minutes to finish work on the engine while it takes only 15 minutes to complete the installation of both the hood and the wheels, the bottleneck is at the engine installation. Hence, cars will come off the assembly line at 20-minute intervals.

Assembly lines don't increase the speed of producing a single unit, but only increases rate when there is a stream of units to be produced. In the example above, a car will still require 35 minutes to be made; however, when there is a stream of cars it will only take 20 minutes to have each new car coming off the assembly line.

14.8 INDUSTRIAL AUTOMATION

An 'Automation' is automati.: operation, it is basically concerned with automatic 'goods' production or manufacturing.

Three Steps Towards Automation

Steps towards automation constitute a history of increasing manufacturing efficiency and labour productivity; three steps have been followed:

- introduction of tools
- mechanization and
- automation.

Use of Tools

It was around four million years ago when human beings appeared on the earth.[1] Since then they have made 'tools' (so they are called tool makers or tool-making animals (B. Franklin, 1978). A tool (or an instrument) is an extention of a hand which performs effective actions. Human beings have utilized tools for productive activities, thus differing from the other animals.[2] In this sense the human being can be called Homo faber as well as Homo sapiens. Thus, the first step in history of increasing manufacturing efficiency is the birth and utilization of tools.

Mechanisation

The second step is the replacement of human physical labour by machine. A machine consists of three components:

1. prime mover converting a variety of energy to mechanical energy
2. transmitter transmitting a driving force to the workplaces
3. machining unit conducting mechanical work.

'Machine tools' play an important role in increasing manufacturing efficiency for the production of industrial products-tangible goods; they are machines which can in turn produce machines, hence often called 'mother machines'. Tools, which played the basic role at the first stage of increasing manufacturing efficiency, are now attached to the machine tools and implement machining operations. With the Industrial Revolution, two hundred years ago in England, metal manufactured machine tools appeared. The mass-production mode, enhanced by 'standardisation' and the 'principle of interchangeability', appeared around the middle of the 19th century in the United States, hence this production mode is often called the 'American system of manufacture', as mentioned in Section 4.2. The development of precision machine tools has contributed to modernization of manufacturing – the conversion of 'manufacture' to 'big industry' (the highest level of capitalism, as well be explained in Section 31.2). This is a phage of 'mechnisation'.

Automation

The third step was the replacement of human mental labour by machines. The setup, operation and control of machine tools came to be operated automatically rather than by skilled workers as in the second stage. This is the movement toward 'automatisation'. Its first realization was the 'automatic lathe' which was invented in 1873. Developments in electronics and control engineering advanced the tendency toward automatisation. *This word changed to 'automation' after World War II, though this term was originated in 1936 by D.S. Harder (Morris Dictionary of Word and Phrase Origins, 1988).*

Automation has developed greatly since the war, being accelerated by the invention of numerical control (NC) in 1952. At that time this tendency was recognized as a revolution and a new philosophy of manufacturing. Automation thus had a great impact on industry, together with 'cybernetics'[3] which was developed by Wiener (1961).

Mechatronics

'Mechatronics', which is a 'Japanese-English' word meaning a unification of 'mechanics' with regard to physical labour and 'electronics' with regard to mental labour, now plays an important role in automation.

Meanings of Automation

Three Theories on Automation

In the beginning, when D.S. Harder coined the term 'automation' in 1936 whilst working for the General Motors Corporation, he meant it as the transfer of workparts between the machines in a production process, without human operation. In 1946 he established the 'Automation' Department when he was a vice-president of the Ford Motor Company.

After the war, a management consultant, J.. Diebold, wrote two books on automation. In his book published in 1952 he defined automation as automatic operation or a process of automatically making tangible goods. In 1955 he mentioned that automation had two meanings:

1. Automatic regulation by feedback
2. Integration of a plural number of machines.

J.B. Bright presented the development stages of mechanization and automation, and D.F. Drucker recognized automation as a conceptual system beyond technology. These three theories are typical in understanding the concept and meanings of automation.

Origin of Automation

'Automation' may be considered as the abbreviation of 'automatisation' or 'automatic operation'; alternatively, automation is a combination of Greek automatos (meaning self-acting) and Latin'-ion' (meaning a state).

Meaning of Mechanisation/Automation

'Mechanisation' is the replacement of human physical labour by machines, but the control of this machine operation is effected by human operator. However 'automation' also replaces this control action by machines as mentioned in the previous section; that is, 'automation' means the replacement of both human physical and mental activities by machines.

Kinds of Automation

Automation implies processes adopting automatic production methods or full automatic production in factories. Hence this term has been concerned with, and confined to, production activities.

Three Kinds of Automation

Three directions have pointed the way toward automation (Department of Scientific and Industrial Research, 1956) :

1. Aautomatic flow-type production in manufacturing industries
2. Automatic control of continuous production in process industries
3. Increase in business efficiency by computers.

These tendencies created the following three types of automation (Inoue, 1978):

1. Mechanical (or Detroit[4]) automation for manufacturing industries
2. Process automation for process and chemical industries
3. Office (or business) automation for office work.

Factory Automation (FA)

Mechanical automation and process automation are concerned with direct production processes which convert raw materials into products; *i.e.*, the 'flow of materials'. This type of automation is now called factory automation (FA). This terminology appeared in 1961.

Office Automation (OA)

On the other hand, office automation (OA) is concerned with management/control of productive activities, namely, the 'flow of information'. Once FA and OA are integrated, 'corporate (or enterprise) automation' emerges.

Low-Cost Automation (LCA)

In general, high-level automation needs a vast amount of capital investment. Low-cost automation (LCA) emerged in 1965 at the Pennsylvania State University, aiming at automation with a small amount of capital by employing standarised automatic equipment such as actuator, sensors, etc. This type of automation is particularly useful for small businesses, because the cost-effectiveness for automatic [mass] production of a single product is great.

Extended Automation

The word 'automation' is attached to other words, such as:

1. Design automation for speedy automatic design and drawing of parts and products.
2. Laboratory automation for automatic measurement, collection and analysis of test data
3. Store (or sales) automation for sales management by computers using POS (point of sale) techniques, and others.

Even 'home automation' is suggested for a pleasant life in the home.

Development of Automatic Manufacturing

Early Automation

In the latter half of the 19th century, conveyors, which were first employed in an automatic flour-producing plant by O. Evans in 1787, were widely used for transfer of goods. Around 1873 C. Spencer invented an automatic screw lathe.

Transfer Automation

A simple type of automation, which has been widely used until now, is the 'transfer machine' for continuous production, combining in one unit the functions of several separate machining units unified with attendant loading and transferring mechanisms.

The first transfer machine was built by Morries Motors of England in 1924 for making cylinder blocks, but it was unreliable because of its reliance upon mechanical controls. The development of transfer machines reached its climax in the mid-1950s, when the main car factories installed this integrated type of transfer line. This is called transfer automation, only oriented towards the mass-production industries.

Flexible Automation

For automation of non-mass (variety of product, small-batch) production, a versatile machine, capable of automating this production mode and equipped with a control device that enables it to read the instructions for specified machining operations and operate automatically according to those commands under the control of an operator, has been developed. Typical of this 'program-controlled automation' or flexible automation, which was mentioned in Section 4.3, is numerical control (NC). In NC the machining instructions are given to the control device by means of punched or magnetic tape or disk.

The idea of a punched-card control system, however, was first introduced in 1725 by B. Bouchon for a loom to weave correctly the desired pattern, and extended to culminate in the perfection of the Jacquard loom in around 1804. It is still being used today. In the manufacturing area the control of a machine tool by means of a punched tape was first carried out in the USA in 1952 when a numerically controlled milling machine was developed for the US Air Force at the Massachusetts Institute of Technology. This type of program-controlled machine tool has been used in industrial workshops since about 1956.

14.9 ADVANTAGES AND LIMITATIONS OF AUTOMATION

Advantages

Advantages as is apparent from the numerous technological innovations occurring, advanced automated systems are available today and further advances seem close at hand. Thus, a former questions, "Is automation possible?" has been displaced by the query, "Is automation profitable?" As is essentially true of all business concerns, automation is welcomed most where it contributes to profit. Of the several dividends yielded by manufacturing processing automation, two are upper most. (*i*) improved productivity and (*ii*) better product quality.

(*i*) Improved productivity of machines and people is an advantage that almost always translates into greater profitability and return on investment. Several factors enter into improved productivity, but two are most important.

(*a*) Increased production capacity. More goods produced per manufacturing floor area, machines installed and the human work force. In terms of machines, automation usually increases the machines thus yielding more machine hours per day.

(*b*) Better inventory control of raw materials. Goods in process and finished goods: There is an axiom – *To automate well means to understand and plan exceedingly well*. Some authorities have observed that just analysis of a plant's operation and procedures, while considering total automation, is very important even though only a limited amount of automation may be immediately installed. For the first time such analysis may cause an in-depth understanding of the intricacies and interrelationship of a given production situation. A number of special techniques, most supported by excellent software for computerized analysis, have been developed in recent years. These include such concepts as group technology, just in time (JIT) systems, material requirements planning (MRP) and Manufacturing Resources Planning (MRPII) these concepts are described in following chapters of this book.

(ii) Enhance product quality: Which improves competitive position and reduces waste and re-works. Improved competitive position naturally translates into higher volume leading to added economic advantages.
Human inadequacies such as dexterity, awareness, cognition, speed and strength, that far exceed the human limitation, in some of the manufacturing and processing operations, among other variables in same manufacturing and processing industries variables such as temperature, pressure, chemical composition flow, weight etc. are not measurable by human operators. These inadequacies and limitations are among two most compiling forces a that necessitated automation.

(iii) Upward shift of workers role, that is from numerous arduous, low skilled duties to higher skilled supervisory and maintenance responsibilities.

(iv) Reduction of personal accidents through the assumption of accident-prone duties by automated machines and processes.

Limitations

Some of the limitations of automation include :

1. High cost of design, manufacturing and maintaining automated equipment. This cost is finding considerable relief because of the continuous increase in cost of electronic components and equipment, although some of these savings are offset by continuing inflated costs of software. Successful efforts, to date, and that will also continue in future, in terms of standardizing equipment, communication networks. Software costs exists as a barrier to processes that one also likely for automation.
2. Vulnerability to down time exists because of increasing complexity of automated equipment. However, this is being reduced at an accelerated rate due to improved equipment, self-diagnostics, fault-tolerance techniques and more economic approaches to design redundancy into automatic systems.
3. Loss of flexibility. This was a very important restraint on automation until relatively recently introduction and refinement of the concept of Flexible Manufacturing Systems (FMS) has largely negated this restraint.
4. More management attention. Actions in highly automated systems occur sometimes at an almost unbelievably high rate and allow little or no time for human decisions. Currently with state-of-the-art technology, a machine or a process can be driven to make quite a lot of off-spec and scrap material before effective supervision can get into control of a runaway situation. Through the assistance of information networks, ranging from corporate to plant-wide to cells and individual machines and processes, managers can be appraised of factory/floor situations on essentially on a second-by-second basis. Thus, more and better management personnel are needed as a plant increases its content of automation.

QUESTIONS

1. Explain the Impact of assembly line and automation on man.
2. What are the advantages and limitations of Automation?
3. What is assembly line and automation?
4. Explain the principles of motion economy and their significance.
5. Explain the various human factors in engineering design.

QUESTION PAPER

B.Tech 2nd Semester Examination

Science Technology & Society

Paper –HU-1002

Time allowed: 3 Hrs Max Marks: 100

Note : Attempt any five questions in all, selecting one question from each section. Section E is compulsory.

Section – A

1. What is interrelationship of technology growth and economic growth? Explain with suitable examples (20)
2. Define Industrial revolution. What are its impacts? (20)

Section – B

3. "Due to rapid technology growth, there is depletion of resources" Do you agree? Why or why not? (20)
4. What are the various eco-friendly technologies? Do they increase the cost of production? (20)

Section – C

5. Define transfer of technology. Discuss the problems of technology transfer. (20)
6. Write a detailed note on human centered technology. (20)

Section – D

7. What are the social and ethical responsibilities of an engineer? Discuss. (20)
8. Explain the importance of social & ethical audit of industrial organizations. (20)

Section – E

9. Explain the following: (2 × 10)
   (*i*) Science as professional activity
   (*ii*) Indian Science & technology
   (*iii*) Environmental degradation
   (*iv*) Aim of Club of Rome
   (*v*) Eco-friendly technology
   (*vi*) Nuclear threat
   (*vii*) Impact of assembly line and automation
   (*viii*) Safety Engineering
   (*ix*) Politics & Technology
   (*x*) Whistle blowing

QUESTION PAPER

B.Tech 2nd Semester Examination

Science Technology & Society

Paper –HU-1002

Time allowed: 3 Hrs | Max Marks: 100

Note: Attempt any five questions in all, selecting one question from each section. Section E is compulsory.

Section – A

1. How are science and technology interrelated? Discuss in detail interrelationship of social and economic growth with advancement of technology. (20)
2. Describe landmark achievements of science and technology in post independence India that have made it one of the modern developing countries. (20)

Section – B

3. Rapid technological growth has led to resource depletion, give your views supported by examples for or against this statement. (20)
4. What is energy crisis? What are the various reasons for energy crisis in the world? What possible steps can be taken to over come this crisis? (20)

Section – C

5. What are various modes of technology transfer? Describe each mode. (20)
6. What is assembly line production? Describe in detail steps of assembly line production of a product which you know is produced in this manner. (20)

Section – D

7. What is a code of professional ethic? Draft a code of professional ethic which you consider should be followed by engineers in their professional dealings with society and colleagues. (20)
8. What is whistle blowing, is it ethical? What are the legal protections available to a whistle blower?

Section – E

9. (*a*) What is the cause of global warming? (2×10)
   (*b*) What is the difference between nuclear reactor and a nuclear bomb?
   (*c*) Give a list of five non-renewable sources of energy.
   (*d*) Why is development of swadeshi technology necessary for a country like India?
   (*e*) What are problems of automation?
   (*f*) List solar powered gadgets that can replace traditional gadgets in a house.
   (*g*) Name sources of air pollution that emanate from industry.
   (*h*) Technology has widened the gap between developed countries and developing countries, comment.
   (*i*) Briefly describe the technology inputs that ushered in the green revolution in India.
   (*j*) Wind energy is a clean source of energy, comment.

QUESTION PAPER

B.Tech 2nd Semester Examination
Science Technology & Society
Paper –HU-1002

Time allowed: 3 Hrs | Max Marks: 100

Note: Attempt any five questions in all, selecting one question from each section. Section E is compulsory.

Section – A

1. Enumerate the differences between Science and Technology and explain how they depend on each other. (15)
2. Science and Technology are tools of social transformation, elaborate. (15)

Section – B

3. (*a*) What led to energy crisis in mid seventics? How did world respond to this crisis and overcome it? (7)
   (*b*) What are various technologically viable renewable sources of energy? Give two ways in which one such renewable source of energy has been technologically harnessed. (8)
4. How does environmental pollution differ from environmental degradation? Discuss ways by which environmental degradation can be controlled. (15)

Section – C

5. (*a*) What is arms race? How it is an impediment in the development of poor people of under-developed countries, explain in detail. (8)
   (*b*) Arms race is a big business for developed nations. Comment. (20)
6. What is the philosophy behind Schumacher's movement of appropriate technology? How is this philosophy advantageous to poor nations in meeting their developmental needs? (15)

Section – D

7. (*a*) What is the importance safety in engineering practice? (8)
   (*b*) Describe in detail safety regulations according to Factories Act of 1948. (7)
8. List social and ethical responsibilities of an engineer. (15)

Section – E

9. (*a*) Comment on the statement," Science is open and subject to criticism whereas technology is closed and costly". (4 × 10)
   (*b*) "Technology is the panacea of all the ills of society," give reasons in favour or against this statement.
   (*c*) Name a manmade non-biodegradable pollutant and explain how it has spoiled the ecological balance.

(*d*) Science is a social activity, explain.

(*e*) How has assembly line and automation affected the production in industry?

(*f*) Difference between Authoritarian and Democratic control of technology.

(*g*) Why professional bodies encourage their members to follow professional ethics?

(*h*) What were the reasons that Indian science remained in dormant state for almost 500 years?

(*i*) Give negative after effects of industrial revolution.

(*j*) Write a brief note on wind energy.

QUESTION PAPER

B.Tech 2nd Semester Examination
Science Technology & Society
Paper –HU-1002

Time allowed: 3 Hrs Max Marks: 100

Note: Attempt any five questions in all, selecting One question from each section. Section E is compulsory.

Section – A

1. Discuss fully how Science and Technology are tools of social transformation. (20)
2. Discuss the inter-relationship of technology growth with social, economic and our cultural growth. (20)

Section – B

3. What is global warming? How does green house effect and ozone depletion contribute towards global warming? Discuss in detail green gouse effect and mechanism of Ozone depletion. (20)
4. Write short notes on: 1. Environmental Ethics
   2. Eco-friendly technologies 3. Energy crisis (20)

Section – C

5. Discuss the philosophy behind Schrmacher's movement of appropriate technology. How is this philosophy advantageous to poor nations in meeting their developmental needs? (20)
6. What is meant by Human Centered Technology? How is it useful for the society? (20)

Section – D

7. Discuss social and ethical responsibilities of an engineer. (20)
8. Discuss in detail safety regulations according to Factories Act of 1948.

Section – E

9. (*a*) Write a brief note on wind energy. (2×10)
   (*b*) Discuss after effects of industrial revolution.
   (*c*) Discuss the main reasons that Indian Science remained in dormant state.
   (*d*) Professional bodies encourage their members to follow professional ethics why?
   (*e*) Differentiate between Authoritarian and Democratic control of technology.
   (*f*) Explain briefly the report of club of Rome.
   (*g*) Arms race is a luxury which developing nations cannot afford.
   (*h*) "Science and Technology share a symbiotic relationship." Comment.
   (*i*) What culture? Does it differ from civilization?
   (*j*) How would you say that sociology is a science?

QUESTION PAPER

B.Tech. 1st Semester Examination, H.P University

Science Technology & Society (Dec. 2008)

Paper: HU-1002

Time allowed: 3 Hrs | Max Marks: 100

The candidate shall limit their answers precisely within the answer-book (40 pages) issued to them and no supplementary/continuation sheet will be issued.

Note: Attempt *five* questions in all, selecting one question each from Sections A,B,C and D. Section E is compulsory.

Section – A

1. (*a*) Describe major technological innovations which paved the way for industrial revolution. (7½)
   (*b*) What were the negative after effects of industrial revolution? (7½)
2. (*a*) What were the achievements of Science and Technology in medieval India? Describe. (7½)
   (*b*) Give reasons for the slow pace of development of Science and Technology in medieval India. (7½)

Section – B

3. (*a*) Write note on any three of the following sources of energy:
   (*i*) Solar energy.
   (*ii*) Wind energy.
   (*iii*) Nuclear energy.
   (*iv*) Hydro energy. (7½)
   (*b*) What are the reasons behind energy crisis ? Suggest remedies to avert energy crisis. (7½)
4. (*a*) What are Eco-friendly technologies? Name five eco-friendly technologies which have improved the quality of life of a common man. (7½)
   (*b*) What are the causes and ill effects of air pollution? What appropriate steps can be taken to stop air pollution? (7½)

Section – C

5. (*a*) What is Assembly line production? List advantages and disadvantages of assembly line production. Give a few examples where assembly line production changed the production scenario. (7½)
   (*b*) What is Automation? What are various strategies to introduce automation in an industrial enterprise? Give advantages and disadvantages of automation. (7½)
6. (*a*) What are various levels of Technology transfer? What are the various modes by which technology transfer can be effected? (7½)
   (*b*) What are the advantages of Technology transfer? State the problems faced by developing countries in the transfer of technology. (7½)

Section – D

7. Write short notes on any two of the following:
   (*i*) Environment ethics.
   (*ii*) Business ethics.
   (*iii*) Social responsibility of a business enterprise. (7½ × 2 = 15)

8. (*a*) What is Whistle blowing? Discuss the protection available to an employee who blows a whistle against an employer, use case study method to elaborate. (7½)
   (*b*) Write short note on any one of the following:
   (*i*) Autoritarian *versus* Democratic control of technology.
   (*ii*) Social and ethical audit of industrial organization. (7½)

Section – E

9. Write short answers of the following:
   (*a*) What is the significance of Ethics in protecting society?
   (*b*) Give two reasons to encourage indigenous technology.
   (*c*) What is Greenhouse effect? How can it be stopped?
   (*d*) What is Carbon credit ? How can this help in maintaining ecological balance?
   (*e*) What is the difference between Environmental degradation and Environmental pollution?
   (*f*) What is the importance of Factories Act?
   (*g*) "Mobile technology is harnessing the third industrial revolution." Comment.
   (*h*) "Pass time industries such as entertainment industry is a drain on energy resources." Comment.
   (*i*) What is the basic difference between Science and Technology ?
   (*j*) List benefits of Information and Communication technology. (4 × 10 = 40)

QUESTION PAPER

B.Tech. IInd Semester Examination, H.P University

Science Technology & Society (June, 2009)

Paper: HU-1002

Time allowed: 3 Hrs — Max Marks: 100

The candidate shall limit their answers precisely within the answer-book (40 pages) issued to them and no supplementary/continuation sheet will be issued.

Note: Attempt *one* question from each section. Section E is compulsory.

Section – A

1. "There is no inter-relationship between Technology growth and Cultural growth." Comment.
2. Discuss the impact of Ancient, Medieval and Modern technology.

Section – B

3. Write a detailed note on Renewable energy resources.
4. Explain the various causes of Environmental degradation and pollution.

Section – C

5. What are various factors which need consideration while making technology assessment?
6. Discuss the problems of man-machine interaction.

Section – D

7. Discuss the ethical issues in Engineering practice.
8. Explain the various parameters monitored in social and ethical audit.

Section – E

(Compulsory Question)

9. Explain the following:
   (*i*) Science and Social activity.
   (*ii*) Indian Science and Technology.
   (*iii*) Club of Rome.
   (*iv*) Eco-friendly technologies.
   (*v*) Environmental ethics.
   (*vi*) Nuclear threat.
   (*vii*) Human centered technology.
   (*viii*) Safety engineering.
   (*ix*) Politics and Technology.
   (*x*) Codes of Professional ethics.

BIBLIOGRAPHY

1. Environment-Problems and Solutions-D K Asthana & Meera Asthana, S Chand & Co, New Delhi.
2. Environmental Science- The Way the World Works - Bernard J. Nebel
3. Appleyard, R.ed. 1989. the impact of international migration on developing countries paris: OECD.
4. Barger, Bernard 1952 science and the social order New York: Free Press.
5. Gaillard, J 1991. Scientists in the third world Lexington: Kentucky University Press.
6. Gaillard, J., V. V. Krishna and R. Waast, eds. 1997. Scientific communities in the developing world New Delhi: Sage.
7. Kornhauser, William, 1962 Scientists in industry, Berkley; University of California Press, price, Derek J. dSolla, 1963 little science, big science New York Columbia University Press.
8. Krishna, V. V. 1993 S. S. Bhatnagar on science, technology and development 1938-54 New Delhi: Wiley Eastern.
9. Rahman, A. 1972 Trimurti: Science, Technology and society - A collection of essays New Delhi; Peoples Publishing House.
10. Storer, Norman W. 1966. The social system of science New York: Holt Rinehart and Winston.
11. UNCTAD/CSIR Case study in reverse transfer of technology: A survey of problems and policy a India Doc. TD/B/C. 6AC.4/6 and Corr. 1, Geneva.
12. Crane, Diana. 1965. "scientists at major and minor universities: A study of productivity and recognition" America sociological review, 30 (5), Pp. 699-714.
13. Coler, Myron A.ed 1963 Essays on the creativity in the sciences New York: New York University Press.
14. Debroy, Bibek. 1996. Beyond the Uruguay round: The Indian perspective on GATT New Delhi: Sage.
15. Gilpin, Robert, and Christopher Wright eds. 1964. Scientists and national policy making New York: Columbia University Press.
16. Kumar, Nagesh and N. S. Siddharthan. 1997 Technology, market structures and internationalization : Issues and policies for developing countries London: Routlege and the united National University.
17. MacLeod, Roy and Deepak Kumar, 1995. Technology and the raj: Western technology and technical transfers to India, 1700-1947 New Delhi: Saga.
18. Merton, Robert K. 1938. "Science, technology and society in seventeenth - century England" Osiris (Bruges, Belgium), 14 Pp. 360-632.

INDEX

A

B

C

Q

R

S

T